[Issued with Army Orders for November, 1921.

[Crown Copyright Reserved.

40

War Office

7329

MANUAL

OF

FIELD WORKS

(ALL ARMS).

1921.

(PROVISIONAL.)

LONDON:

PUBLISHED BY HIS MAJESTY'S STATIONERY OFFICE.

To be purchased through any Bookseller or directly from
H.M. STATIONERY OFFICE at the following addresses:
IMPERIAL HOUSE, KINGSWAY, LONDON, W.C. 2, and 28, ABINGDON STREET, LONDON, S.W. 1;
37, PETER STREET, MANCHESTER; 1, ST. ANDREW'S CRESCENT, CARDIFF;
23, FORTH STREET, EDINBURGH;
or from EASON & SON, LTD., 40 & 41, LOWER SACKVILLE STREET, DUBLIN.

1921.

Price 2s. Net.

By Command of the Army Council,

THE WAR OFFICE,
    November, 1921

CONTENTS.

(B 14783)Q Wt. 11084—PP 3115/387 62¼M 12/21 H. & S., Ltd. G.S.133

4

CHAPTER V.

MACHINE GUN EMPLACEMENTS.

CHAPTER VI.

OBSTACLES.

CHAPTER VII.

DETAILS OF TRENCHES, FIRE POSITIONS, AND TRENCH ACCESSORIES.

CHAPTER VIII.

A DEFENSIVE SYSTEM.

CHAPTER IX.

CAMOUFLAGE.

CHAPTER X.

THE ORGANIZATION OF WORKING PARTIES AND THEIR TASKS.

PART II.—BRIDGING.

CHAPTER XI.

KNOTTING AND LASHINGS.

CHAPTER XII.

BLOCKS, TACKLES AND USE OF SPARS.

CHAPTER XIII.

ROAD BRIDGES AND THE PASSAGE OF GAPS.

PART III.—ACCOMMODATION.

CHAPTER XIV.
CAMPING ARRANGEMENTS.

CHAPTER XV.
SHELTERS AND DUG-OUTS.

PART IV.—COMMUNICATIONS.

CHAPTER XVI.
CROSS-COUNTRY TRACKS, ROADS AND TRAMWAYS.

PART V.—DEMOLITIONS.

CHAPTER XVII.
EXPLOSIVES AND DEMOLITIONS.

CHAPTER XVIII.
LAND MINES, TRAPS, &c.

LIST OF APPENDICES.

LIST OF PLATES.
(*At end of book.*)

MANUAL

OF

FIELD WORKS (ALL ARMS).

GLOSSARY OF TERMS.

Batter.—The slope of the face of any earthen, stone or masonry structure which is not vertical.

Baulk.—A timber beam, or road bearer of a military bridge.

Bay.—The distance bridged by one set of baulks or road-bearers.

Berm.—The distance between the edge of an excavation and the parapet formed of the excavated earth—in a defence work.

Bight.—The portion of a rope used double when the ends are not available.

Bisect.—To divide into two equal parts.

Bivouac.—A camp without tents or huts.

Borrow-pit.—An excavation from which earth is taken for a particular purpose, *e.g.*, building a breastwork.

Breastwork.—A defence work of which the greater portion of its height is above ground level.

Butt.—The thick end of a spar.

Calibre.—Diameter of the bore of a gun in inches not counting the depth of the grooves.

Case.—A structure used in underground work consisting of a top sill, bottom sill and two sides, usually made of 2-inch or 3-inch planks.

Chess.—A specially prepared plank which forms the decking of a pontoon bridge.

Cleat.—A small piece of wood fixed to another to form a support or stop against movement.

Command.—The vertical height of the crest of a work above the ground level or above the crest of a neighbouring work.

Communications.—Roads, railways, waterways and air routes.

Consolidation.—Making ground secure against attack by careful organization of the troops and by provision of protection.

Cover.—Concealment from view or protection from enemy projectiles.

Crest.—The highest point of a parapet, usually the intersection of the superior with the interior slope.

Dead ground (or *water*).—Ground or water over which aimed or observed fire cannot be brought to bear.

Dead load.—A stationary load which is applied continuously to a structure.

Defensive system.—An area of ground which embraces the complete organization of defence across any front.

Defilade.—The adjustment of the levels of the crest and interior of a work to secure cover for the defenders.

Detonator.—A small amount of very high explosive in a container which can be fired by ignition to explode a charge.

Direct laying.—The method of laying a gun by looking at the target over or through the sights.

Dog.—A bar of iron of which the ends are pointed and bent inwards at right angles—used for fastening heavy timbers together.

Dug-out.—An underground chamber or passage.

Embrasure.—The aperture in the wall of an emplacement through which a gun fires.

Enfilade fire.—Fire which sweeps the position from a flank.

Field of fire.—An area of ground exposed to the effective fire of a given number of troops or group of guns.

Fire bay.—A length of trench from which it is intended to deliver rifle fire.

Frame.—A structure used in underground work consisting of a top sill, bottom sill and two legs.

Frontage.—The extent of ground covered laterally by a body of troops.

Glacis.—The ground round a work within close rifle fire, sometimes formed artificially. A term used to describe an even natural slope.

Gradient.—A slope represented by a fraction, *e.g.*, 1/30 represents a rise or fall of one unit measured vertically for every 30 units measured horizontally.

Grazing fire.—Fire which is parallel or nearly so to the surface of the ground.

Groundsill.—The bottom member of a frame, sett or case used in work underground.

Headcover.—Protection against frontal or oblique fire for the heads of men when firing over a parapet.

Helve.—Handle of pick-axe or felling-axe.

High angle fire.—Fire at angles exceeding 25°.

Intelligence post.—A post occupied by unit, brigade, divisional or corps, observers for watching the enemy and ground on their front.

Keep.—A work within the system of defence of a defended post or locality, and distinct from it. Commanding a field of fire within the outer defences with a view to assisting in their recapture in case of need.

Lead.—Pronounced LEED. Any wire used to convey an electric current.

Live load.—A load which is suddenly applied to a structure or part of a structure with an impact producing stresses in excess of those due to its weight when at rest (*see* Dead load).

Look-out post (L.O.P.).—A post from which the progress of the battle can be watched.

Observation post.—A position whence the fire of a battery or a smaller unit of artillery is observed, corrected and controlled. (The term is similarly used with reference to machine gun units.)

Outpost zone.—The part of a defensive position nearest to the enemy.

Overhead cover.—Protection by means of a roof, against splinters, shells or bombs.

Parados.—A bank of earth constructed to give protection against reverse fire, and the back burst of high explosives.

Primer (except artillery).—A specially prepared nature of high explosive which acts as the medium of detonation between the detonator and the demolition charge.

Profile.—The outline of the section of a work at right angles to the crest line.

Rear position.—Part of a defensive position.

Redoubt.—A field work entirely enclosed by a parapet giving all round fire.

Relief.—The length of time that men have to work before being relieved, or a party of men who are on duty or who work for a given length of time.

Retrenchment.—A system of trenches sited to form a second, though not necessarily a separate, line of defence, usually to reduce the number of rifles required in the front line.

Reverse fire.—Fire directed against the rear of the position.

Sangar.—A parapet composed of dry built stone wall.

Sanitation.—The practical application of certain well-established laws with regard to the preservation of health and the prevention of disease.

Sap.—A trench dug by men working at the bottom and constantly extending the end towards the enemy.

Searching power.—The power of a projectile to reach a target behind cover; it varies with the range of the weapon and the angle of descent of the projectile.

Sett.—A term used by miners for frame or case.

Shelter.—A roofed area giving cover from weather, splinters, or shells.

Slope :—

    Interior.—The inner slope of a parapet extending from the crest to the fire step.

    Exterior.—The outer slope of the parapet from the exterior crest to the ground level.

    Superior.—The slope of top of the parapet immediately in front of the crest.

Span.—The horizontal distance between the supports of a bridge. The length of a bridge from shore to shore is called the total span.

Spitlock.—To mark out a line on the ground with the point of a pick-axe.

Spoil.—The material resulting from any excavations.

Spreader.—A piece of timber nailed along the top sill or ground sill of a mining frame to prevent the sides from being forced inwards.

Tankodrome.—A place where tanks are parked.

Tasks.—The amount of work to be completed by a man, or party of men, during one relief.

Trace.—The outline of the plan of a work on the ground.

Traverse.—A buttress of earth provided between two adjacent fire-bays for protection against enfilade fire, and to localise the burst of shells.

Templet.—A pattern, guide or model used to indicate the shape any piece of work is to assume when finished, *e.g.*, wood laths nailed together to outline the section of a trench or parapet used to check the accuracy of the work.

Topsill.—The top member of a frame, sett or case used in work underground.

Transom.—A transverse bearer or support on which the baulks or road bearers of a bridge rest.

Trenails.—Pegs of hard wood used for jointing heavy timbers instead of dogs or spikes.

INTRODUCTION.

Field engineering consists of those elements of the science of engineering without which an army in the field cannot live, move or fight.   Just as the training manuals of the several arms of the service aim at the training of leaders to lead and the soldier to fight, so this book endeavours to assist in the training of the leader to direct and the soldier to supply intelligently, the labour, which is required to carry out any field engineering work, however elementary.

The subject has been divided into five principal parts, of which a general knowledge is most necessary in war, namely :—

    i. Field fortification.
    ii. Bridging.
    iii. Accommodation.
    iv. Communications.
    v. Demolitions.

Part I—**Field fortification.**—Has been arranged to give a graduated course of instruction.   It begins with the definition of field fortification and general principles ; it passes through training in the use of tools and materials and the construction in detail of the actual defence works ; and ends with the application of the principles and details, already learnt, to the organization of a defensive position on the most comprehensive scale.

Part II—**Bridging.**—Includes the use of ropes and spars and the details of "light" bridges. The necessity for calculation has been avoided as far as possible by giving standard examples of these bridges. The more technical information on this subject is given in Military Engineering, Vol. III.

Part III—**Accommodation.**—Is devoted to the construction in camp or trenches of all shelters, dug-outs, including cooking and sanitary accessories.

Part IV—**Communications.**—Deals with tracks, roads and tramways, and the information is confined to that which would be required for work in the more forward areas.

Part V—**Demolitions.**—Includes the use of the common explosives supplied and found in the field. Accidents have occurred in the past which have been due to ignorance and the mistaken idea that any work, which involves the use of explosives, is the prerogative of experts. All ranks must be trained in handling the many and various explosives used by all arms.

Further information on the above subjects can be found in Military Engineering, Vols. I to IX.

Night work.—No special notes have been given as regards work at night. It must be recognized that work at night in the forward area is the rule and not the exception, and that both officers and men require the most careful training, so that loss of time and waste of labour are reduced to a minimum.

PART I.—FIELD FORTIFICATION.

CHAPTER I.

GENERAL PRINCIPLES OF FIELD FORTIFICATION.

1. *Definitions.*

1. **Field fortification** includes all measures taken to strengthen a position by means of works constructed in the face of the enemy, or in anticipation of his immediate approach to the scene of action.

These works are called field works or field defences.

2. The main objects of field defences are :—

    i. To place the occupant in the best position for using his weapons,

    ii. To protect the occupant from the weapons of the enemy, and thus to reduce casualties,

with the general effect that, by skilful use of field defences combined with fire effect, a commander may be able to reduce the strength of his force in actual combat with the enemy to a minimum ; and thus to form a reserve for the decisive action either in his own sector or for transfer to some other part of the theatre of war. The stronger the defences, and the more work put into their construction, the greater will be the economy of strength in holding them.

3. Although field defences presuppose a defensive attitude locally, they play a most important part in all offensive operations, and in this connection they must be most carefully studied.

Whether on the offensive or defensive, it must be remembered that the guiding principle remains the same, namely, that the defences are to be constructed to conform with the tactical plan of operations and that they are only a means to an end and not an end in themselves.

4. Field defences are of two classes :—

    Hasty.

    Deliberate.

5. **Hasty defences** are made actually on the battlefield, by the attacking troops, prior to another bound forward, to secure the ground they have taken ; or by the other side to hold up the progress of the enemy, while fresh dispositions of troops are being made in rear.

The design of hasty defences follows generally the rules laid down for deliberate works, but the conditions under which they are made do not allow the same accuracy of line, dimensions, &c., to be maintained.

The aim of a commander who has been ordered to make good the ground he holds, is to get his men under cover in the quickest possible way.

In an encounter battle the tools available will be the entrenching implements and such entrenching tools as it may be possible to bring up during the night. The amount of digging that can be done is comparatively small, so that to get the protection required every use must be made of the cover, which actually exists on the battlefield.

This cover may consist of hedges, walls, buildings, banks, sunken roads, railway embankments and cuttings, woods and shell-holes. All of these are easily convertible into strong defences if intelligently treated; they have all played important parts in battles, and it has been proved that troops well trained in adapting natural cover to defence are extremely difficult to eject when once they have dug themselves in.

The works are described in detail in Chapters IV, V, VI and VII.

6. **Deliberate field defences.**—Under this head come all defences which are not included under hasty defences, but which are not so imposing as to be called permanent fortifications. The only considerations which limit their scope are :—

    i. The time, material and labour available.

    ii. The industry and skill of that labour.

    iii. The ability of the enemy to interfere with it.

Deliberate field defences would be employed in :—

    i. The gradual development of a defensive system, when once the opportunity of manœuvre has ceased.

    ii. The later stages of consolidating an objective taken during the period of position warfare.

    iii. The preparation of rear defensive systems.

They are described in detail in Chapters IV, V, VI, VII and VIII.

2. *Effect of modern war equipment on the design of field fortifications.*

1. In order that field fortifications may be designed to the best advantage, it is necessary to study the characteristics of the various forms of war equipment which may be used against them.

The power and nature of the equipment employed affect the design of field fortifications in two ways :—

    i. Their penetration, searching power and destructive effect govern the amount, disposition and nature of the protection necessary for security.

    ii. The range, rate of fire and radius of activity influence the siting of the works and obstacles.

2. **Classification of fire.**—Fire is termed frontal, oblique, enfilade or reverse, according to its direction relative to the object fired at.

It is called :—

Frontal, when the line of fire is perpendicular to the front of the target.

Oblique, when the line of fire is inclined to the front of the target.

Enfilade, when it sweeps the target from a flank.

Reverse, when the rear instead of the front of the target is fired at. A line of troops or defences thus attacked is said to be "taken in reverse."

As regards the angle of descent of the projectile, fire is said to be **high angle** when it is delivered from guns or howitzers at any angle of elevation exceeding 25°.

3. *Rifles and machine guns.*

1. **Rifle fire.**—Modern military rifles are sighted to about 2,800 yards. Their maximum range may be taken as about 3,700 yards. The slope of descent of the bullet varies from about 1/120 at 600 yards and 1/30 at 1,100 yards, to 1/4 at 2,200 yards.

The following table gives the maximum penetration of a **single** pointed rifle bullet into various materials. It does not allow for a number of bullets hitting on or near the same spot. To be bullet-proof under service conditions, the thickness of loose materials, such as earth or shingle, must be about 50 per cent. greater than that given in the table :—

Material.	Maximum penetration.	Remarks.
	Inches.	
Steel plate, ordinary mild or wrought iron	¾	
Shingle	6	Not larger than 1-inch ring gauge.
Coal, hard	9	
Brickwork, cement mortar	9	
Brickwork, lime mortar ...	14	
Chalk	15	
Sand, confined between boards, or in sandbags	18	
Sand, loose	30	
Earth, free from stones (unrammed)	40	Ramming earth reduces its resisting power.
Sawn timber :—		
Hardwood, *e.g.,* oak ...	38	In timber, the penetration is much less in round logs than in scantling, owing to the deflection of the bullet, but care must be taken to fill the interstices.
Softwood, *e.g.,* fir ...	56	
Green timber :—		
Logs 12 inches diameter and over	24	
Poles 4½ to 6 inches in diameter	36	
Clay	60	Varies greatly. This is maximum for greasy clay.
Dry turf or peat	80	
Snow...	—	Varies greatly. 3 feet of rammed frozen snow, well consolidated with water, will stop a bullet, but the power of resistance will decrease as the temperature rises. Soft snow unrammed has little power of resistance.

2. **Machine gun fire** is a concentrated form of rifle fire, capable of being directed with great accuracy on to a small area. It can also be traversed when a less concentrated stream of bullets will be delivered against a target, such as a trench or a line of troops in the open. In this way it can be used to support an attack by keeping down the heads of firers in a trench. Machine guns can be used to fire indirect, that is from behind cover, the fire being directed generally with the aid of maps.

The machine gun attack from aeroplanes will take the form of direct fire at point blank range, and the rule of slope of trajectory will not apply in this case.

At ranges beyond 300 yards the penetration of machine gun fire may be taken to be equal to that of concentrated rifle fire. At distances under 300 yards, owing to the cumulative and shattering effect produced by a number of shots striking rapidly in succession over a small area, penetration is effected more rapidly and with fewer number of rounds than by rifle fire.

4. *Artillery.*

1. The **natures of artillery**, any or all of which may accompany an army in the field, consist of guns, howitzers and mortars.

Guns and howitzers are classified thus :—

Classification.	Guns.			Howitzers.		
	Nature.	Weight of shell.	Approx. maximum range.	Nature.	Weight of shell.	Approx. maximum range.
LIGHT ARTILLERY.		lbs.		ins.	lbs.	
(a) Pack artillery	2·75-in.	12½	5,800	3·7	20	5,800
(b) Horse artillery	13-pr.	13	8,500	4·5	35	7,000
(c) Field artillery	18-pr.	18	9,500	4·5	35	7,000
MEDIUM ARTILLERY.						
Horse and tractor drawn ...	60-pr.	60	15,500	6	100	10,000
HEAVY ARTILLERY.						
Tractor drawn or on railway mountings	6-in.	100	19,000	{ 8 { 9·2	200 290	12,300 13,000
SUPER HEAVY ARTILLERY.						
Tractor drawn or on railway mountings	9·2-in. 12-in. 14-in.	390 850 1,400	24,500 28,200 35,600	} 12 } 18	750 2,500	14,300 23,000 (calculated).

2. The principal characteristics of " **guns** " are their high muzzle velocity (1,500 feet per second and upwards) which permits of them firing at long ranges; the flatness of the trajectory of their shells which, when using shrapnel, gives deep searching effect combined with small angles of descent (about 12° at medium ranges); and in the case of light artillery the rapidity of their fire.

Pack guns are peculiarly suited to broken, hilly country and are comparatively easily concealed on the move.

Horse Artillery guns (13-pr.) are the most mobile form of artillery and are primarily intended for use with mounted troops. They may, however, also be employed to support the combined action of other arms, similarly to field guns.

Field guns (18-pr.) form the bulk of the artillery with a force. Their essential task is to assist the infantry in every way to close with the enemy. They are specially suited for covering fire, repelling attacks in the open and raking communications. They are the most effective artillery weapon against tanks.

Medium guns (60-pr.) are employed in counter-battery work, for raking communications and covering fire beyond the range of light guns.

Heavy and super heavy guns are used for raking distant communications and for shelling camps and dumps, &c., beyond the range of other artillery.

3. The principal characteristics of " **howitzers** " are their steep angles of descent (about 25° at medium ranges) and of departure, their great shell power and their accuracy.

As compared to guns, they can be placed in positions offering greater opportunity for concealment and cover from shell fire.

They are especially adapted for the destruction of strong works and for engaging entrenched troops and batteries. With instantaneous fuzes, they are very effective against troops in the open and for wire cutting.

As a class, their range is not as great as that of guns.

Pack howitzers are particularly useful for the close support of infantry in hilly or very enclosed country.

Field howitzers (4·5-in.) are employed for covering fire and the bombardment of the weaker defences; for neutralization with gas shell and for counter-battery work generally in the war of movement.

Medium howitzers (6-in.) are used for covering fire, the destruction of defences and the neutralization of hostile batteries. In the war of movement, they are particularly suitable for counter-battery work.

Heavy and super heavy howitzers are employed in counter-battery work against batteries provided with good cover and in the destruction of strong defences.

4. The principal characteristic of " **mortars** " is their power of developing destructive and accurate fire at short ranges.

Their utility is restricted not only by their range but also by difficulties of ammunition supply.

Light mortars are an infantry weapon used for close support and barrage purposes ; they are also of great value, during an advance, in dealing with machine guns and defence posts.

Medium mortars are employed for wire-cutting and in a preliminary bombardment, or to support local actions.

5. **Shells** consist of time shrapnel shell, high explosive shell and special shell.

They are classified thus :—

Nature.	Fired by.	Action.	Employment.
Time shrapnel	Light and medium guns	**TIME SHRAPNEL SHELL.** The shell is burst in the air short of the target by the "time" arrangement of the fuze and a shower of bullets is projected from the mouth of the shell along the line of flight on to the target in the shape of a cone. The bullets form an ellipse covering an area which varies with the range and the type of equipment. The beaten area of 18-pr. shrapnel is 70 yards long at 3,000 yards, and 40 yards at 6,000 yards. That of the 60-pr. is naturally much longer. Time shrapnel becomes ineffective at 6,000 yards with the 18-pr. and 15,000 yards with the 60-pr.	Against troops in the open or under very light cover ; for covering fire or barrages. Its forward effect makes it especially effective in enfilade.
H.E. shell with instantaneous fuze	All natures	**HIGH EXPLOSIVE SHELL.** The shell bursts immediately on impact, the body breaking into fragments which fly outwards mainly at right angles to the axis of the shell. The shell is intended to detonate (a far more violent action than explosion) ; detonation produces violent concussion. Unreliable at angles of descent less than 6°.	Against troops and guns in the open or under light cover ; for covering fire or barrages, harassing fire and wire-cutting. For concussion effect against works constructed of concrete or masonry. Against tanks.

Nature.	Fired by.	Action.	Employment.
H.E. shell with non - delay fuze	All natures	The action is similar to that of H.E. shell with instantaneous fuze, except that the splinter effect is reduced owing to a proportion of the fragments burying themselves, due to the burst taking place an appreciable time after impact. The crater effect is good and shell will burst at all angles of descent.	For destruction of buildings and strong cover other than concrete. For destruction of trenches by the crater effect produced.
H.E. shell with delay fuze	All natures	The shell has two actions, depending on the angle of impact and the surface of the ground. i. *Ricochet effect* (guns only).—If the shell ricochets into the air during the time of burning of the fuze, the shell will burst in the air; fragments fly laterally and downwards. Concussion effect is good. No crater. Burst may be as far as 70 yards from the point of impact, and 30 feet, or more, in the air. ii. "*Burst after penetration*" *effect.*—The shell buries itself and bursts under the ground. A terranean crater is produced with concussion effect.	Against troops in trenches, prepared shell-holes and sunken roads; against the personnel of shielded guns and for covering fire or barrages with light artillery at ranges not exceeding 4,500 yards. Against strongly protected field works and permanent fortifications.
		SPECIAL SHELL.	
Gas shell ...	Light and medium artillery	The shell is burst instantaneously on impact by a sufficient charge to break up the case, causing the liquid contents of the shell to be scattered on the ground. The liquid evaporates at once or slowly (depending on the nature of the filling), producing poisonous or noxious gases.	To cause death or disablement to personnel. To depreciate the enemy's fighting efficiency by forcing him to hamper himself by anti-gas appliances. To deny a given area to the enemy.

Nature.	Fired by.	Action.	Employment.
Incendiary shell	... ...	The shell is similar in action to time shrapnel, except that in place of bullets it is filled with discs or short cylinders which, on being ignited by the bursting charge of the shell, are projected forward as a shower of flaming "stars."	To set buildings, farms, hayricks, &c., on fire and to burn crops or undergrowth.
Smoke shell ...	... ...	The chemical contents of the shell, having been ignited by the bursting charge of the shell, are scattered in the same manner as the contents of a gas shell. The ignition of the filling produces dense clouds of smoke.	To increase the screening effect of a creeping barrage, to mask hostile fire and deny observation to the enemy.

5. *Gas, smoke, mines, tanks and aeroplanes*

1. **Gas.**—The design of defences will be affected only in regard to the sites selected for the trenches and in the accessories of trenches. It has been found that trenches in woods and in the bottom of valleys are particularly affected by gas bombardments ; the same applies to villages which have not been totally destroyed. The protection of dug-outs against gas is effected by means of curtains, the details of which are described in Chapter VII.

2. **Smoke** can be produced in the form of a cloud from projectors, smoke cases, candles, tanks, and aircraft, or by a concentrated bombardment by artillery or mortars using smoke shell. It is used to cover movement and to effect surprise. It can be used to screen the movement of troops, working parties, &c., but not to cover stationary troops or working parties.

3. **Mines.**—These influence the design of a trench system in the provision of special dug-outs for listening, arrangements for shaft heads, &c. Mine warfare is dealt with in Military Engineering, Vol. IV.

4. **Tanks.**—Tanks influence the design in the provision of emplacements of guns, machine guns and mortars to destroy them, and obstacles to hold up them or the infantry accompanying them under fire. The details of these measures are discussed under their own headings.

5. Aeroplanes.—The nature of attack by aeroplanes on troops occupying trenches will be generally machine gun fire from low flying planes, *i.e.*, planes at a height of 200 feet. They will use every endeavour to increase the effect of their attack by surprise, by diving rapidly from a height, or by swooping suddenly over the skyline where the field of view is limited. Protection against this form of attack can only be afforded by shelters, slit trenches as described in Sec. **33**, para. 2, or deep fire trenches with deep traverses.

In special cases bombs may be used ; the protection against these would be similar to that against artillery fire. Otherwise aeroplane attack by bombing will be usually confined to the attack on horse lines, camps, strategical centres, headquarters, railway and store depôts, ammunition dumps, &c. In such cases protection must be afforded for personnel by deep or bomb-proof dug-outs, as described in Chapter XV.

Aerial bombs vary in weight from 25 lbs. up to 660 lbs. or even heavier. They can be fitted with either an instantaneous or a delay action fuze. When fitted with an instantaneous fuze they are intended for use against personnel. Splinters from these bombs fly very horizontally and they have a powerful man-killing effect. A splinter from a 660-lb. bomb has penetrated a 14-inch brick wall. The crater formed is generally shallow.

When fitted with a delay action fuze the bomb may penetrate two or three stories of an ordinary dwelling house before bursting. The crater formed on grass land will vary from about 3 feet in diameter and 18 inches deep for a 25-lb. bomb, up to 30 to 40 feet in diameter and 20 to 25 feet deep for a 660-lb. bomb. One of the latter formed a crater 40 feet in diameter and 10 feet deep in a roadway paved with wooden blocks on 9 inches of good concrete.

Splinters from bombs fly very horizontally ; consequently their effect can be limited by traverses of a reasonable height. Protection can be given in this way to troops in tents and huts, men in hutted hospitals and to horses in stables. The methods are described in Chapter XIV. A 9-inch brick wall or a revetted earth wall, 2 feet thick, will stop splinters from bombs hitherto used against troops in camps or horse lines.

CHAPTER II.

TOOLS AND MATERIALS.

6. *Tools.*

1. Every soldier must be able to march and dig. Careful training is necessary to ensure that the correct method of using the tools available is employed.

2. The tools generally used in field fortification are grouped under the following heads :—

Entrenching tools.
Cutting tools.

3. Entrenching tools.—These are the pick-axe, the shovel, the spade, the crowbar and the entrenching implement.

The most important are the pick-axe and shovel ; to obtain a satisfactory result with them, men must be trained to use them as methodically and thoroughly as they are trained to use a rifle.

While being trained, men must not be allowed to dig except under proper instruction, so that they may acquire the correct action from the start, and exercise the right muscles. They must be taught to use the pick and shovel with either hand leading, in order that they may be able to get out and throw the earth on either side of a trench, without changing position, and to work their tasks from front to rear to avoid risk of injuring their neighbours.

The **pick-axe** is intended for loosening soil ; the pointed end is for use in hard, the chisel end in soft ground.

To use the pick left hand leading, advance the left foot, hold the pick with the handle horizontal, the left hand in front of the right and near the head (Plate **1**, Fig. 1).

Carry the pick head back to behind the left shoulder without letting the pick move through the hands.

Swing the pick forward and upwards with the left hand ; at the same time allow the helve to slide through the left hand until the hands meet (Plate **1**, Fig. 2) and guiding it to the desired spot allow the pick to fall under its own momentum (Plate **1**, Fig. 3).

Break up the soil by raising the hands thus using the pick head as a lever.

Pull the pick towards the body with the right hand and again raise it by picking it up with the left hand near the head. For picking right hand leading the position of the feet and action of the hands are reversed.

The **shovel** is the tool for clearing away the soil loosened by the pick.

To use the shovel right handed (for throwing to left or front only) advance the left foot, hold the shovel with the right hand on the handle, left hand below it, bend down and fill the shovel by thrusting it horizontally into the loosened earth, using the inside of the right leg (Plate **2**, Fig. 1). Push the left hand to the blade of the shovel and straighten the legs (Plate **2**, Fig. 2).

Swing the shovel back taking the weight on the right foot, then swing forward at the same time transferring the weight to the left foot and allow the shovel to slide through the left hand without jerking so that the earth leaves the shovel in a compact mass (Plates **3** and **4**).

For shovelling left handed (for throwing to right or front only), position of the feet and action of the hands are reversed.

In easy soil, the pick should not be used, and **digging** should be done in " spits " in a straight line across the trench, so as to leave a face to work to (Plate **5**, Fig. 1). The point of the shovel is driven vertically into the ground (Fig. 1) and pushed home with the foot by using the weight of the body (Fig. 2), not by kicking which only wears out the boots and hurts the feet. The soil is broken by pulling the shovel down into a horizontal position, and thrown as when shovelling.

Where picks and shovels are to be used spare helves for picks and spare shovels must be provided. If the ground is hard special arrangements must be made for sharpening picks, *e.g.*, field forge and spare steel points.

4, The **entrenching implement** or **grubber** consists of a head and a helve. One end of the head is a hoe shaped like a shovel and the other end is similar to the chisel end of a pick-axe. Men must be taught to use the entrenching implement lying down, and to start the intended excavation from the rear. Hard ground is more easily broken up by this method and a hollow for the disengaged arm is provided, which gives cover for the digger.

5. The **spade** is used for cutting sods, and trimming slopes.

6. The **crowbar** has many uses ; all ranks should be instructed in its use as a common lever for raising weights.

7. **Cutting tools.**—The service cutting tools are the felling-axe, hand-axe, bill hook, cross-cut saw, hand saw, and folding saw.

The felling-axe, cross-cut and folding saws are used for felling ; the hand-axe, bill hook and hand saw for clearing brushwood, hedges, &c., and for trimming.

The **felling-axe** can be used with effect only by a man trained to use it. An axe in the hands of an unskilled man is a source of danger to himself and his neighbours.

The **cross-cut saw** is safer, easier and quicker in the hands of men unskilled in the use of the axe and is worked by two men who pull the saw in turn across the timber. No pushing is required. When used for felling, wedges are required to prevent the saw from jamming.

Where cutting tools are in use means to keep them sharp must be provided, *e.g.*, grindstones, saw sets, files and honing stones.

8. **Other tools in common use.**—Mauls, sledge hammers, augers, field levels. Men require practice with these to get the best value out of them. Mauls and sledge hammers must be used with a full arm swing. Augers must be screwed in vertically to the surface on which they are worked so that the hole may be true. The field level is described in Chapter III.

7. *Materials.*

1. A list of materials commonly used in the construction of field defences is given in Appendix III, and their use is described in Chapter VII. The characteristics of some of the more common materials are given in the following paragraphs.

2. Earth.—Of all the materials used in the construction of field defences, earth is the most valuable and the one most generally available. In combination with timber, steel, concrete or masonry, it can be used in a great variety of ways for giving protection against projectiles. Earth slopes, when freshly cut, will stand nearly vertically for a short time, but quickly disintegrate and crumble after exposure to air, sun, rain and frost, and in time the earth loses its cohesion and will stand only at the slope of excavated earth. Alternate frost and thaw are particularly destructive to earth slopes.

Excavated earth stands naturally at a slope of about 1/1 to 2/3. To make earth stand at a steeper slope, it must be revetted (Chapter VII).

The weight of earth varies from about 80 to 100 lbs. per cubic foot.

3. Stones in a parapet stop bullets, but cause damage from splinters and increase casualties if subject to artillery fire. In countries where there is no earth, stones are used to build defences of loose stone walls called sangars. Only the largest stones which can be handled should be used for such walls.

4. Sods are pieces of turf cut 18 by 9 by $4\frac{1}{2}$ inches; they are used for revetting and are used like bricks in brick work.

The best tools to cut sods are spades and handaxes, the worst are shovels as their blades are curved and it is, therefore, difficult to cut the edge of the sods square.

5. Timber in round logs, squared beams, scantlings or planks is used in nearly every detail connected with field engineering such as shelters, bombproofs, gun platforms, stockades, bridges, huts, &c.

Before felling timber, the trees should be carefully inspected to see that the most suitable are taken for the purpose, and the direction in which they are to fall should be carefully planned, so that time is not lost by trees falling into those which are standing, or on those which have already been felled.

To fell a tree in a required direction, it should be strained in that direction by a rope. It is then cut into as far as the centre on that side and finished off by a cut about 4 inches higher up on the opposite side.

6. Brushwood is used for roadmaking, hutting and revetting. Willow, birch, ash, Spanish chestnut and hazel are the most suitable kinds, and work best if cut when the leaf is off. It should be cut as low down as possible.

As a rough rule it may be taken that 1,000 square yards of brushwood, up to 2-inch diameter, make up three G.S. wagon loads.

If brushwood has to be carried any distance it should be tied into bundles, weighing about 50 lbs. If nothing else is available these may be bound with pliable rods called "withes," which should be well beaten and twisted before use (Plate **6**, Figs. 1 and 2). Brushwood can be made up into fascines or hurdles.

7. A fascine is a long bundle of brushwood tightly packed and bound, used for drains from trenches, foundations of roads in marshy sites, and occasionally for making steps, &c. The usual dimensions are 18 feet long and 9 inches in diameter. It is made in a cradle of trestles placed at a uniform level as shown on Plate **6**, Figs. 3, 4, 5.

8. Hurdles are chiefly used for revetment and unless for a special object, are usually made 6 feet long and 2 feet 9 inches high in the web (Plate **7**, Figs. 1 and 2).

To make a hurdle a line 6 feet long is marked on the ground and divided into nine equal parts, and a picket (about 3 feet 6 inches long and from 1 inch to 2 inches in diameter) driven in at each division, the two outside ones being somewhat stouter and longer.

The web is then constructed by a process called randing which consists in working with single rods commencing from the centre. Each rod is taken alternate sides of the pickets, twisted round the end pickets and woven back towards the centre. A fresh rod must overlap by several pickets the one which it supplants.

Pairing rods (Plate **7**, Fig. 2) are used in the centre and at both ends of the web, which is usually sewn top and bottom in three places.

A rougher type of hurdle is shown in Plate **7**, Fig. 3. This is made much more quickly than the ordinary type and is equally efficient for most purposes.

Hurdles can also be made of expanded metal as shown in Plate **7**, Fig. 4. For convenience of carrying, the vertical stiffeners can be temporarily removed and the X.P.M. rolled up.

9. Gabions are hollow baskets or boxes, made of almost any material capable of being bent or woven into a cylindrical or square form, such as brushwood, expanded metal, stout canvas, wire netting, &c. Two patterns are shown on Plate **8** from which the general method of construction can be obtained.

The most convenient form is the square gabion made of expanded metal (Plate **8**, Fig. 5). It is made by bending a piece of X.P.M. round a stout rectangular framework 3 feet high with four 18-inch faces. The joint where the ends overlap is secured by a hurdle lacing of plain wire or other device (Fig. 6).

Gabions are used for revetments, hasty cover and repair work.

10. Sandbags.—The service pattern of sandbag measures 33 inches by 14 inches empty. It is made of canvas and issued in bales of **250**, weighing 96 lbs.

Sandbags are used for revetments, loopholes, spoil bags for mining or dug-out work.

11. Sacks.—Grain sacks or bags which may be available on service can be substituted for sandbags.

They usually contain about 2 bushels (2½ cubic feet) of grain—if used for field defences they should not be more than half filled, otherwise they are too heavy to handle easily.

It is not necessary to close or tie up a sack if the mouth is carefully folded under it when it is being placed in position; the weight of the sack will prevent loss of earth.

12. **Frames revetting** are wood frames specially made for revetting or repair of trenches (Plate **9, Fig. 1**). Their use is shown on Plate **41**, Fig. 3.

Trenchboards are wooden gratings of the dimensions shown on Plate **9**. They are used to give a firm footing in trenches in combination with the short revetting (Plate **41, Fig. 3**) frame or on overland tracks.

8. *Method of distribution of tools, stores and materials.*

1. All tools and materials in excess of those included in the Mobilization Store Tables of units, A.F. G 1098/9, required for field engineering, are supplied through the engineers. In each formation from the base to the front, engineer parks, stores or dumps as they are more generally called, are organized, and the channel of supply is from the base park through the army and corps parks to the divisional engineer dumps.

2. In stationary warfare the divisional organization for the distribution of engineer stores is as follows :—

The main divisional dump is situated, if possible, on a light railway by which it is fed by the corps dump, and is in such a position as to be reasonably immune from interference by the enemy's shelling.

An advanced divisional dump is formed in advance of this, usually on a road where transport can deliver by day and can be loaded and sent forward by night; it should also be on the light railway.

In front of this the brigade dumps are situated as far forward as horse transport can go at night, or at the light railway railhead and adjacent to the tramway, existing or projected.

During offensive operations it is frequently necessary to form special (trench) dumps, which are stocked with articles required for the work of consolidating the objective, and these stores are not drawn upon for ordinary and current work. They are under the control of the infantry commander.

3. In open warfare this organization must be modified according to circumstances.

4. A list of principal tools, materials and stores suitable for use in field engineering is given in Appendix IV.

CHAPTER III.
FIELD LEVEL AND FIELD GEOMETRY.
9. *Measurement of slopes and laying-off angles.*

1. **Slopes** are usually described in field works by fractions, in which the numerator expresses the height, and the denominator the base of

the slope. Thus a slope, described as 4/1 (*i.e.*, four in one) is one in which the vertical height is four times the base (Plate **10**, Fig. 1); whilst that expressed by 1/6 (*i.e.*, one in six) is, on the contrary, one in which the base is six times the vertical height (Fig. 2).

To convert angles of slope, given in degrees, into fractions as used in field works, a rough rule is to divide 60 by the number of degrees in the angle. Thus 6°= 1 in 10 roughly. This rule should not be used for angles greater than 30°.

2. Field level.—In laying out field defences certain simple geometrical operations may be necessary. The only special instrument employed for the purpose is the field level, which is used for setting off angles on the ground, and for gauging slopes.

The level is shown on Plate **11**.

The limb C, which contains the spirit level, must be opened out first, and afterwards the limb B ; these are then joined by a catch at X.

The level is used in the following ways :—

    i. As an ordinary spirit level, as used by a carpenter or a pavior; for this purpose it need not be opened at all (Fig. 4).

    ii. As a square for setting off right angles. The limbs B and C form the right angle (Fig. 1).

    iii. As a protractor, for laying out an ang'e from a given point on a given line. The limb A is made to coincide with the line, the point of the arrow head being at the given point. The required angle can then be laid out by stretching a tape from the arrow head over the angle as numbered on limb B or limb C (Fig. 1).

    iv. For setting off any slopes from the horizontal to the vertical as a mason's level. For this purpose the plumb-bob, kept in a recess of limb C, is required. The plumb-bob must be suspended from the brass socket in limb C (near the end remote from the spirit level), and allowed to swing freely, and the level moves until the string coincides with the required angle $\frac{1}{4}$, &c. ; the edge of the limb A will then be at the required slope (Figs. 2 and 3).

    v. One side is graduated in feet and inches and can be used as a four-foot rule (Fig. 4).

3. The **improvised level** shown in Plate **12** may be found useful. By means of it a series of points, A, B, C, D, E, F, G, H, I, on the same level, can be fixed and marked by pegs. Suppose AI is 40 yards and the difference of level of the drain or trench between A and I is found to be 1 foot, then the fall is $1\frac{1}{2}$ inches each length of 5 yards ; the bottom of the drain or trench can be graded by measurement from A, B, C, &c.

10. *Field geometry.*

1. In some of the more technical operations of field engineering, such as the construction of bridges and in road and camp work, a

knowledge of the following applications of simple geometry in the field will often be of use. No special instruments are required for this purpose.

2. To lay out a right angle. Let X be a point in a given straight line AB (Plate **10,** Fig. 3), from which it is required to set off a right angle.

From X measure off a distance of 4 units XC along AB. Take a piece of line or tape 8 units long, apply one end to point X, and the other to point C ; find a point in the tape 3 units from X, and, seizing it at this point, draw the bight out to D, till the line is taut, then CXD is a right angle. For example, if 1 foot is the unit—XC = 4 feet, XD = 3 feet and CD = 5 feet. The longer the sides of the triangle, the more accurate will be the right angle, and it will be found that when laying out long lines, such as a parade ground, or football ground, the sides of the triangle should not be less than 16 feet, 12 feet, and 20 feet.

3. To trace a perpendicular to a given line from a point outside. Let X be the point outside the line AB (Fig. 4), from which it is required to draw a perpendicular to that line. Take a tape or cord longer than the perpendicular will be ; fix one end at X, and stretch it taut, so that the other end shall cut AB in C. Drive in a peg at C, find D, the middle point of CX. With D as centre, swing DX or DC round to the position DE, cutting AB in E. Join XE, then XE is at right angles to AB.

4. To lay off an angle of 60° or 120°. Let X be the point in the line AB (Fig. 5) from which it is required to lay off an angle of 60°. Take any point C in AB towards that end of the line from which the angle of 60° is to be drawn. Take a tape or cord twice the length of XC, and fasten the ends to X and C. Seize it at the middle and draw the bight out taut to E. Then the angle EXC is 60° and AXE is 120°.

5. To bisect a given angle. Let ABC be the angle which it is required to bisect (Fig. 6). In AB take any point D. Fasten the end of the tape at D, and take it round B and back again to D. With the length thus found mark E in BC and make the loose end fast at E. Take the centre of the tape from B and stretch it tight in the position DFE. BF will bisect the angle ABC.

6. To lay out an angle equal to a given angle. Let X be the point in the straight line AB (Fig. 7), from which it is desired to lay off an angle equal to the angle DEC. In the bounding lines of the angle DEC take any two points DC, and from X measure XG equal to EC. Take a tape equal to CDE. Put the ends at C and E, and make the tape cover CDE. Holding the tape by the point above D, transfer the ends which were at E and C to X and G respectively, and pull the tape taut. Then the point which had been at D will be at some point F, and the angle FXG will equal the angle DEC.

7. To find the distance between any two points A and B when it cannot be measured directly. From B (Fig. 8) lay off the line BD as nearly at

right angles to AB as possible, D being at any convenient distance. In BD select a point C so that BC is some multiple of CD. From D lay off the angle BDF equal to the angle ABD, and on the opposite side of the line BD. Make DE of such a length that the point E is in line with A and C.

Then AB : BC :: DE : CD,

$$\text{or } AB = \frac{BC \times DE}{CD}.$$

CHAPTER IV.

THE SITING OF TRENCHES.

11. *Reconnaissance.*

1. The siting of trenches is an operation which can be carried out satisfactorily when conditions allow of certain facilities for reconnaissance before work is commenced.

2. When the encounter battle crystallizes into position warfare, the front system of trenches of either side register the mean high water mark of the attack, and are actually the places at which the foremost fighters have dug themselves in. The conditions allow of little latitude for siting of trenches. The individual pits, convenient shell-holes, remains of hedges, debris of walls and buildings, are gradually merged into a system, until at last there is a semblance of a continuous line.

3. Behind this, if the circumstances demand it, other lines or systems will be developed, where preliminary reconnaissance can be made. Their reconnaissance will be carried out according to the principles given in Field Service Regulations, Vol. II, and will determine the general siting of the whole defensive system, which is described in detail in Chapter VIII.

12. *The detailed siting of infantry trenches.*

1. The detailed siting of infantry trenches requires a close study of the ground in order to make the best use of its possibilities. It is rarely possible to grasp the whole of the possibilities of the features of the ground at the first attempt at siting the trenches, and the junctions of the different sectors of the defensive system will demand adjustments of the siting as first determined. Unless these adjustments are made before the trenches are traced (Sec. 71), time and labour will be wasted; the labour will become disheartened and the completion of the trenches will be delayed.

When the conditions admit, the siting should be marked out with small flags, so as to allow of alteration without waste of time and labour, and the position of the flags should be determined finally before the tracing parties are set to work.

2. The infantry trenches must be sited so as to secure the observation posts of the position from capture by the enemy in a minor operation (Plate **13**, Figs. 1 and 3), and must cover positions from which the artillery and machine guns can break up an enemy attack and can afford adequate support to the infantry manning the trenches.

3. Machine guns form the framework of any defensive system. Infantry trenches must, therefore, be sited in close co-operation with the machine gun defence and in such a manner as not to interfere with the machine gun field of fire. Suitably placed obstacles (Chapter VI), will force the attacker to adopt lines of approach which can be swept by machine gun fire.

4. The guiding principles in the siting and design of fire trenches are :--

> i. *Field of Fire.*—A fire-trench should admit of the fullest possible development of the power o the weapons used by the defenders.
>
> ii. *Protection.*—A fire-trench should restrict to the fullest possible extent the power and effect of the weapons of the attackers.
>
> iii. *Mutual support.*

5. **Field of fire.**—The ground which a system of fire-trenches is intended to cover must be swept by fire, either frontal or enfilade as the local conditions permit.

The distance within which a determined defence can stop an equally determined attack, has been sensibly reduced by the improvements in the rate of fire and accuracy of rifles, machine guns and artillery. A minimum field of fire of 100 to 150 yards is accepted as satisfactory on positions, which it is intended to hold to the last, provided that artillery observation of the enemy's advance can be obtained from some points within the position.

In positions which are lightly held, such as outpost positions, if the conditions admit of any choice of site, a field of fire of from 400 to 600 yards should be aimed at. When deciding the field of fire from a proposed trench, the eye should be at the level of the top of the parapet.

Enfilade fire is most demoralizing to the attacker and most heartening to the defender, since the attacker comes under the fire of a defender with whom he is unable to close. The alignment of the trenches, therefore, should be very irregular, following the lie of the ground, forming alternate bastions and re-entrants, running forward on spurs and back in the valleys.

6. **Protection.**—Protection is best provided by concealment of the trenches, which in addition affords opportunities for surprise.

Concealment of trenches. — The improvement in fire arms has necessitated more attention being paid to concealment of trenches, and although now, systems of trenches cannot be hidden from air photography, they can be concealed, to a great extent, from direct observation by correct design and careful siting, so that an enemy can be kept in doubt as to the portions of the position which are occupied, and the strength in which they are held.

In design, the first step towards this was the abolition of the high command parapet, and the introduction of the deep fire trench with low command. Later experience has confirmed this change of design, even though the field of fire is more readily affected by minor undulations, because the rifle or machine gun is brought nearer the ground.

In siting, trenches are concealed by using folds in the ground and natural cover, such as hedges, banks, crops, &c. Even when the general line to be held is on a forward slope, much may be done to hide individual lengths of trench by siting them on the reverse slopes of undulations of ground, while still retaining the requisite field of fire and observation of the enemy.

From the point of view of concealment the worst position for earthworks is on the sky line, or with a distant background when seen from the attackers' observation posts. Trenches when placed even well down the slope of a hill will sometimes be found to be on the sky line, when viewed from the enemy's position (Plate 13, Fig. 2). **Whenever possible, therefore, siting must be examined from the enemy's point of view.**

When it is not possible to conceal earthworks they may be sited so that it is difficult for the enemy to observe the burst of his shells, as, for instance, on a low ridge with depressions to front and rear. These depressions will render it difficult for the enemy's ground observers to see where his shells fall.

7. **Mutual support.**—Fire-trenches must be sited so that they give mutual support. By this means, dead ground in front of one trench may be covered by one to a flank ; enfilade fire can be obtained and, should any portion of the position be penetrated, the enemy may be prevented from reinforcing or exploiting the penetration.

13. Forward and reverse slope positions.

1. A forward slope position is one in which the trenches are on the slope of a hill nearest to the enemy so sited that they give the defender, from his trenches, a clear uninterrupted view of the enemy's trenches and the ground over which he must advance to the attack.

A reverse slope position is on the side of a hill furthest from the enemy and the defenders' trenches are hidden by the contour of the ground from direct ground observation by the enemy (Plate 13, Fig. 1). Before a reverse slope position can safely be taken up, positions in rear or on the flanks must be found, from which the enemy advance can be observed.

It is impossible to find a position of any extent in which the slopes are even and uniform. All irregularities of ground present either a convex or a concave surface. These irregularities offer temptations either of going too far forward on a convex slope for a good view, or of drawing back too much on a concave slope to escape enemy observation, with the result that pronounced and therefore inconvenient salients are formed in the general lines of a position.

In order to avoid these salients and to make use of those features of the ground which offer the best facilities for defence, it may be necessary to site trenches in one place on a forward slope and in another on a reverse slope.

Therefore possibilities of both forward and reverse slopes must be considered.

2. **Forward slope positions.**—When trenches can be placed some way down the forward slope, it is generally easy to site them so as to protect observing posts, giving a good view of the enemy's trenches and the ground over which he must advance to the attack, but such trenches must not be sited so far down the slope that they cannot be supported by artillery within effective range (Plate **13**, Fig. 3 and Plate **14**). Also, when siting the front line, the position of support and reserve trenches must be considered. These trenches may be concealed from ground observation by the enemy by skilful use of minor undulations. When these conditions can be fulfilled and adequate communications between the trenches is provided, a position well down a forward slope is generally difficult to attack successfully.

There is a natural tendency to place trenches on high ground : such ground is not always the best. The advantages of high ground are, that the defenders instinctively feel greater confidence, that communications are more easily concealed, that a better view of the enemy is obtained and that trenches, generally, are more easily drained. The disadvantages are that the defenders' fire is more plunging than grazing, that the position of the trenches can be located more easily by the attack when at a distance, that the assaulting infantry can be supported by the attackers' guns until a later moment, and that the enemy may work round the position and take it in flank and reverse.

3. **Reverse slope positions.**—When the slopes of the summit of a hill are gradual on the defenders' side and the crest is broad, it may be necessary to place the trenches of the battle position some distance on that side of the crest. Under these conditions the crest of the hill will screen the trenches from ground observation by the enemy's artillery observers (Plate **15**), but it is often difficult to provide the necessary field of fire and observation, and, should the enemy succeed in establishing himself between the crest of the hill and the defender's trenches, the advantage will lie, generally, with the enemy. The defender must have observation over the front slopes either from some position in rear, or from the flanks, and he must be able to bring effective fire on them.

14. *Communications and drainage.*

1. **Communication trenches** require as careful siting as fire trenches ; they must not be laid out in stereotyped zig-zags and waves. They should be sited with the main object of affording concealed approaches, but selected portions should be sited as fire trenches for flank defence, so as to form pockets in which an enemy attack penetrating the front line can be held up under fire until he can be annihilated by artillery fire or be dealt with by counter-attack. A complicated system of communication trenches should be avoided. They should provide one " up " and one " down " route per company front between the front line and support trenches, and one " up " and one " down " route per battalion front between support and reserve trenches. One communication trench per battalion up to the reserve trenches will usually suffice.

2. **Drainage.**—Drains must be dug at the same time as trenches which they are to serve, so that it is necessary to consider the drainage plan when the trenches are sited. The slopes of the ground must be used to carry off the water to the natural drainage channels. Sumps should be necessary only in very flat country, and should be considered a last resort, when no modification of the siting will induce a natural flow.

15. *Considerations summarized.*

To sum up :—

    i. Infantry trenches must be sited so as to cover artillery observation posts and battery positions. and so as not to interfere with the siting of machine gun positions.

    ii. A field of fire of 100 yards in front of each fire trench is necessary. When deciding this, the eye should be at the level of the top of the parapet.

    iii. Enfilade fire is the most effective form of fire. Provide for it, remembering that a man armed with a rifle always fires at right angles to his parapet.

    iv. Trenches on a forward slope must be within supporting distance of covered artillery positions. When siting the front line, consideration should also be given to the position for support and reserve trenches and to concealing them from the enemy's ground observers.

    v. The artillery must be able to get good observation over the ground immediately in front of the fire trenches. Where this can be obtained from higher ground, in rear or from flank positions in the neighbourhood, advantage should be taken of reverse slopes in siting trenches so that, while the enemy is under observation, the defending troops are concealed from view.

    vi. Communication trenches should be sited so as to form adequate but simple means of communication between the different fire positions. Selected portions also should be sited for flank defence.

    vii. Drainage must be considered when siting trenches.

16. *Improving and clearing the field of fire.*

1. Preparation of the foreground.—In order to comply with the condition that a field of fire of at least 100 yards is required, even in the most open countries, it will be found that a certain amount of clearing will have to be done.

This must be done in such a way as to give no assistance to the attackers in their advance or in the use of their weapons. At the same time the possibility of adapting and improving any existing cover for the use of defenders should be borne in mind. Natural obstacles, which may be left, should be such as will not interfere with counter-attack troops or screen the enemy from fire. It should be remembered that concealment of the works of the defence is a vital factor in holding them against an enemy equipped with powerful artillery.

It will be advisable first to improve the field of fire near the position and work forward as time permits ; but in case of a delaying action where fire effect at long ranges is required early, it is better to prepare for bringing fire to bear upon points at some distance from the position.

Before commencing any work, a rough estimate of the time, labour and tools required should be made so that the result aimed at may not be too ambitious. A field of fire only partially cleared may provide more effective cover than there was in its original state.

When clearing the foreground, it is frequently of advantage to leave a natural screen, concealing some portion of the position from the enemy's view. For instance, a line of trees may be left standing when clearing a wood ; these will obstruct the enemy's view, whilst offering very little hindrance to the fire of the defenders.

Hedges impede the attack and can be converted into very effective obstacles. They should seldom be entirely cleared. Thick hedges should be thinned and entangled with barbed wire, gaps being cut at intervals to give a clear view.

2. Trees.—Large scattered trees give less cover when standing than when cut down, and may sometimes be useful as range marks. Unless they can be removed, only their lower branches should be trimmed off.

3. Brushwood.—Thick brushwood, especially in the case of some tropical growths, forms a very effective obstacle. In place of clearing it altogether, portions may be left to deny special points to an enemy, to break up his attack, and to compel him to adopt particular lines of advance.

Thin brushwood, however, unless cut and entangled, can generally be easily traversed by infantry without great loss of order, and if left standing may serve to screen an advance.

4. Walls.—Walls must be dealt with on the same principles as hedges. When it is required to demolish them, they can frequently be knocked down by a party of a dozen or more men, using a trunk of a tree, or a rail, as a battering ram.

Low buildings may be similarly treated. Houses and buildings should be burnt and left standing—so that there may be no access to the upper floors, which might be useful as observat'on posts ; the entrance to the cellars must be blocked. The debris of a house or wall forms very good cover for a machine gun emplacement.

If it is decided to blow them down, it must be remembered that the amount of explosive carried in the field is limited, and that the debris of the buildings will be more valuable as concealment and protection for the cellars against artillery fire than the buildings themselves.

5. **Woods and orchards.**—It is rarely possible to undertake the wholesale clearing of a wood—the work is usually restricted to the thinning of the undergrowth and removal of lower branches—arrangements being made to deal with the enemy just after he has emerged from the wood by holding him under fire with suitable obstacles.

Wide rides may be cut if time permits. These rides are like peep-holes cut through the wall of a house into the rooms beyond. The rides combined with a wire obstacle run obliquely through the wood may often assist in recording and checking the progress of the enemy (Plate **16**, Fig. 1).

If the wood is heavily indented on the side of the defence, the indentation may be exaggerated ; by this means the enemy advancing through the wood may be induced to " bunch " at the salients " A " before emerging, and losses can be inflicted if the defenders are alert (Plate **16**, Fig. 2).

6. **Crops.**—Grain crops must be treated in the same way as woods. There is never time to clear the ground entirely, but with the help of cutting machines, rides and indentations are quickly made.

Clearing crops with sickles and scythes is a very slow process, and requires skilled reapers.

Range marks should be provided, and should be placed on that side of large trees, houses, banks, &c., which is only visible to the defence. The simplest arrangement consists of one white object per 100 yards range ; 500 yards may be denoted by the sign V, made with two boards, poles, &c., and 1,000 yards by the sign X, intermediate hundreds being indicated by single objects in addition, as above described.

Every soldier should, in addition, know the ranges to points under fire from his post, which are likely to be traversed by the enemy. These points should not be selected merely because they are prominent.

CHAPTER V.

MACHINE GUN EMPLACEMENTS.

17. *General principles.*

1. The importance of the machine gun, and its great fire power, are now so fully realised, that one of the first cares

of a commander, who offers or accepts battle, will be to destroy or neutralize the hostile machine guns opposed to him. It is essential, therefore, that, both in attack and defence, machine guns are provided with the most complete concealment and protection, which the circumstances of the case will allow.

2. The degree of protection provided will vary between concealment from ground and air observation, which in open warfare will be often the only protection possible, and the concrete emplacement or deep dug-out, such as can be constructed in position warfare, or when ample time and material are available.

3. Machine gun emplacements may be, therefore, classed under two headings :—

    i. Hasty emplacements.

    ii. Deliberate emplacements.

4. The responsibility for siting machine gun positions rests with machine gun commanders, acting under the orders of the commander of the force of which they form a part. It is the duty of the commander to define the task which he wishes the machine guns to perform, or in the case of a defensive action, the ground which he intends to defend ; it is the duty of the machine gun commander to select the best positions available for carrying out his commander's intentions, and to arrange for the construction of emplacements most suitable to the circumstances of the case. The type of emplacement selected will depend on the rôle of the gun, the lie of the ground, the time and labour available, the nature of the soil, and the proximity of the enemy.

Close co-operation between commanders and their machine gun commanders is particularly important when siting new field works or new wire, in order to ensure that the best use possible is made of the available ground from a machine gun point of view, and that the minimum number of machine guns are used for its defence.

New trenches dug without considering machine gun positions may not only blind the machine guns and limit their field of fire, but also afford covered approaches to the enemy leading to the machine gun positions.

5. The considerations governing the siting of machine gun positions are given in detail in " Machine Gun Training," Chapter XVIII, which must be studied in conjunction with this chapter. These considerations may be summarized as follows :—

    i. The site and type of the emplacement selected must give the field of fire necessary to fulfil the tactical requirements.

    ii. They must be concealed from observation, from the ground and air, so as to secure surprise, and to avoid discovery and subsequent destruction by hostile fire.

    iii. Protection, in the form of shelters, dug-outs, &c., is required for the gun and personnel against their destruction by fire not specially directed against the position, such as area bombardments.

        These conditions are difficult to reconcile, and the choice of site will be a compromise.

6. In open warfare there will seldom be time for elaborate emplacements, and guns will have to rely entirely on concealment for their protection.

7. In position warfare more elaborate emplacements can be provided. The greatest care must be exercised to prevent these being discovered, either in process of construction or when they are in occupation by the machine gunners.

In the forward area, owing to the difficulty of constructing and concealing strong works, the type of emplacement will be that which can be concealed most easily, and, owing to the likelihood of such positions being surrounded in the case of determined hostile attack, emplacements for all-round fire should be provided as described in Sec. **20** with a shell slit for the personnel (Plate **46**). Such an emplacement should be sited generally away from the trenches, to avoid the fire directed on the trenches. Access should be obtained by means of a carefully camouflaged trench or subway, to avoid overland tracks, which are very conspicuous from the air. If it is impossible to avoid making tracks, these should be beyond the position occupied to a dummy position, or to other trenches.

Further back, it will be possible to bring up special materials and construct strong emplacements. In most cases there will be also more cover, such as woods, hedges, buildings, in which the emplacements can be concealed. It will be possible, therefore, to construct splinter-proof emplacements, emplacements proof against light shells up to 4-inch (Plate **19**, Fig. 3), a shell-proof concrete emplacement, or an elaborate nest connected by underground passages. Such a position must not be surrounded by belts of high wire which would show up the presence of the work to the airman. If tactical wire is used (Sec. **21**) to force the enemy to advance in a particular direction and to bring him into the belt of machine gun fire, machine gun emplacements should not be placed in the angle of the wire, where the enemy is bound to suspect their presence ; dummy emplacements may be made at these points if time permits ; but the real emplacements should be sited in concealed" ground to a flank or in rear.

8. In most cases, positions for harassing fire, or for covering an attack, are chosen for one operation only, and need not be of such strength as is necessary for emplacements of a more permanent nature. These may consist of shell slits with open platforms for firing ; or emplacements with light splinter-proof cover and wide loophole, such as that described in Sec. **19**. In wet weather, when sustained fire is required, and when there is no opportunity for an elaborate work, a single sheet of galvanized iron, or other cover against rain, will be invaluable, and will greatly assist belt-filling.

18. *Concealment and drainage of emplacements.*

1. Many machine gun emplacements depend for their security on concealment, and in all of them it is of first importance. The subject of concealment is dealt with in Secs. **60 to 64.**

The following paragraphs, however, show how a hasty machine gun emplacement may be concealed, when no materials except those found locally are available.

2. A **camouflage screen** should be improvised, if possible, large enough to cover the whole emplacement. For an open emplacement this should be about 8 feet by 6 feet and can be made of strips of sandbags or canvas woven into wire netting. The netting should be fastened to two light poles, one at its centre and the other at one end. The screen is thrown over the emplacement with the pole at the centre over the rear edge of the platform. The other pole which is placed across the front of the emplacement can be raised to fire (Plate **18**, Fig. 2). The screen can be carried rolled round the poles.

3. Tracks are very conspicuous from the air. It is impossible to avoid making them, but they will not disclose the position of the emplacement if some procedure such as the following is adopted (Plate **19**, Fig. 2).

Lead the team past the position chosen, to another 50 or 100 yards away. Dump stores there, and occupy this as a temporary position. Meanwhile, prepare the chosen position for occupation, using the tracks already made. When the proper position is completed and occupied, make all carrying parties, relief, &c., proceed past the position to the dump, round which tracks may be multiplied, and then move back carefully on their tracks to the occupied position. By this means the tracks will appear to lead past the position to the dump, in which dummy works can be made.

4. Excavated earth is conspicuous, especially from the air. Small excavations may be thrown into a shell-hole, but any considerable excavations should be removed in sandbags well away from the position and dumped round a dummy position. Approaches to a position should be constructed by sapping towards it so that the earth can be carried away along the sap.

5. Work must be concealed during its progress. For instance, if a platform or chamber has been dug out during the night, and it has not been possible to roof it over, a camouflage cover should be thrown over the work before daylight, and all tools and materials hidden under the cover.

6. Concealment of deliberate emplacements includes the concealing of all traces of work and occupation, and requires to be carefully planned on the lines laid down in Chapter IX, before any work is started.

7. Drainage and revetment of machine gun emplacements must be carried out on the same lines as the drainage and revetment of trenches described in Chapter VII.

19. *Hasty emplacements.*

1. These consist generally of an open emplacement of the dimensions shown on Plate **18**, Fig. 4.

The platform should be cut well into the bank or parapet, so that when covered the work will not appear to break the continuity of the bank. Cover for the personnel should be provided in the form of shell slits (Plate **46**).

If indirect fire is to be used, a **T**-base (Plate **21**), must be placed in position on the platform.

When further time is available, a light roof can be built over the emplacement to give cover from weather. The roof will consist of about 2 sheets of corrugated iron or boarding, supported on 3-inch by 3-inch rafters, about 7 feet long, resting on light poles or 4-inch by 4-inch scantling about 4 feet long.

The inside of the emplacement should be revetted (Chapter VII), and a box or slit loophole will be required for the gun to fire through.

Only enough earth should be thrown on to the roof to hide it. If more than a few inches of earth are used, the emplacement will collapse, when a shell bursts near it, the occupants will be buried and the gun put out of action.

2. Emplacements proof against light shell up to 4-inch, can be made if one of the shelters described in Sec. **102** is available (Plate **19**, Fig. 3), but it is generally better to rely on concealment for protection unless one of the emplacements described in Sec. **20** can be made.

The minimum dimensions of machine gun emplacements are given on Plate **17**.

The roof covering over the shelter should be :—

    i. Two feet thickness of earth.
    ii. A burster course of 1 foot to 1 foot 6 inches of hard material,
        e.g., stones, brick, &c., in sandbags.
    iii. Enough earth for concealment.

A double course of logs wired together as used in Sec. **106** is a useful addition.

A box loophole must be provided to fire through. The emplacement should be built as low in the ground as will admit of the required field of fire, otherwise it will form a very upstanding target.

3. Hasty emplacements will often be made in shell-holes. They are made on the lines indicated above, but should be as simple as possible, so that they may be concealed. Plate **18**, Figs. 1, 2 and 3 shows a type of this kind of emplacement; in Fig. 3 the camouflage cover has been removed to show the framing of a light weather-proof roof.

Drainage in this case is best effected by carrying the water off to a deeper shell-hole, but the drain must be camouflaged.

20. *Deliberate emplacements.*

1. These may be classified under the following types :—

  i. Open emplacements.

  ii. Champagne type, which is also an open type, but which gives
  dug-out accommodation for the team in addition.

  iii. Moir pill-box, which is a concrete emplacement for the gun,
  and has cover for the team in a shelter or dug-out close
  by.

  iv. Reinforced concrete pill-box.

2. Open emplacements are the same as those already described,
except that better provision can be made for the personnel.

These are required in forward areas, when it is impossible to construct
either the Champagne type or the concrete emplacement, and when
the Moir pill-box cannot be erected. When constructed in isolated
positions away from the trenches, the emplacement should be in the
form used in the Champagne type. When located in a trench position,
all that is required is a platform on which the gun mounting can stand.
In both cases, the emplacement must be camouflaged from overhead
observation.

3. **Champagne type emplacement.**—The emplacement itself
(Plate **20**) is merely a rectangular pit with revetted sides. A base is
provided over a well, giving access to the dug-out below. Emplacements of this type are generally constructed in pairs communicating
with a single dug-out. But, in order that the gun detachment may
reach the emplacement without delay, the distance between the
emplacements constructed in pairs should not exceed 35 yards.

This type of emplacement can only be used in localities where the
water level admits of the construction of dug-outs. It is especially
suitable for employment in an area normally liable to shell fire,
more particularly in sites under direct observation from the enemy's
observing posts, as it is invisible to ground observation and easily
camouflaged from the air. When under direct observation, it is most
undesirable that any splinter-proof cover should be provided, as it
may disclose the emplacement and render it more difficult to remove
casualties. A wounded man lying in the emplacement would seriously
impede the work of the gun detachment and might block the exit
from the dug-out, while the splinter-proof cover may be destroyed
by shell fire and so render exit from the dug-out difficult or impossible.

In the area not normally liable to shell fire, Champagne emplacements are also most valuable, but their construction involves the use
of much material and labour, and they should only be used to protect
points of special tactical importance. They should be used in preference to concrete emplacements wherever possible.

Details of the T-base for the open machine gun platform of a Champagne emplacement are shown on Plate **21.**

4. Moir Pill-Box.—This emplacement (Plate **22**) consists of a circular concrete block wall, supporting a dome from which is suspended a revolving steel bullet-proof shield. The gun is carried by a special mounting attached to the shield. By the addition of concrete on the dome and round the block wall, the emplacement can be made practically proof against field gun fire, and completely proof against rifle or machine gun fire. It cannot be made in any way proof against 6-inch shells, but will not, if suitably protected, be affected by a shell of that calibre bursting 3 feet or more from it.

This pill-box is suitable for use in an area not normally liable to shell fire, where it will, in the first instance, be exposed only to fire from field guns and machine guns. Against these it provides practically complete protection. The question of provision of accommodation for the detachment in this area will depend upon whether the pill-box will normally be occupied or not. In fixing sites it is essential to avoid localities, such as positions in the neighbourhood of trenches or houses, which are likely to be shelled.

In an area normally liable to shell fire, this pill-box can only be used in localities not under direct observation from the enemy's observing posts. It is further undesirable to use it in localities within the normal zone of concentrated bombardment or near places which are likely to be shelled, unless circumstances preclude the construction of a form of emplacement, such as the Champagne or concrete emplacement, which is better suited for use in such a zone. Whenever used in an area liable to shell fire, cover for the detachment must be provided. This will usually take the form of a splinter-proof shelter (Sec. **105**), either connected with, or immediately in the vicinity of, the emplacement.

The value of this emplacement lies largely in the fact that it is easily concealed from air observation and photography, and possesses an all-round traverse. It is essential, therefore, that the greatest care be taken to camouflage the site before erection, and to maintain the camouflage later. This applies equally to any shelter erected in connection with the emplacement, and especially to entrances both to the emplacement and the shelter. Concealment from ground observation can frequently be obtained by locating the emplacement in a hedge, or by the transplantation of bushes, grass, &c., to form a screen. Dummy pill-boxes can be easily constructed.

It is not desirable to use the Moir pill-box in conjunction with a dug-out as in the Champagne type. If it is proposed to provide accommodation for the detachment in a dug-out, the exit from the dug-out should not lead directly into the emplacement, but should be connected with it by a short length (about 6 feet) of trench, which must be most carefully camouflaged.

5. The more permanent types of emplacements made of reinforced concrete are not dealt with in this book ; they are described in Military Engineering, Vol. II.

CHAPTER VI.

OBSTACLES.

21. *Siting of obstacles.*

1. Obstacles are used to check or direct into certain channels the movements of enemy troops advancing to the attack and to hold them under fire as long as possible.

Obstacles are of two kinds :—

    i. Tactical.
    ii. Protective.

2. **i. Tactical obstacles** are intended to :—

    (*a*) Break up an enemy's attack formation.
    (*b*) Restrict his power of manœuvre.
    (*c*) Force his troops into positions in which they are more easily dealt with by fire, particularly machine gun fire.

These obstacles are, therefore, sited in conjunction with the machine-gun defence, and in lesser degree with the artillery defence (Plate **23**).

They usually take the form of irregular blocks of entanglement, or wired areas, such as small woods and stream beds.

3. **ii. Protective obstacles** are intended to hold the attackers under close rifle fire of the defenders.

They must be sited, therefore, in conjunction with the infantry defences. Enfilade being the most effective form of fire, the obstacles should be so sited that their outer edge is under enfilade fire from some portion of a fire trench.

They should be far enough from the trench to prevent the enemy from bombing the occupants with hand grenades, but not so far that they can be cut under cover of darkness or mist. The trace must be irregular not parallel to the trenches, but arranged in bold zig-zags, so that the obstacles are not destroyed by the same artillery barrage as the trenches. Generally, these conditions will be fulfilled by keeping the obstacle a minimum of 30 yards and a maximum of 100 yards from the trench, and they must not afford any cover to the enemy.

4. In spite of the great improvements introduced for the destruction of obstacles by artillery, experience shows that a well sited and well constructed obstacle has always some value even after the most severe bombardment.

5. Obstacles should be hidden from direct observation as far as possible in hollows and folds in the ground, behind and in hedges and ditches, below banks, or in brushwood, woods and crops.

In special cases it may be desirable to sink the obstacles in trenches, but labour is rarely available for this heavy work, and this method of concealment is usually confined to those portions of the front where breastworks have to be built—the borrow-pits are then laid out and dug with this end in view (Plate 24).

6. A sunken obstacle is of great value where defence against tank attack is required. A narrow sunken obstacle will not check the tank, but the tank will not be able to crush the wire to form a passage for the accompanying infantry. The obstacle will thus have the effect of breaking up the co-ordination of tanks and infantry and will go far towards checking the attack.

7. Obstacles should be difficult to remove or surmount, and are more effective if, in order to destroy them, the enemy is forced to carry special equipment. Special attention should be paid to their anchorages.

8. An obstacle covering a considerable area is less easily crossed or destroyed, and is less visible on aeroplane photographs than one made with the same quantity of material concentrated in a narrow belt. A uniform thickness and depth of obstacle should be avoided. Irregularity tends to break up an attack. Every opportunity should be taken to form pockets, in which the enemy will be held up under fire, by making obstacles along communication trenches (see Sec. 14).

22. Gaps.

1. The only gaps required in front line wire are a few small concealed exits for patrols ; in rear lines, however, it is important to have plenty of well-marked gaps for counter-attacking troops to advance through or in order that troops and guns retiring may not be hindered and delayed under the enemy's fire. All such gaps must be provided with knife rests, wire concertinas, &c., so that they may rapidly be closed after the passage of our rear-guards and not form an easy approach for the enemy. If a gap is to be closed with knife rests, the ends of the entanglement on each side must be square, so that a complete block is effected (Plate 25). The knife rests must be securely anchored to the entanglement or to stout pickets driven into the ground, their inner ends being provided with loops of plain wire with which they can quickly be connected together when in position across the gap. Gaps for infantry should be provided about every 100 yards ; to avoid additional gaps, they should coincide with communication trenches, where such exist. They should be zig-zagged through the obstacle zone, but should not be too complicated for a mounted man to pass.

2. Road gaps are extremely important, and must be carefully prepared for blocking (Plate 26), as they are the weakest points in the obstacle zone. Where they are numerous, the infantry gaps should be made to coincide with them as far as possible, with the exception that, in the case of important main roads, it is better to make an infantry gap 20 to 30 yards to one side, so that the passage of infantry may not interfere with the traffic.

3. Where sufficient roads and tracks do not exist, special artillery gaps must be made every half-mile ; they should go straight through the obstacle zone, should be well marked, and arrangements must be made to leave the trench undug opposite them or to bridge existing trenches ; ramps into or out of trenches should not be made, as they become impassable in wet weather.

The obstacles must be protected from damage by guns and vehicles by rows of strong posts on each side of the gap.

4. Special gaps must be made for counter-attack and no obstacle should be made without reference to the commander of the sector of defence.

5. All gaps should be well marked either by :—

    i. " GAP " boards.

    ii. Painting the posts at the sides of the gap white on the defender's side.

Every effort must be made, however, to conceal these gaps from hostile ground and aerial observation.

23. *Order of priority of work.*

1. The obstacles of a defensive position should be made in the following **order of priority** :—

    i. A continuous defensive obstacle will be made throughout, except for such gaps as are described above.

    ii. This will be deepened and thickened, the depth and thickness being varied along different portions of the front.

    iii. Tactical wire will be erected.

24. *Types of wire obstacles.*

1. The various forms of wire obstacles used in military operations are described below.

2. Barbed wire obstacles are at once the most effective and the most rapidly made.

The construction of wire obstacles is the duty of the troops holding the position to be defended, and, in order that this duty may be performed with efficiency and despatch, all ranks must be thoroughly trained in the use of the materials which may be available.

The following are the ordinary types of wire obstacle :—

    i. Emergency obstacle (French wire).

    ii. Belts of concertinas.

    iii. Low (or knee high) wire entanglement.

    iv. Double apron fence.

    v. Simple 4-strand fences for spider wire.

3. Standard French wire (concertina plain wire) is the most rapid form of entanglement. It must not be regarded as a permanent obstacle, but merely one that can be rapidly put up, and is capable of being strengthened afterwards. It is a standard to be adopted on emergency, and every man should be trained in its erection.

The pattern selected consists of two belts of French wire, one yard apart in the clear, with a horizontal barbed strand along the top of each belt; a trip wire windlassed on the front of the enemy belt; and loose wire thrown in between the belts.

The essence of a French wire entanglement is rapidity, and its chief use is in a situation when rapidity is essential. The addition of loose wire and a trip wire certainly make the entanglement more efficient, and can be made as quickly as the French wire itself can be erected (Plate **27**, Figs. 1, 2).

4. Concertina wire.—A very rapid entanglement consisting of concertinas of barbed wire, fixed by pickets, and with one horizontal wire along the top of the pickets is shown on Plate **27**, Figs. 3 and 4. It has two rather serious disadvantages, in that it requires a good deal of preparation beforehand, and entails large carrying parties.

At least two rows of concertinas should be erected (one yard apart in the clear) to form an effective entanglement. One row is not sufficient.

5. Low wire entanglement consists of a series of rows of medium pickets not less than 2 feet 6 inches high, one horizontal wire along the top of each row, one diagonal wire in each of the two bays formed by each set of three rows, and loose wire thrown into the bays (Plate **28**, Figs. 1 and 2). This entanglement depends for its efficiency on its width, which should not be less than 30 feet.

It is less conspicuous than any other entanglement, but is more difficult to put up at night.

Any form of entanglement is of little use if it is less than 2 feet 6 inches high, unless concealed in water, high grass, &c.

6. Double apron fence consists of four horizontal strands on the fence, and three, including the trip wire, on each apron.

Taking into consideration the following points :—

    i. Effectiveness.
    ii. Amount of preparation required beforehand.
    iii. Size of carrying party.
    iv. Rapidity and simplicity of erection.

The double apron fence is the best pattern of entanglement, and stands up against shell fire as well as any other. For very rapid work over long lengths, the back apron may be omitted ; the entanglement thus modified is sufficient to stop the most determined enemy attacks for a time, but it is easily damaged. The value of the entanglement lies chiefly in the front apron, which should never be omitted (Plate **28**, Figs. 3 and 4).

Belts of double-apron fences form an excellent framework for a wide obstacle ; concertinas, or loose wire can be thrown in between the bays for thickening purposes (Plate **29**).

7. Spider wire.—The spider wire shown consists of a series of cattle fences placed according to Plate **30**, so as to divide up the ground into compartments. In laying out, care should be taken that not more than two or three fences meet at one point. The method of construction is the same as for the apron fence, omitting the aprons.

8. The visibility of wire obstacles from the air depends upon the length of time the wire has been erected, because, after a short time, the difference between the surface of the ground within the wire entanglement, which has been protected from traffic and the effects of the weather, will show as a dark shadow, and this shadow will be accentuated by the light lines across it wherever there is a track through the wire. It is not the wire entanglement which shows, but the difference in the surface, e.g., increased length of grass, untrodden ploughed land, &c.

9. In eastern countries, where mirage occurs, the presence of a wire entanglement is frequently betrayed by its mirage at a height above the earth's surface.

25. *Preparations for rapid wiring.*

1. The rapidity of the work of making an obstacle depends very largely on careful preparation beforehand. The following points are essential :—

    i. The line of entanglement **must** be taped ; if this is not done the party is sure to lose direction, the natural tendency being to come nearer and nearer to one's own trench.

    ii. Dumps of wiring material should be made in convenient positions close to the work. This enables long lengths of wire to be erected in one night, and prevents the infliction of casualties by the enemy seeing progressive wiring being done night after night.

    iii. Tapes should be laid from each of these dumps to the flank of the tasks concerned.

    iv. Coils of wire should be prepared for use (para. 4).

2. Wire-cutters.—It is very seldom that there are enough wire-cutters to give a pair to every man in a wiring party. If stores have been prepared properly beforehand, there is no necessity for anybody, except the officers and N.C.Os., to have them, and the issue of wire-cutters should be strictly limited.

3. Windlassing sticks.—Every man of the wiring party should carry the helve of the entrenching implement, a short 2-foot stake, or iron rod ($\frac{1}{2}$ inch diam.). These are necessary for :—

    i. Screwing in pickets.

    ii. Running out coils of barbed wire.

    iii. Windlassing wire.

Jumping bars are only necessary when working in hard ground. They should be bound with whipcord, or a double thickness of canvas, to prevent noise.

    4. Handling of material.—Rapidity in wiring depends very largely on the ability of the men to handle wire. Men must be trained to use it with confidence and not be afraid of it. It is like a stinging nettle ; if a man is not frightened of it, and treats it as if it were a rope, it will not hurt him. If gloves are used they should be fingerless, as the fingers, especially the little ones, are apt to catch in the wire. The best sappers and men, who have had long experience in wiring, never use gloves.

The plain wires securing a coil of barbed wire must be cut and a piece of sandbag or white cloth tied to the running end of the coil in order that there shall be no difficulty in finding it at night ; the pieces of tin on the wooden drums must be broken off to prevent noise. All this should be done before material is taken forward for work.

Any temporary lashing that may be required for the transport or carrying of materials should be of twine, so that it can be cut easily in the dark. Binding wire must be reserved for permanent lashings. This is a most important point in the manufacture and use of barbed wire concertinas.

    5. Screw pickets.—The following rules should be adopted for all work with screw pickets, the standard sizes of which are given on Plate **31**.

        i. **Laying out pickets.**—Pickets must be laid so that the point of the screw faces the enemy, and indicates the spot at which the picket is to be screwed in.

        ii. **Screwing in pickets.**—It is important that the eyes of all screw pickets should face the same way, as it is then much easier to fix the wire in the eyes. Pickets must be screwed in so that the eyes are parallel to the length of the entanglement and the cut end of the loop forming the top eye faces the direction from which the men are working, i.e., the head of the task. It should be carefully explained to the men that the top eyes of some pickets are in the form of a loop and those of others terminate in an upright point. In these latter, the cut end of the loop has been straightened out, and in applying the above rule, it must be imagined that it has been bent down again.

    6. Fixing wire.—For fixing wire on the screw pickets, the following rules should be adopted (Plate **32**) :—

    i. Men fixing the wire must always work facing the enemy.

ii. To fix wire in the top eye of a long picket or the loop of an anchorage picket (Fig. 1) :—

Pull the standing end taut and slip the wire up into the eye ; continue the upward movement in a circle coming down between the body of the eye and the point (the wire is now through the eye). Then take a turn with the running end round the picket below the eye, working counter-clockwise.

iii. To fix wire in the lower eye when there is already a wire in the top eye (Fig. 2) :—

(a) Pull the standing end taut and slip the wire up into the eye. Then take the bight on the running end, pass it round the picket above the eye, then finish off by taking a turn with the bight on the running end.

(b) In the long picket, if one eye is on the opposite side of the picket to the other three, the wire must, in this case, be forced down into the eye and the bight on the running end passed round the post under the eye.

iv. All horizontal wires of an apron must be fixed to the diagonal stays by windlassing (Fig. 3).

If these rules are carried out, the wire will be firmly fixed in the eye and cannot slip up or down the post ; also, if one bay is cut, the wire in the bays on either side remains taut and does not slip through the eyes. They apply whichever way the wirers are working—from right to left or left to right.

These methods of fixing wire are found to be far more satisfactory and rapid than employing short lengths of plain wire. The latter method is slow, and the plain wire almost invariably runs short, or is forgotten or lost at night.

7. Holdfasts.—Screw anchorage pickets must be screwed in in the direction of the stay wire, or they will be drawn in the direction of the strain. In sound earth " Hair-pins " or French wire entanglement staples can be used in lieu.

26. *Drills for making wire obstacles.*

1. Drills.—Many drills have been evolved by which long lengths of good wire entanglement can be erected rapidly by well-trained squads. In practice, such squads are seldom available.

Any drill which is to be of value must be so designed that :—

i. It is easily carried out by partially trained or untrained men in the dark or under fire.

ii. Casualties can be replaced as they occur, without disorganization of work or duties.

The drill must, therefore, be as simple as possible, the ideal solution being " one man one job."

It may be found sometimes that this is not possible, owing to the necessity for keeping the party small enough to be supplied by a platoon under normal front-line conditions.

The following additional points have been considered in working out the drill given below :—

 i. Men should work in pairs or groups of three.

 ii. No one group should ever cross another in the course of its work.

 iii. All groups should work in the same direction, from one flank of the task towards the other flank.

 iv. Groups should work at intervals so that the men are not bunched.

 v. The pattern of the entanglement and method of erecting should be such that no group has to step over the wire previously erected by another group.

2. Drill for Double Apron Fence.
(9 horizontal wires).

Party : 1 N.C.O. and 10 men (no more are likely to be available from a platoon holding a front-line post).

Fall in and number : 1, 2, 3, 4, 5, 6, 7, 8, 9, 10.

Stores for 50 yards double apron fence.

20 long screw pickets	5
40 short screw pickets	5
8 coils barbed wire (approximately 100 yards each)...	8
Man loads	18

First Duty (Stores).

1, 2, 3, 4, 5, 6, 7, 8, 9, 10 ...	All numbers carry out all stores and dump at the end of task (two journeys).

Second Duty (Pickets).

1, 2, 3, 4, 5, long picket Nos. ...	1, 2, screw in long pickets three paces apart (7 ft. 6 in.) along the tape, being fed by 3, 4, 5.
6, 7, 8, 9, 10, short picket Nos.	6, 7, screw in short pickets opposite the intervals between long pickets, and 6 ft. from the fence on each side, being fed by 8, 9, 10.

THIRD DUTY (WIRE).

| 1, 2, 3, 4, 5, 6, horizontal wire Nos. | 1 and 2, 3 and 4, 5 and 6, run out and secure three horizontal fence wires, then three horizontal front apron wires, then three horizontal back apron wires. |
| 7, 8, 9, 10, ... | 7 and 8, 9 and 10 run out and secure front diagonal wire and rear diagonal wire respectively. N.B. —The horizontal fence wires must be given a start by 9 and 10. |

27. *Methods of thickening a framework of apron fence obstacles.*

1. Various means of thickening a framework of apron fences are shown on Plate **29**. The method of preparing the material is given below.

2. **Barbed wire concertinas.**—Draw a circle 4 feet in diameter. Place nine posts equally spaced round this circle and drive them in, leaving a height of 5 feet above ground ; angle iron pickets are better than wooden ones. Make a framework to fit over the top of pickets to prevent them from being forced inwards (Plate **33**). One coil is required per concertina with short lengths of plain wire for binding. Three men make the concertina. No. 1 works inside the framework, Nos. 2 and 3 run out the coil.

Construction :—

i. Take two complete turns round the nine posts with No. 12 plain wire or four turns with No. 14 wire, and bind these turns together at each interval between the posts so as to form a secure end for pulling the concertina out.

ii. Fasten the end of the barbed wire to the plain wire and take 24 turns round the posts in a spiral form, binding two consecutive turns together at every other interval.

iii. Make two turns with plain wire and make fast as in i. above.

The time required to make one concertina is 20 minutes.

The best method of preparing a concertina for carrying is shown on Plate **33**. The 5-foot laths must be lashed together tightly with twine. A man must use both hands to pull the concertina out, holding the plain wire turns at the end of the spiral.

A barbed wire concertina can be extended to a length of 18 to 20 feet, and requires to be pegged down with staples or hair-pins in the same way as French wire. To stiffen it, screw pickets can be used (as for French wire) ; the pickets are screwed in first of all, at 4 yards interval ; the concertina is then extended, dropped over the pickets and pegged down.

3. **Method of preparing loose wire.**—The task of throwing loose wire into an entanglement from a coil is a long and tedious one. It is made very much easier and quicker if the wire is coiled in a spiral form beforehand.

To do this, drive in two 3-foot stakes, 3 feet apart, and two more at right angles to them 1 foot 6 inches apart. Then wind 100 yards of barbed wire round this diamond shaped framework, gradually working it up the stakes in a spiral. Finally tie the spiral together in four places with twine and take it off the stakes.

A spiral thus made can be easily carried by a man on his shoulder in a trench.

To use it as loose wire, cut the bindings, carry the spiral on the left arm and walk along, throwing two or three coils at a time into the entanglement.

One spiral supplies enough loose wire for a bay 2 yards wide and 25 yards long. It takes two men 5 minutes to make one of these spirals, and a man can throw it in as loose wire almost as fast as he can walk. If spirals are needed in large quantities, a winch is useful and saves time and labour.

If time and opportunity to make spirals are lacking, loose wire can be placed as follows :—Uncoil a 50-yard length on the ground, cut it, pick it up with a long forked stick, twisting it to and fro, and throw it on the entanglement. Press it well down and secure it to the wires already in position by windlassing.

4. **Knife rests.**—Forms of knife rests are shown in Plate 34. They can be readily improvised. Sufficient lengths of the distance piece must be left at each end for carrying.

28. *Man loads.*

The following are found to be convenient man loads of various materials used in wire entanglements. The numbers have been worked out not only as fair loads for the average infantryman, but also that they may be in the proportions required for wiring.

Material.	No.	Average total weight.
Light screw (long) pickets, 5 feet 7 inches long, with four eyes ...	6	36 lbs.
Light screw (medium) pickets, 3 feet 9 inches long, with two eyes	8	36 ,,
Light screw (anchorage) pickets, 2 feet 1 inch long, with two eyes	16	40 ,,
Angle iron pickets, 5 feet 10¼ inches long	4	43 ,,
,, ,, 3 feet 6 inches long	6	37 ,,
Wooden posts, 5 feet long, 3½ inches diameter	4	—
Wooden pickets, 2 feet 6 inches long, 2½ inches diameter ...	16	—
Coil barbed wire, 130 yards long	1	35 ,,
French wire coils	3	42 ,,
Concertinas	1	40 ,,

29. *Miscellaneous obstacles.*

1. **A tree entanglement** may be formed by cutting trees, brushwood, the strongest timber in overgrown hedges, &c., nearly through, about 3 feet above the ground, bringing the upper parts down to the ground and interlacing and securing them by pickets. Large trees thus treated form obstacles specially useful for blocking roads ; the ends of thick branches should be pointed, and all weak places strengthened by ordinary abatis.

This is the best method of entangling the edge of a wood to prevent the enemy troops from rushing trenches behind it. Vines or hops woven together with their tops picketed to the ground form good entanglements.

The tools and time required for this class of obstacle vary according to the material of which it is formed. Axes, saws, billhooks, mallets and ropes are generally necessary.

2. **Barricades** are used to close streets, roads and bridges against a rush of enemy troops, motor machine guns, armoured cars, &c.

As a rule, they should not close the road completely, but should be made in two overlapping portions or placed where a house standing back from the general line allows a passage round them (Plate **35**, Fig. 1).

They will be rarely prepared as defensive parapets, their defence being affected by machine guns and rifle fire from hidden positions in front and in rear of the barricade (Plate **35**, Fig. 2).

They can be made of nearly any material but have the disadvantage of being opaque and thus giving the enemy cover from view.

A useful form of movable barricade against rushes by motor machine guns and armoured cars is shown on Plate **35**, Fig. 3. Carts filled with stones, &c., have been used for the same purpose. They are kept in a side road, until required, when they are run into place.

3. A method of converting the railing on an esplanade wall into an obstacle against an attempted landing (Sec. **59**), is shown on Plate **36**.

4. **Inundations** can be made in the broad flat valleys of slow running rivers or streams by damming the stream. It is important to do this at points where the greatest effect can be produced with the least labour, *e.g.*, bridges.

If the valley is a shelled area it is rapidly made an impassable obstacle for even if the water is only 6 inches above ground level it prevents the troops from avoiding the shell-holes. Loose barbed wire adds to their difficulties.

5. **Mines.**—Surface mines are used to inflict casualties on the enemy and lower his *morale*, and are usually some form of trap set off by pulling a string or cutting a wire. They are dealt with in Chapter XVIII.

6. **Tank obstacles**.—Tank mines are the best obstacle against tanks. These would be provided and laid by the engineers of the formation responsible, and are dealt with in Military Engineering, Vol. IV.

A tank cannot surmount an obstacle with a nearly vertical face of 6 feet height; a ditch 10 feet wide and 6 feet deep is an effective obstacle, but the labour of making this is generally prohibitive.

It may happen that it is possible to scarp a road bank on a hill side in such a way as to form an obstacle which may hold up a tank attack under fire.

7. The illumination of obstacles may be effected by the use of Very lights and parachute rockets.

8. Passage of obstacles.—The destruction of obstacles such as abatis wire entanglements, and barricades, prior to an assault by the infantry, is usually undertaken by the artillery with fire from guns, howitzers, and trench mortars with instantaneous fuze.

Failing this the Bangalore torpedo has been found effective. The Bangalore torpedo is an explosive charge contained in a cylindrical case—the effect of the charge is to make a gap in the obstacle the length of the torpedo. Details of this torpedo are given in Chapter XVII, Sec. **121.**

The passage of other obstacles such as inundations, streams, ditches and ravines, is dealt with in Chapter XIII under Bridging, and in Military Engineering, Vol. III.

CHAPTER VII.

DETAILS OF TRENCHES, FIRE POSITIONS, AND TRENCH ACCESSORIES.

30. *General remarks.*

1. All fire positions, trenches and works intended for occupation by troops must be designed to give the most efficient protection possible against the effect of the enemy's projectiles from all directions. This protection is afforded by :—

    i. A bullet-proof parapet against frontal and oblique fire.

    ii. Traverses, against enfilade fire, and to limit the effects of shells which burst directly in the trench.

    iii. Parados, or parapets on the rear side of the work, against reverse fire and the back blast from high explosive shells and bombs fitted with instantaneous fuze.

    iv. Trenches not less than 6 feet 6 inches wide to minimize the risk of men being buried by collapse of sides under bombardment.

    v. Shelters and dug-outs, which are described in Chapter XV.

2. The efficiency of any design depends upon the combination of trace and profile to meet the tactical and physical conditions of the ground and the probable nature of the enemy's attack.

The trace of a work is the general plan on the ground, the profile is its cross section.

31. *Fire trenches.*

1. The **trace** must not contain long straight lengths of open trench, which will be exposed to enfilade fire, except where protection against bombing is necessary (*see* Sec. **37**). The length of any one bay should, therefore, not exceed 30 feet. In special circumstances where a trench system has to be completed quickly the length may be increased to 50 feet.

Traverses must not be less than 15 feet thick, and they must overlap the rear edge of the fire bay by not less than 5 feet at ground level, so that in trenches of the trace shown on Plate **37**, Fig. 1, the fire-bay must be at least 28 feet long, if the trench is 6 feet 6 inches wide.

Besides being irregular in itself, the general line of the trace must be laid out in bold curves, so as to increase the enemy's difficulty in organizing bombardments and barrage fire.

A berm 18 inches wide should be left clear from the top edge of the trench to the toe of the parapet or parados, to prevent the collapse of the sides of the trench from the weight of the earth.

The forms of trace in general use are :—

i. The " square " trace which consists of a series of fire-bays separated by traverses at right angles to the fire-bays (Plate **37**, Fig. 1).

This type gives the best protection for all the angles are well closed in, but it is slightly extravagant in time and labour.

ii. The " bastion " trace (Plate **37**, Fig. 2) is similar to the square trace but the sides of traverses are set at about 120° with the fire-bays. This type gives good protection but is more open at the angles. It is more difficult to trace, but does not involve quite so much work over a given length of line, and is easier for traffic and fire control.

iii. The " zig-zag " trace (Plate **38**, Fig. 1) which is a number of fire-bays laid out in a series of zig-zag, of which no angle should be greater than 135°.

This trace is simple to lay out, quickly constructed, but depends for protection on its irregularity of line—for there are no traverses.

Some alternative traces based on combinations of the above are shown in Plate **38**, Fig. 2 and Plate **39**, Figs. 1, 2 and 3. The dog-leg trace (Plate **40**, Figs. 1 and 2) is very useful for a continuous line across a valley with steep sides.

2. The **profile** or section of a trench must be designed so that the trench provides :—

i. **A** position from which men firing can use their rifles effectively.

ii. A passage or communication trench, which should be deep enough and wide enough to allow of the safe passage of stretcher bearers, &c.

A typical section with the names of the parts of a trench is given on Plate **41**, Fig. 1. The height from the fire step to the top of the parapet for fire standing is shown as 4 feet 6 inches. This, however, must be modified according to circumstances :—

i. The height of the men var'es, and a first condition must be that every man must have a parapet as high as he can fire over conveniently, but no higher. Men must be trained to test the height of the parapet immediately they occupy a trench, and to add or reduce the height to suit themselves.

ii. The slope of the ground on which the trench is sited will vary. If the trench is sited to fire up hill, the parapet may be slightly higher than that of a trench on level ground ; while if sited to fire down hill, the parapet **must be lower,** if the men are to cover the ground in front with effective fire.

When consolidating a captured position or entrenching under fire, the trench will be of a section as shown in I (Plate **41**, Fig. 2), and the earth will be thrown up to form the parapet only. This section gives a trench, the bottom of which is the length of a pick helve below the original surface of the ground. **The trench must not be deepened further until it has been widened, as shown in II** (Plate **41**, Fig. 2). If an attempt is made to deepen the trench before it has been widened, the fire step disappears, so that it is impossible to fire out of the trench, and the trench becomes impassable and collapses under shell fire.

Trenches behind the front line, which are generally dug by working parties, should be dug to the full width from the beginning, provided that the tasks can be so arranged (Chapter X), that a depth of 3 feet can be dug in the first relief. No working party should be allowed to leave a trench which may have to be used as a fire trench, until that trench has been dug throughout to such a depth that it gives good cover to men firing standing in it. The advantages of digging a trench to the full width from the start are :—

i. That a proper fire step is assured.

ii. The labour of digging is lessened, because the bulk of the excavation is finished before water can collect in the trench and make the digging difficult.

iii. Trenches dug in this way to the proper slopes last much longer than narrow trenches, which rapidly disappear under the combined effect of weather and shell fire.

The completed section of the trench should be of the minimum dimensions shown in Plate **41**, Fig. 2. The parapet must be bullet-proof at the top, and the top should be as irregular as possible, provided that it does not interfere with the firer. Slopes should not be steeper than 4/1, and the fire steps should be at least 2 feet wide. The back of the trench should provide a passage at least 2 feet wide at the bottom, which should be a minimum of 6 feet below the top of the parapet.

The interior slope of the parapet should be revetted, if possible, so as to provide a firm support for the forearm of the firer.

A revetted section is shown on Plate **41**, Fig. 3. The fire step must be revetted first in all cases, since when men are firing the whole of their weight is thrown on to the rear edge of the fire step, and unless the step is wide and the edge revetted, it is very quickly destroyed and the fire bay becomes useless. Bricks, rubble, trench boards or boarding may be laid on the fire step, so as to provide a hard standing at the correct level.

The remainder of the trench should not be revetted if it will stand without revetment (Sec. **41**).

When the trench is revetted, the interior slope may in good ground be cut at a slope of 6/1, but not steeper.

The parados should be irregular at the top and 2 to 3 feet high, so as to form a background for the heads of the men in the trench.

3. Parapets must not be under cut to form shelters, recesses for ammunition, Very lights, &c. ; this practice invariably results in the collapse of the parapet and many casualties from men being buried by shell fire ; if recesses of this nature are required they must be properly lined with steel shelters, corrugated iron, timber, boxes, &c. Fire trenches must be provided with frequent exits, consisting of well-revetted steps, both to the front and rear : exits are required for the use of patrols, to facilitate the reinforcement of trenches in front, and to enable men to get out to effect repairs, engage in new work, and carry out conservancy.

4. Trenches for firing lying down should be made obliquely to the line of fire. The height over which a man can fire in this position is from 9 to 12 inches. The legs of the firer are very exposed to shrapnel, and the trench should be deepened as soon as possible.

5. The adaptation of hedges, walls and embankments as fire positions are described in Sec. **42.**

6. The construction of a fire position among shell-holes is described in Sec. **43.**

32. *Communication trenches.*

1. To afford protection from enfilade fire and to minimize exposure to shrapnel, communication trenches must be irregular in line, zig-zagged, or traversed. The winding trace (Plate **42**, Fig. 1) is best, but the curves must be sufficiently pronounced to give real protection against enfilade fire. When it can be avoided traverses should not be made in communication trenches, as they make the movement of carrying

parties difficult. If traverses are made, the best pattern is an island traverse with the trench going round it on both sides. The corners of traverses should be rounded to enable loaded men and stretchers to pass ; they are easier to revet when rounded than when square. The minimum curve in winding communication trenches so that a stretcher can be carried round it, is 16 feet radius in a trench 3 feet wide.

2. Except in such soil as solid chalk, communication trenches which are required to remain serviceable for a long time or to stand wet weather must be revetted. A berm of 18 inches must be left between the edge of the trench and the parapet. The minimum width at the bottom should be 2 feet 6 inches, and 3 feet is better. Increasing the width reduces the protection afforded, and the width of 3 feet at the bottom should seldom be exceeded. The revetted sides must be sloped at between 4/1 and 3/1. The depth of the trench from top of parapet to bottom of trench or floorboard should be 7 feet, if possible, the proportion of depth of trench to height of parapet depends on the site and facilities for drainage (Plate **42**, Figs. 2 and 3).

3. **Passing places.**—The communication trenches may be the only means of effecting reliefs in the trench system. Instances have occurred when relieving troops have stuck fast in the trenches and been unable to proceed.

Passing places, and in a long trench occasional sidings, should be arranged ; sign-posts should always be placed at the entrance to communication trenches, and at any branches off them, to show where they lead.

4. **Defence of communication trenches.**—Special arrangements must be made to prevent the enemy's bombers working down a communication trench to attack the lines behind. Any communication trench entering a support or back line trench from the front must be made straight for the last 45 yards, and Lewis gun or rifle fire provided down the straight portion (Plate **43**, Fig. 1). A dog-leg trench will do, if proper arrangements can be made for enfilading both reaches of it. Provision must be made for blocking this last 45 yards of the trench at both ends. Chevaux de frise (" knife rests ") or other wire obstacles are placed in a recess at the point where the block is to be made, so that the last man to retire can quickly pull them down into position (Plate **43**, Figs. 1 and 2). The straight length must be well wired on both sides.

5. **Communication trenches prepared for use as fire trenches** are of the utmost value for flank defence when the enemy has succeeded in penetrating the front line. T-heads or D-heads should be dug off the trench so as to form fire-bays facing in the required direction, or fire trenches should be cut across a re-entrant angle in the trench (Plate **44**): the occupants of these trenches must be protected from rifle and machine gun fire from positions in rear. A communication trench prepared for use as a fire trench should be protected on both sides by a good wire entanglement.

Trench junctions.—A communication trench should enter and leave a fire or traffic trench as shown in Plate **45**, Fig. 2 : the entrance and exit are separated by a space of 30 yards, so that one shell cannot block the communication both ways. Well revetted steps must be provided on either side at intervals of 100 to 200 yards to serve as exits.

Overland tracks, with all trench crossings properly bridged, on either side of a main communication trench, relieve congestion of traffic at night or by day when conditions are favourable.

33. *Support and reserve trenches.*

1. **Support and reserve trenches** should be similar in design to traversed fire trenches. Protection against shell fire in the form of tunnelled dug-outs or concrete shelters (Chapter XV) should be provided.

2. **Slit trenches** afford very good protection from a hostile bombardment. They are 1 to 2 feet wide and 7 feet deep, dug at right angles to and on either side of the communication trenches. They must be strutted at the top to prevent collapse, and exit steps must be provided at the end away from the communication trench. Each " slit " should be wavy in plan and long enough to hold 10 to 12 men (Plate **46**). These trenches are used also for cover for reserves, machine gun crews, and, in artillery positions, for the personnel of the guns.

34. *Breastworks.*

1. **Breastworks** are made when it is impossible to obtain cover by digging trenches ; for instance, in rocky country where there is no earth, and in marshy country where the water lies on or close to the surface. Their construction is slow and laborious. In spite of their being more conspicuous than trenches, well built earth breastworks are not damaged unduly by artillery fire, and are more easily repaired than trenches.

The trace and profile of breastworks follow the same general rules as for trenches. The parapet must be 10 feet thick at the top, the exterior slope between 1/2 and 1 3, and the borrow-pit, from which the earth for the parapet is obtained, must be traced so that a berm of 3 feet is left between the toe of the exterior slope and the edge of the pit (Plate 45. Fig. 1).

2. Breastworks must be constructed with traverses in the same manner as fire trenches, and must have a fire step, to allow of every man using his rifle over the top. The necessary amount of cover for free movement along the line (6 feet 6 inches as a minimum) can be obtained, either by building up the parapet to this height, when a raised firing step will be required, or by having the firing step at ground level and digging a narrow shallow trench immediately behind it and round the ends of traverses. A parados must be constructed to protect

the garrison from the backblast of high explosive shells. This parados should be bullet-proof (4 feet thick) at its top, and strongly revetted on both faces. It should be as high or slightly higher than the parapet. A path paved with brick or trench boarded just behind the parados is a great convenience ; it should communicate with the fire bays by openings through the parados behind at least every other traverse. The space between the breastwork and parados should be trench boarded, and drainage must be provided.

3. The labour of moving the earth, required to make a breastwork, by shovelling only, is so great that some special arrangements must be made to reduce it. Wheelbarrows, hand barrows, baskets, wheeling planks, trench boards for tracks, and horse scoops should be employed as required.

4. A breastwork once begun should be completed as quickly as possible for, while incomplete, it is very vulnerable to artillery fire. It is also important to complete the work in dry weather for the borrow-pits are likely to fill with water and progress is then very slow if not impossible.

5. A breastwork may be constructed as follows :—

> Put up two revetments of gabions or hurdles—or if using sandbags build two sandbag walls—10 feet apart ; fill in between with earth ; build up a bursting course in front ; finally make a very gentle slope to the front.

6. Breastworks constructed of sandbags are much more vulnerable to artillery fire than earth breastworks. Sandbags are used when silent work is required. A sandbag breastwork must be built in the same manner and with the same precautions as laid down for the sandbag revetments (Sec. **41**, para. 5).

35. *Sapping.*

1. **Sapping** consists in constantly advancing a trench in the direction of its length by a party, who work standing on the bottom of the trench and, by throwing up a parapet on the exposed flank and end of the trench, keep themselves under cover.

The width of a sap is just wide enough to allow one man at the face to use his tools (Plate **47**, Fig. 1).

Sapping is the method of making trenches when the fire of the enemy is too accurate to do ordinary trench work or when it is necessary to establish communications between listening or other forward posts with the front trenches.

The average rate of progress is from 2 to 3 feet an hour. The man at the face must be constantly changed. Saps should be wired in on both sides to prevent the enemy from raiding and capturing the occupants.

2. **Russian saps** are tunnels driven from 2 to 3 feet under the surface in the same way as described for dug-outs and subways in

Chapter XV. They are made to establish concealed communications between the front trenches and the forward posts, or to provide exits from the former for raiding and assaulting troops.

In position warfare, prior to an offensive, Russian saps are driven forward towards the enemy's trenches so as to enable communication to be established as quickly as possible, when these trenches have been captured. A Russian sap is converted into a communication trench by removing the top sills of the frames and allowing the supported earth to fall in. This earth must be cleared away at once, otherwise the sap soon becomes impassable and the sides of the sap must be prevented from collapsing. The side timbers may be kept in place by screwing home short screw pickets in the side of the sap about three-quarters the height of the sap every 5 feet and passing an iron rod or pipe through the eyes (Plate **47**, Fig. 3).

36. *Drainage.*

1. **Drainage** of trenches and fire positions is of the greatest importance ; if neglected, trenches collapse and disappear in bad weather. Apart from the question of convenience and health, failure to provide it, therefore, may have disastrous results on operations. More trenches are destroyed by neglect of drainage than by the enemy's fire.

The question of drainage must be carefully considered when trenches are sited. Drains should be put at the lowest point of each fold in the ground, and the bottom of the trench graded so as to fall towards them without any intermediate depressions.

Excavation of drains should be done uphill and the bottom of the trench graded before work ceases each day, so that pockets, formed by unfinished tasks, are not left to collect water.

2. **Pumps.**—Every scheme for keeping a trench system clear of water must include an ample supply of pumps : the necessary pumping parties are supplied by the garrison.

3. **Sumps** or soakage pits (Plate **48**) should not be relied on unless natural drainage is impossible. The only part of a sump which is effective is that below the level of the bottom of the trench ; unless the sump reaches a permeable stratum, it must be pumped or baled out. If a sump ceases to absorb water, it is probable that the pores of the permeable stratum have become choked with particles of mud ; if the sides of the sump are shaved off it will again absorb water. •

Sump pits must be revetted above water level with a skeleton revetment, kept in position by bracing across the sump : below water level, the pits must be revetted with brushwood, X.P.M., or corrugated iron.

When constructing a trench system, until the main sumps can be provided, it will be necessary to provide small sump pits in the trench itself : these must be well revetted, and kept clear by pumping.

4. In ground where the water level is close to the surface, the depth of the trenches must be reduced accordingly and cover obtained by increasing the height of the parapet up to a full breastwork, if necessary. In such soil, sumps are of no value.

5. In occupied trenches, the mud which is churned up by traffic will make drainage impossible, unless trench boards are laid with a clear space for the water to flow beneath them.

Trench boards should be laid as soon after digging as possible, even in dry weather for, after a heavy shower, traffic will quickly convert the bottom of the trench into a slough.

6. The **maintenance** of a drainage system is essential and must be carried out by the troops in occupation. Special trench wardens must be detailed for communication trenches, so that blocks caused by falls or shell fire may be removed without delay.

37. *Traverses.*

1. **Traverses** are strong buttresses of earth butting out from the front or the rear face of the trench, so as to split it into a series of compartments. They give protection to the garrison against enfilade fire and localize the effect of a shell or bomb bursting in the trench. For both these purposes they must be strong and solid, and not less than 15 feet thick. The top of the traverse should be higher than the parapet, so as to protect the heads of firers from enfilade and traversing fire of machine guns. The earth forming the top of the traverse should be thrown well forward and occasional forward traverses provided (Plate **39**). The traverses, however, should never be higher than the parados, or the fire bays will be marked out by them. Traverses add to the length of trench necessary to accommodate a given number of rifles, and, if they are too frequent, make supervision and control difficult. They facilitate bombing attacks along the length of the trench, as grenades can be thrown from under cover of a traverse, generally into the next bay but one. As a protection against this there should be, at intervals in the line, straight lengths of trench in which the distance between two adjacent traverses is beyond the range of a bomb thrown by hand, *i.e.*, 45 yards. The traverses at either end should be loopholed for fire inwards (Plate **49**, Fig. 1).

2. Traverses will often have to be made in a completed trench which is insufficiently traversed. To do this cut out a " **D**," the inner trace of which is 15 feet wide, and sufficiently deep to give an overlap of 5 feet. When this **D** has been dug to depth, drained and trench boarded, build two revetment walls (Sec. **41**) across the old trench and fill in the space so formed with the spoil taken from the **D** (Plate **49**, Fig. 2). Two parties of shovellers will be required, one for filling the space between the revetment walls and the other for reforming the parados.

3. **Bridge traverses** are traverses built across a trench, but which allow traffic to pass below. They are used to screen trenches which are enfiladed by the enemy, and the effect is very much the same as that of the flies in the scenery of a theatre (Plate **50**).

38. *Overhead cover, head cover, loopholes.*

1. **Overhead cover** is never used in any trench which is to be occupied as a fire trench. Overhead cover for shelters and dug-outs is dealt with in Chapter XV.

Beyond this the only cases in which overhead protection is required are behind defended walls (Sec. **42**, para. 2), or as a protection against rifle grenades in posts on the lip of a crater.

2. **Head cover.**—Hasty head cover may be provided by placing large stones or bags of shingle in the parapet ; the firer must fire obliquely across the parapet in order to get protection from hostile fire. More deliberate head cover is provided by loopholes.

3. **Loopholes.**—All firing by night, and to meet an attack whether by day or night, must be over the top of the parapet. A certain number of loopholes are necessary, however, in all trench systems for the use of snipers to inflict casualties on the enemy whenever opportunity offers, to annoy him, interfere with his work, keep him under cover, and keep down the fire of his snipers (Sec. **39**), and for observation.

Various types of loopholes are shown on Plates **51**, **52** and **53**.

The art of building loopholes so as to make them secure, invisible and convenient for firing at definite points requires most careful study and training. The chief points to be observed are :—

i. They are usually made at night and, therefore, the alignment must be sited and marked out by day.

ii. The work must be completed in one night and all signs of new work must be obliterated by daylight.

iii. The maximum amount of protection must be given to the firer. A service steel loophole plate with a metal flap to close the aperture is the best form of loophole at close quarters.

iv. The recess must give sufficient room to allow the firer to use his rifle **obliquely** through the loophole from either side (Plate **51**).

v. Loopholes must be concealed from the front, they are therefore set obliquely in the parapet.

vi. Curtains must be hung at the back of the loophole, so that its position is not established by light showing through it (Plate **97**).

vii. Concealment may be facilitated if the extreme slope of the parapet is made irregular combined with beams, timber and rubbish of all sorts thrown over it(Plate **53**).

The methods of concealing the deep shadows thrown by the loophole are described in Chapter IX.

39. *Snipers posts.*

No definite rules can be laid down as to the best position for snipers. It must be left to the ingenuity and enterprise of the snipers to discover suitable places and to utilise them skilfully. Many excellent places will be found for observation and sniping in rear of the firing line. The best time to reconnoitre for such points is during the evening light, when the enemy cannot see very far, but while it is still possible to see whether they command the view required.

A tunnel through the parapet, if the opening is carefully concealed, may prove a good sniper's post (Plate **53**). Sniper posts should be made for two men, one to observe and another to fire or make notes.

40. *Observation posts, intelligence posts, look-out posts.*

A good system of observation is of the utmost importance to the artillery and infantry in any form of warfare. Observation and intelligence posts are the eyes of the artillery and infantry commanders respectively, and the enemy will spare no pains to blind them ; any building, feature or point which is suspected as being used as an observation post will certainly become a target for his artillery (Chapter VIII), betraying the place as an observation or intelligence post.

For good observation work the observer must be comfortable. A shelf rest in front for his elbows, field glasses, &c., is required.

The post must not be too dark, otherwise the eyes of the observer are strained whenever he turns his head from the bright daylight outside to the darkness within. The rear of the post must not be open to full daylight whenever anyone enters or leaves it, otherwise the enemy can see **through** the slit and observe such movements as may take place behind it, and the daylight showing through betrays the place as an observation or intelligence post.

The bottom of the slit should be 5 feet 6 inches from floor level, to enable a tall man to use it. A small platform can easily be placed for a short man.

The observation slit should be of irregular shape and not less than 6 inches deep. If the slit is less than this, the field of view is too small unless the observer keeps his eye close up to the slit, which attitude is much too fatiguing for prolonged observation. Headroom above the slit should not be less than 6 inches.

The slit should be about 3 inches long or according to the field of view required.

Observation posts have often become useless from having been built with the slit too near the local ground level ; when the grass and weeds grow, observation is obscured, and cutting down is quite impossible in many cases owing to the proximity of the enemy and the amount of clearing required which would naturally betray the observation post.

The floor area must be as small as possible in order to reduce the labour of construction and to facilitate concealment. Thirty-six square feet is the minimum area in which work can be done properly. (Plate **54**).

For artillery it should be large enough to accommodate one observer, two telephonists, and maps ; the telephonists can be accommodated with advantage below the observer (Plate **55**).

41. *Revetment of trenches.*

1. **Slopes.**—The amount of revetment in trenches can be very much reduced if the sides of the excavations are carefully sloped. There is no difficulty about this provided the work is properly set out and explained to the men. Unless the slopes are cut smooth and uniform, rain lodges on the uneven surface and soon soaks into the earth and makes it disintegrate and fall.

A trench A,B,C,D, with well cut slopes and badly cut slopes is shown on Plate **56**, Figs. 1 and 2.

Slopes should never be cut at the same time as the excavation is being dug out ; the general principle of cutting slopes is shown on Plate **56**, Fig. 3.

Taking the side A B at a slope of 2/1 the horizontal distance from A to B, is 3 feet. It will be convenient to dig the full depth of 6 feet in two stages of 3 feet depth.

First stage, leave a step 18 inches wide at A and dig from a, vertically down for 3 feet to b.

Second stage, leave a step of 18 inches wide at b and dig from a vertically down for 3 feet to B.

Third stage, clear the steps by digging out carefully the triangular blocks, Aab and ba^1B.

This is done best by first of all digging out narrow slits W,X,Y,Z, at intervals (Plate **56**, Fig. 3) as guides and then clearing the remainder, using spades or flattened shovels.

The width of the step aA is arrived at in this way. The slope Ab is 2/1 ab = 3 feet, therefore aA = 18 inches ; similarly for the step $a^1 b$.

The side CD can be done in exactly the same way only as the slope of DC is 3/1, the width of the steps would be 1 foot because the slope of DC = 3/1 and cd = 3 feet, therefore dD = 1 foot.

2. **Selection of type of revetment.**—The principal objections to revetments in a trench are the great amount of time, labour and material required for their construction, and that, should the trench be blown in, the revetting material is difficult to clear away and obstructs traffic, this is particularly the case with corrugated iron, expanded metal and brushwood hurdles.

The upper part of a trench is most exposed to damage by shell fire and should not be revetted unless absolutely necessary ; sandbags and brushwood are most suitable as they can be cleared away with ordinary cutting tools and shovels.

The lower part of a trench is less exposed and it is convenient to revet it with some more permanent form of material such as corrugated iron, expanded metal, hurdles or brushwood. This provides a firm foundation on which to build the sandbag wall, facilitates drainage and greatly assists in clearing the trench by providing a hard surface to clear to.

Firesteps should be revetted as soon as the digging is finished. This may be done by using the short revetting frames or pickets with revetting material, such as corrugated iron, expanded metal, brushwood or fascines. Sandbags should never be used for making or revetting firesteps ; they become very slippery in wet weather and men cannot get a secure footing to fire from.

Sandbags and gabions are most useful for repair work.

3. **Revetments** are of two types :—

(a) Those which consist of a " skin " held in position against the face of the earth by fixed uprights, *e.g.*, corrugated iron, expanded metal, brushwood, hurdles supported by pickets or frames.

(b) Those which are built up like a retaining wall or dam and which hold back the earth by their own weight, *e.g.*, sandbags, sods, gabions.

4. **Type** (a).

i. **Pickets**.—If pickets are used as uprights, their feet must be driven well into sound ground at a slope of 4/1 and their heads securely anchored back so that the pressure of the earth may not force them out of position. The whole efficiency of the revetment depends on this anchorage.

Stout anchorage pickets at least 2 feet 6 inches long should be driven in sufficiently far back from the face of the revetment to be well beyond the angle of repose of the earth (Plate **57,** Figs. 1 and 2) roughly twice the height of the revetment from the face. The revetment pickets should be 2 to 3 feet apart and wired back to the anchorage pickets by at least eight strands of 14 S.W.G. wire twisted together and windlassed tight. These wires should be fastened to the anchorage picket at ground level and to the top of the revetment picket, except in the case of breastworks, when the wire should be attached to the revetment picket at a point about one-quarter of its exposed length from the top. (Plate **57,** Fig. 2). The anchorage wires must be perfectly straight.

In bad ground a second anchorage should be driven in 3 or 4 feet behind the first and the head of the latter anchored back to it.

Anchorage should, as a rule, be driven in or laid at right angles to the line of pull.

Screw pickets when used as anchorages should on the contrary be screwed in, in prolongation of the line of pull.

ii. **Revetting frames**.—(Plates **9** and **69**) provide the supports for trench boards with a drainage channel below.

In using these revetting frames the revetment must commence at the bottom of the frame (Plate **41,** Fig. 3). The trench must be dug deep enough to allow this to be done. Trenches in which it is intended to use these frames should be checked by means of templets during excavation, 3 inches clearance being allowed on each side of the frames, earth must be tightly packed against the revetting material, especially below the struts of the revetting frames.

The frames must not be fitted into slots cut in the bottom of the trench.

The frames must be upright and properly aligned so that each takes its share of the earth pressure.

The distance apart of revetment pickets and revetting frames depends on the stiffness of the revetting material used ; in ordinary ground they should be from 2 to 3 feet apart.

iii. **Corrugated iron** is the strongest and most durable revetting material. The sheets should overlap each other by 3 inches, an upright being placed at each overlap and opposite the centre of each sheet.

In waterlogged ground it is advisable to make weep-holes in the sheets to assist in the drainage of the earth behind.

iv. **Expanded metal** should be used in the form of hurdles (Plate **7**) or gabions (Plate **8**). If hurdles are not available, 4 inches by 1 inch longitudinal battens can be nailed to the expanded metal to give it some measure of rigidity : in either case the expanded metal and not the battens should be placed next the earth. When revetting frames are used as uprights, these battens should be nailed or wired to them ; this helps to keep the frames in position and strengthens the revetment. For convenience in carrying the X.P.M. may be rolled up, and the battens nailed on at the site where they will be used. Plate **8** shows a wire hook which is useful for lacing together plain X.P.M. sheets in revetments or gabions.

If expanded metal is used for revetting firesteps, a picket or plank must be fixed along the edge of the firestep to prevent it from being trodden down.

v. **Brushwood.**—Remove the leaves and twigs and pack the brushwood in behind uprights spaced at about 2 to 3 feet intervals. The brushwood need not be woven between the uprights. Brushwood is very bulky and requires a great deal of transport ; it is most useful for work near woods, where it will be obtained when clearing the field of fire.

If brushwood hurdles are used, they must be held in position by proper uprights ; it is not sufficient to anchor back the pickets of the hurdle itself.

vi. **Planking.**—Forest planking or half-round waste outside cuts of logs may be used for revetting in positions not exposed to artillery fire.

vii. **Wire netting and canvas** are almost useless—they bulge excessively under the pressure of the earth.

5. Type (b.)

i. Sandbags.—Sandbags should be three-quarters filled with earth or sand so that when beaten with a shovel to a rectangular shape they measure about 20 by 10 by 5 inches. Hard ground, gravel, chalk, bricks, &c., must be broken small so that when the sandbag is filled the material can be shaken into a compact, pliant mass. In this case the sandbags must not be beaten.

A filling party should consist of three men, two holding and tying and one shovelling : building parties should work in pairs. The size of the carrying party connecting the filling and building parties depends on the distance that the bags must be carried. Three men should fill and two men should lay 60 bags per hour, so that the carrying party should be sufficient to deal with this number of bags.

Sandbags rot quickly and should not be used where the revetment is required to stand for a long period if other material is available. They are used chiefly in the repair of damaged parapets and for quiet work close to the enemy.

The revetments must take the form of a properly built and bonded retaining wall, with the thickness at the base proportional to the height ; it must not be a mere veneer or skin of bags. The common faults in building sandbags are shown on Plate 58.

The most important part of a sandbag revetment is the foundation ; this must be in sound ground and must be excavated so as to be perpendicular to the slope of the face of the revetment. The " batter " (slope) for a sandbag revetment is 4 in 1 ; the foundation must, therefore, be cut to a slope of 1 in 4.

In unsound ground the foundation of the sandbag must be revetted ; short revetting frames are best for this purpose.

A bag is said to be a " stretcher " when it is laid with its longest side parallel to the face of the wall, and a " header " when at right angles to the face. The bond used in sandbagging is known as English Bond, *i.e.,* alternate course of headers and stretchers (Plate 58). The first course should be headers. Headers should be laid with the chokes (tied ends) towards the parapet ; if a stretcher has only one seam, this also should be turned towards the parapet.

ii. Sods.—Sods should be laid in the same way as sandbags, grass downwards ; if available a split picket should be driven through each sod to hold it in position and strengthen the revetment. Bundles of heather and grass can be used in the same way for temporary work.

iii. Gabions.—Gabions should be set at a batter of 4 in 1, on a foundation as described in sub-para. i above. They should be filled solid and kept steady by earth thrown up against them at the same time.

If sandbags are used above the gabions, they must be set back behind their edge to prevent the expanded metal cutting the bottom layer of sandbags.

42. *Defence of hedges, walls, &c.*

1. **Hedges.**—It is most important to conceal the fact that the hedge is occupied ; for this reason, the back of the hedge must be cleared so that the upper branches may form a screen against aeroplanes, and the front of the hedge must be cleared so that the defender can see and fire through without being seen, and the foliage or branches hides the earth excavated and thrown to the front to make the parapet. The front of the trench must be close to the centre of the hedge, so that its thick stems may interfere with the firer as little as possible (Plate 59).

If the ditch is on the enemy's side of the hedge, excavated earth can be thrown into it and then covered with the trimmings of the hedge.

Hedges should be trimmed in front with a jack knife and not in a wholesale manner with billhooks or hand-axes.

The roots of hedges will make the work of excavation difficult.

2. **Walls.**—It is rarely advisable to occupy walls if the enemy's artillery is efficient—machine gun fire would usually enforce the use of loop-holes. In any case, men occupying walls or buildings should be protected from falling debris by overhead cover (Plate 55, Fig. 5).

3. **Embankments and cutting.**—Fire positions in these features are easily made by cutting " D " and " T " heads into the bank ; the chief point to be remembered is that protection from the back burst of shells must be provided as shown on Plate 60.

D-heads should be 30 feet long, so that both entrances cannot be destroyed by one shell.

4. **Blockhouses** are small isolated defensible buildings suitable for occupation if the enemy has no artillery or machine guns.

They are usually made of a combination of wood or corrugated iron and shingle (Plates 61 and 62).

The dwarf rubble wall or bank of earth supports two corrugated iron skins 6 inches apart, packed with hard shingle ; loopholes are provided as shown. The roof is composed of corrugated iron or a tent, the whole structure being supported on a wooden frame. The entrance is partially underground and protected by a traverse (Plate 61, Fig. 2).

A circular fire trench should be provided for the sentry, and a wire entanglement constructed round the completed blockhouse ; provision must be made for the storage of water (Plate 61, Fig. 3).

A log blockhouse, suitable in a wooded country, is shown on Plate 62.

5. **Stockades** are improvised defensive walls ; their design and thickness will depend upon the nature of the weapons of the enemy.

Their loopholes must be arranged so as to bring flanking fire on to each face, and at such a height that the enemy cannot use them from outside, *i.e.*, 6 feet 6 inches from ground level (Plate 63, Fig. 1).

6. **Sangars** have the same limited use as other walls and stockades (Plate 63, Fig. 2).

43. *Shell-hole and crater defences.*

1. In heavily shelled ground the shell-holes can be quickly converted into a hasty defensive position. These positions should be organized in depth to afford material support by flanking fire.

It is almost impossible to conceal organized shell-holes from low flying aeroplanes, and they are easily detected in air photos, but they can be concealed from ground observers by carrying out the following instructions :—

No fresh earth is to be thrown up.

The lip of a shell-hole is not to be disturbed.

Excavated earth is to be dumped in neighbouring shell-holes if not required for cover.

Routes to occupied shell-holes are to be constantly changed.

Connecting trenches must be narrow and camouflaged.

Drainage of shell-holes, though a difficult problem, is of vital importance. Small shell-holes may be connected by drains to deeper holes or it may only be possible to dig a sump in the bottom of each hole covering it with a trench board (Plate **64**, Figs. 1 and 2).

In sodden ground fresh shell-holes are drier and easier to work in than old ones, but in drier soils the sides of old shell-holes are more settled and are free from gas.

Where the shell-holes are not waterlogged large deep ones can be selected and rapidly organized for defence. Fire positions should be made first, either by cutting away the front face or, if the soil is much disintegrated, by digging slits outwards. Labour is saved if the cutting line be taken about half-way down the slope of the shell-hole ; a deeper cut can be made in a shorter time than if the forward edge of the shell-hole be taken as the cutting line. If the firestep is to stand unrevetted it must be dug in the more solid earth beyond the radius of rupture ; but revetting is necessary in any ground which has been subjected to heavy shelling. In all work it is most important to avoid undercutting, unless the soil is properly supported.

Later, when further work is possible, the position can be made stronger either by digging out a **T**-head in front or by widening the first firing position into a small crescent-shaped trench (Plate **64**, Figs. 3 and 4).

Where the shell-holes are contiguous they should be selected in pairs to accommodate a section, and the rear faces of the pair joined up, thus making the ground between them into a traverse (Plate **65**, Fig. 1). Plate **65**, Fig. 2, shows a shell-hole position. Details of a Lewis gun emplacement are shown on Plate **66**.

Owing to difficulties of command and communication, the organization of shell-hole defences can only be considered a temporary expedient, during the construction of a trench system in ground which has been as little damaged by shell fire as can be found on a suitable alignment.

2. Defence of craters.—When two opposing forces settle down into the period of position warfare it is possible that one or other of them will commence to mine. If craters are formed it is important to occupy them at once, because of the increased observation which is usually obtained from posts on the lip of the crater.

The occupation of these posts should be planned in consultation with the officer in charge of the mining of the sector.

Two schemes are shown on Plates **67** and **68**, which show the nature of the work required. Each post must be carefully protected with wire (Plate **68**); shelters must be provided for the men of the post giving protection against rifle grenades and light bombs.

The inside of the crater must be watched as well as the outside of the lip; this is done by observation tunnels (Plate **67**). The posts on the crater must be connected with the trench system by communication trenches.

The construction of a crater position absorbs a large amount of labour, especially in carrying parties, owing to the heavy material required to ensure that the revetments withstand the strains brought about by the settling of the debris. It is usually impossible to obtain good holding ground for anchorages and frames consisting of two uprights and a ground sill, or special deep revetting frames (Plate **69**) must be used: ' pit props, 6 inches in diameter, are the most suitable material.

44. *Defence against tank attack.*

Field defences in themselves present no difficulties to tanks and the labour required to make artificial obstacles such as pits and ditches is prohibitive.

Field works constructed for defence against attack by tanks will be, therefore, confined to the laying of tank mines dealt with in Chapter XVIII and the improvement of natural obstacles, such as sunken roads, banks and streams and marshes which is dealt with in Chapter VI, and the construction of shell-proof gun positions described in Military Engineering, Vol. II. .

Sunken wire obstacles will be of value against tank and infantry attacks (*see* Sec. 21, para. 6).

45. *Defence against gas.*

Defence against gas will be confined to rendering gas-proof, shelters, dug-outs and cellars by means of specially designed curtains and air filters.

The entrances to all dug-outs, shelters and mine shafts within the alert and ready zones should, if possible, be provided with gas-tight doors or with curtains of anti-gas material, fitted so as to give a good joint at the sides and bottom of the doorway, thus stopping all draughts. If two curtains are used with a space between them complete protection is obtained, and it is possible to enter or leave the dug-out without introducing appreciable quantities of gas.

A frame of 4-inch by 1-inch timber, covered with anti-gas material, is fixed flush with the wall, sloping outwards at an angle ot 20° from the vertical. Anti-gas material is cut to the required size, so that when fastened to the top of the frame it will close the entrance completely with about 9 inches resting on the ground. Three pairs of laths are nailed horizontally to the curtain to keep it stretched. The lath on the underside must be 1 foot shorter than the one on the front, so as to clear the frame (Plate **71,** Fig. 1). The lowest of the laths should be 4 inches from the floor. Two curtains should be provided, as shown in the diagrams. The frame for the inner curtain should slope inwards, as shown on Plate **70.** All wires and pipes must pass through the frame, which may be widened on one side to allow of this, and the hole through which they pass must be made gas-tight. They must not interfere in any way with the adjustment of the curtain (Plate **71,** Fig. 2). The curtains should be not less than 3 feet apart, so as to allow a man to stand between them and adjust one before raising the other. The distance must be increased for dressing stations to allow stretcher cases to be carried in.

Frames for gas curtains should be built into the entrances of pill-boxes and other shelters while the entrances are in course of construction. Machine gun loopholes in pill-boxes should be lined with wood on the inside edges, so that they may be closed with frames covered with anti-gas material. Openings in the sides or roofs of shelters and cellars must be provided with curtains or closed with sandbags, so that no gas can enter. Care must be taken to provide means for closing ventilating shafts and flues.

When not in use curtains must be kept rolled.

46. *Cover for anti-aircraft guns and searchlights.*

Experience has shown that anti-aircraft defences are liable to repeated attacks by aircraft and they must, therefore, be provided with suitable protection.

For the personnel shelters or dug-outs, as described in Chapter XV, will be constucted. The lorries will be best protected by being run into a cutting in the bank of a sunken road.

Plate **72** shows a type of emplacement for a 90 c.m. or 120 c.m. anti-aircraft searchlight, suitable for skew gear pipe control with telescope.

47. *Field defences for artillery.*

1. In the following paragraphs are described only those measures for protection which apply exclusively to the artillery, and are generally defences suitable for position warfare, which must be modified to suit varying conditions.

The construction of any artillery position should be carried out in the following order of importance :—

    i. Concealment.

    ii. Cover for headquarters and telephone.

    iii. Cover for personnel and ammunition.

2. Before any work is begun, the site must be camouflaged on a sufficient scale to conceal every indication of work. The position will be located if work is started before the camouflage is complete, and time, labour and material spent on camouflage subsequently will be wasted. The methods to be employed are given in Chapter IX. The cover should be progressive, depending on the time, labour and material available, from weather-proof, splinter-proof, to shell-proof.

3. The command post and dug-out for the wireless operator must always take precedence of cover for the gun personnel and ammunition.

A command post, including the telephone, can be accommodated in a space 9 feet by 9 feet by 6 feet high (Plate **73**), but a separate chamber for the telephone is a great advantage.

A convenient position as regards the battery is from 20 to 60 yards in rear and to one flank of it. It should be provided with an entrance on the sheltered side and a pulpit from which to megaphone to the guns.

When the battery is split up into sections, each section commander will require a similar post.

The wireless chamber, when provided, should be sited clear of the battery position, with separate inter-communication to it.

4. **Shelters for personnel.**--The following instructions apply specially to shelters for artillery personnel.

Neither officers nor men are to be accommodated *en bloc* in any dug-out which is not shell-proof.

Dug-outs for cooks, men off duty and spare telephones may be made well clear of the battery.

Dug-outs near the gun must not lead direct into a closed gun pit, owing to the risk of gas poisoning from the carbon monoxide produced during firing.

The dug-outs near gun pits must be protected as described in Chapter XV.

5. **Shelters for ammunition.**—There are no fixed sizes for ammunition shelters, and any available shelter can be used, provided that not more than 50 large and 100 small shells are stored together in one shelter, and that large quantities are divided up by traverses not less than 4 feet thick. The important point is that the cartridges, other than those of fixed ammunition, should be separate from the shell.

Recesses for 18-pr. ammunition may be made in the gun pit. These can be made with wooden uprights and shelves made of angle iron pickets. In the gun pit, not more than three shelves should be placed in each recess, and not more than two layers of ammunition on each shelf. The shells must be kept from contact with the ground. The recesses must, therefore, be floored with planks, trench boards, brushwood, &c., and lined with canvas or boarding.

A type of shelter for dry ground is shown on Plate **74**. It should not be constructed in the communication trench between guns. On wet ground a shelter can be built of the type shown on Plate **75**.

Ammunition for field guns can be also stored in a trench as described below for ammunition for medium and heavy artillery. Ammunition is not to be stored on the berm of a trench ; this will make the sides collapse.

Recesses for 60-pr., 6-inch, 8-inch and 9·2-inch howitzer cartridges may be made :—

    i. In banks.

    ii. In trenches specially dug for them.

They must not be stored in the emplacements. Recesses in banks should be made at 6 feet interval and only sufficiently deep to allow one row of 8-inch cartridge cylinders lying on their sides, or one row of 6-inch metal-lined cases—the height depends on the number of cartridges to be stored which may be as follows :—

    8-inch, 30 to 40 per recess.

    6-inch, 60 per recess.

Cartridges should be kept about 12 inches above the ground. If wood is used for this purpose, it should be covered with tin or corrugated iron as a precaution against fire. Cartridges must be screened from the sun's rays.

If no bank is near the battery, the cartridge recesses must be made in a similar manner along the sides of a trench. The trench must be deep enough to allow a man to walk upright in it. Entrances must be provided at each end to ensure a through draught and they must be stepped.

The trench and recesses must be covered with a weather-proof roof, and the trench must be drained.

When cartridges are kept in metal-lined cases these should be laid on their sides to prevent entry of damp and rain when the luting is removed from the lid, and to facilitate the extraction of the cartridges. A splinter-proof ammunition recess is shown on Plate **76**. Shell-proof protection for ammunition, if required, will be provided on the same lines as that for personnel described in Chapter XV.

6. **Gun emplacements** may be classed as :—

    i. Camouflaged emplacements without any protection.

    ii. Camouflaged emplacements with splinter-proof protection.

    iii. Camouflaged emplacements with shell-proof protection.

The last case requires special material and skilled labour, and is dealt with in Military Engineering, Vol. II.

The extent to which it is possible to sink a gun below ground level depends on the nature of the ground and the minimum range at which the gun has to fire. All emplacements must be made so that the gun can be run in and out without difficulty. Drainage must be provided for (Sec. **36**).

Diagrams showing the minimum vertical and horizontal dimensions of emplacements of various types of guns, &c., are shown on Plate **77.** These may be modified according to the traverse required.

Embrasures and entrances of emplacements can be protected by the methods shown on Plate **78.** The protection can be adjusted according to the switch required. A wooden framework hung of 9-inch by 3-inch timber, with six layers of wire netting nailed on in front and four layers behind, will also prevent splinters entering the embrasure.

The embrasures and entrances of covered gun pits and splinter-proof screens must be covered with light removable screens to hide the shadows which are invariably cast (Chapter IX).

Splinter-proof overhead protection can be given to the lighter natures of guns and howitzers provided the span of their emplacements does not exceed that of the large steel shelter.

For the larger guns and howitzers the span of the emplacement is so large that very heavy steel joists are required to carry the weight of the roof, and the effect of a direct hit on a splinter-proof roof of this nature would do more damage to gun and personnel than if the emplacement were open.

Splinter-proof protection at the sides and rear of all emplacements should be provided whenever possible.

7. **Reverberation.**—Gun pits constructed of elephant shelters, or any gun pit, the roof and walls of which are curved, are much more noisy than rectangular pits.

The gun, being along the centre line of the curvature, is the centre to which the sound returns after striking the sides. The reverberation which is set up is distressing to the gun detachments and especially to the gun layers.

Reverberation can be reduced by making the forward portion of the pit curved and the rear rectangular.

8. **Platforms.**—A platform for any nature of gun or howitzer consists of two parts—a support for the trail and a bed for the wheels. The former is the more important and work should always be done on this first. Both are essential if prolonged firing is to be carried out from the same position and, for equipments which do not carry their own platforms, must be improvised from the material available.

Trail-support.—This should consist of two parts:—

i. A fixed support, which may be of concrete or pit props firmly fixed with pickets. It should be circular in shape so that the thrust is always at right angles to the tangent to the curve at the point of support.

ii. A cushion, which may be made of fascines or sandbags. The former are more satisfactory in action and do not require renewal so often (Plate **79**).

Wheel-bed.—This may be made of any of the following materials :—

(*a*) Natural earth.
(*b*) Rubble or brick well rammed.
(*c*) Wicker.
(*d*) Wood.
(*e*) Concrete.

(*a*) **Natural earth,** under the best conditions, will only stand a limited number of rounds, dependent on the nature and weight of the gun.

(*b*) **Rubble or brick,** well rammed, forms an excellent wheel-bed (Plates **80** and **81**). It should be at least 1 foot deep, and may be extended across the pit, or may be packed in wooden boxes to form a bed for each wheel, if there is a difficulty in obtaining material.

(*c*) **Wicker** or fascines are useful as a temporary measure in very wet ground, but they are not durable and are unsteady. " Mats, gun, wheel " answer the same purpose.

(*d*) **Wood** wheel-beds facilitate the man-handling of the gun, are portable and can be quickly put together.

They should be made of 9-inch by 3-inch planks, dogged together not nailed (Plate **82**).

Wheel guides may be added to keep the gun in its line of fire and prevent " frogging" (Plate **82**).

(*e*) **Concrete** forms a good wheel-bed if time is available to allow it to set; it is useless unless well mixed. The concrete may be extended across the pit or a separate bed for each wheel may be made.

9. Details of platforms for the various natures. (*a*) **Light artillery.**—The cushion for the trail support may be fixed with wire in the angle between the spade and trail-eye.

With the 18-pr. Mk. II equipment, if the length of time it is expected to occupy the position justifies it, better results will be obtained if two or three trail supports at varying heights are prepared for use at different angles of elevation (Plate **80**).

(*b*) **Medium artillery. 60-pr.**—Pit-props make a suitable trail support. As a temporary measure, fascines laid under the trail in front of and behind the spade are of great assistance in preventing the trail from burying itself.

The Mk. I carriage is steadier on a brick rubble wheel-bed, but wood or brick is equally suitable with the Mk. II carriage.

If whole bricks are obtainable, they may be placed on edge on a layer of expanded metal with brick rubble on top. It is advisable to enclose the whole in a wooden box held in place by pickets.

6-inch howitzer.—Brick rubble is the best trail support with a fascine placed in the spade as a cushion, as the trail is liable to buckle if the support is too firm. The rubble requires constantly replenishing. In soft ground it may be necessary to place a baulk in the spade to prevent the trail burying itself.

In wet ground a bed constructed of one layer of 9-inch by 3-inch timber is necessary (Plate 83).

(c) **Heavy artillery. 8-in. howitzer.**—Mark VI and upwards carry their own platform, but for the earlier Marks of carriage a double decked platform of wood is necessary (Plate 84).

9·2-inch howitzer.—This howitzer carries its own platform in the form of firing beams, but in many cases this must be supplemented by a bed of hard material.

CHAPTER VIII.

A DEFENSIVE SYSTEM.

48. *General remarks.*

1. In order to conduct a successful defence of any locality or area, the defender must fight in front of, and not on, those features of ground which are vital to the defence. Any ground which affects the security of the position must be denied to the enemy.

2. A modern defensive system on a front of any importance consists of :—

    i. A battle position.
    ii. An outpost zone.

3. The general line followed by a defensive position depends upon broad considerations such as the defence of a frontier ; a vital line of communications ; a sea base ; a railway centre ; an area of country containing national resources such as mines, factories, &c. These form the objective to be covered.

4. The principles affecting the selection of a position are given in F.S.R., Vol. II, Sec. **129** The following considerations affect the more detailed siting of the position :—

    i. *Co-operation.*—The ground selected must afford facilities for all arms to co-operate in its defence. The framework of the defence consists of machine gun positions sited independently of the trenches. Due weight must be given to facilities for artillery support, both as regards observation of fire and positions for the guns.
    ii. *Observation.*—Observation over the immediate foreground and the distant approaches is necessary.
    iii. *Concealment.*—Concealment of defences from observation is vital, but against reconnaissance from the air, made either direct or by photography, is extremely difficult. It is

therefore of the greatest importance to devise surprise dispositions such as well concealed machine guns, and to make use of all forms of cover such as hedges, reverse slopes, &c.

iv. *Depth.*—The depth of the defences must be such as to prevent the enemy's field artillery destroying by bombardment the defences of the battle position and to prevent the enemy adequately supporting attacks on the rear of the battle position without moving forward his artillery.

v. *Communications.*—Good communications of all sorts covered from hostile view where possible, are essential to the defence. Trench tracks and roads are necessary for the movements of reserves from rear to front and laterally, while roads, tramlines, and railways facilitate supply of ammunition and materials necessary for the construction of defences. These facilities should, where possible, be denied to the enemy.

vi. *Localities.*—The position must be organized in defended localities protecting the important tactical features. Intermediate defences are only required for inter-communication purposes or for night dispositions : and to deceive the enemy's artillery as to the exact posts held by the infantry of the defence.

vii. *Obstacles.*—The position must enable obstacles to be sited so that the enemy are checked by them under the fire of the defence. Anti-tank obstacles are of especial importance.

5. **Reconnaissance.**—Reconnaissance and siting of a defensive system is carried out by representatives of the general staff, the artillery, the engineers and the machine guns working together.

The co-ordination of the reconnaissance is the responsibility of the general staff.

It is necessary to study a map of the area in question before going on to the ground ; a layered map 1/100,000, or $\frac{1}{2}$ inch to 1 mile is suitable. A plan or several alternative proposals should be prepared after consideration of the important tactical features, observation, lines of approach, and ground that must be denied to the enemy. The approximate battle position proposed should be marked on the map.

It will save much time if flags can be erected beforehand in prominent places along the proposed battle position, in order to assist the reconnoitring officers in picking up the general direction of the position from time to time.

Consideration should be given at this stage to the division of the position into sectors in accordance with the proposed allotment of formations (corps, divisions and brigades) for its defence.

The more exact alignment of the battle position and the siting of the defended localities for its defence will be the first task on the ground. It is important to ensure that the siting of the defences is co-ordinated at the junctions of formations.

Wooden pickets or small flags should be used to mark the machine gun positions, the wire obstacles, and the trenches of the important localities, as all these items of the defence must be carefully co-ordinated.

49. *Organization of the battle position.*

1. The battle position is organized in a series of defended localities which form the fighting areas of the troops allotted to their defence. Positions for local reserves are organized in the rear of these defended localities, and for sector reserves further in rear again. In this way the battle position is organized in sufficient depth to prevent penetration by the enemy in one rush. The defended localities will be selected to protect the main tactical features in the position. The intervals between these localities will vary with the nature and importance of the ground.

2. After the machine gun positions have been decided upon and adequate observation ensured for the artillery, the infantry defences will be sited. These will usually consist of a front and support trench some 150 to 200 yards apart linked up by communication trenches which form the fighting positions of the unit allotted to the locality or area concerned. In rear of this network of trenches there will be a reserve trench some 500 to 1,000 yards behind the support trench. This reserve trench will be organized to form a further series of defended localities for occupation by the local reserves in the event of a penetration of the battle position.

Should time permit a further set of trenches and dug-outs will be constructed for the accommodation of infantry sector reserves, and these will be organized into defended localities to deepen still further the battle position, and localize any break through.

The three principal lines of trenches—front, support and reserve trenches—will be linked up by communication trenches and the defended localities will be connected by lateral communication trenches. The latter are necessary to facilitate communication, and to conceal the groups of trenches forming the defended localities. In short, the principal defences of the battle position will be organized in three lines in which, however, the trenches are not necessarily continuous.

3. The most efficient defence is obtained by the organization of the trench system into a series of defended localities of sufficiently large area for a definite unit to be allotted to the defence of each. The internal organization of these localities should be in the form of a series of defence posts mutually supporting each other for the forward sections or platoons of infantry, together with provision of cover and fire positions for the platoons or companies told off for immediate counter-attack or in reserve.

The defence posts may be sited at trench junctions or wherever the necessary field of fire can be obtained (Plates **86, 87, 88**).

50. *Organization and siting of the outpost zone.*

The outpost zone (Plate **85**) will occupy all the ground in the defender's possession, which lies between the battle position and the enemy ; especially ground over which it is necessary to maintain observation, in order to get immediate information of any hostile movement. It should form a buffer zone, organized and held in such away as to absorb the shock of the attack, to deprive it of its momentum, and to break up the enemy's organization as far as possible. The depth of the zone must depend upon the natural features of the ground and the distance between the enemy and the main battle position, but it should be of sufficient depth to prevent bombardment of the battle position by the enemy's field artillery. It will usually consist of three lines of trenches, a front and support trench for observation and fighting and the outpost line of resistance. The organization and siting of the outpost zone will be carried out on the same lines as that of the main battle position, subject to certain modifications as follows:—

i. The position of the line of observation will usually be decided as the result of fighting, except in the case of rear systems deliberately sited.

ii. With a view to the resumption of the offensive, it will often be desirable to maintain the outpost zone further forward than would be necessary if a purely passive defence were contemplated.

iii. As the outpost zone will only be lightly held, it is not necessary that the trench systems should be so highly organized, nor need the various lines be continuous.

iv. The line selected as the outpost line of resistance should, however, be strengthened as much as possible, being organized into defended localities, and the whole zone should be sufficiently complete to prevent those parts which are occupied and dug to full depth from being conspicuous targets, to facilitate communications and command, and to conceal from the enemy where the outpost zone ends and the battle position begins. In this connection, particular attention to the employment of camouflage is necessary.

The first line of the outpost zone may be the final infantry position after a hard fight, and the men are probably occupying shell-holes, or hastily dug rifle pits. These can be improved as shown on Plates **64, 65,** and **66.** Further in rear, trenches can be dug and made into efficient fire positions ; communications must be made as early as possible and organized for fire ; entanglements must be erected both parallel to and at right angles to the front, so that by degrees the whole zone is organized for a step by step defence, upon which an enemy attack will break.

51. *Organization of a defensive system.*

The line of trenches throughout the whole system must be connected by communication trenches, fire-stepped and wired.

Frequent communication trenches are necessary to link up the various trench lines within the outpost zone and within the battle position.

Each zone must be connected to the next one by means of switches.

Switches are trenches or trench systems, connecting two trench systems, and sited so as to provide a defensive flank ready made, should part of the forward line be taken by the enemy. In this way they limit the development of a successful attack, and form pockets in which the enemy can be destroyed by artillery fire or counter-attack.

The siting of the trenches and works of each position must follow the general rules laid down in Chapter IV, but in a defensive system which extends over many miles of country, it will be found that the physical features of the terrain impose many serious difficulties which are not experienced in considering individual trenches. The final selection must be a compromise between the most desirable and the best attainable as regards the system as a whole.

If, after the compromise has been effected, the position is still unsatisfactory, the defences at the point of danger must be made in greater depth, or arrangements made so that its weakness is covered by artillery or machine gun fire from the flanks or from the rear.

The construction of headquarters for infantry, and artillery formations and units must be regarded as an essential part of the organization of a defensive system.

Priority of work.—The priority of work on the construction of all the various parts of the defensive system is a matter which will be decided by the commander according to the labour, time and material likely to be available, but generally will be :—

i. **Constructing observation posts, machine gun and Lewis gun positions** to get the greatest fire effect in the shortest possible time, with the exposure of the least possible number of men.

ii. **Clearing and improving the field of fire** so that it may be difficult for the enemy to approach the position unseen and that it may be easy for the defender to shoot him in the open.

iii. **Creating obstacles** so that the enemy may be hindered in his assault, and that the defender may shoot him while in difficulties.

iv. **Digging trenches or fire positions** so that it may be difficult for the enemy to see and shoot at the defender while the defender is hidden and in a comfortable position to shoot the enemy.

v. **Improving communciations** so that it may be easy for the defender to transfer his troops from one part of the position to another, under cover, and difficult for the enemy to know the real strength at any point.

vi. **Constructing dug-outs** for the protection of the garrison under the heaviest shell fire, so that the greatest number of men may be available for counter-attack,

52. Rear systems.

1. To cover areas of great importance or where very heavy hostile attacks are expected, it is necessary to construct defences in greater depth than that afforded by one defensive system in order that the enemy may be prevented from penetrating all organized defences in one rush or as the result of one offensive operation.

In such cases one or more rear systems of defence will be constructed at such a distance behind the system in front that the enemy will be forced to move his artillery and organize a fresh offensive.

Such rear systems will invariably be organized as described above with outpost zone and battle position.

2. Switch trenches or switch systems may be constructed to connect one defensive system with those in rear in order to localize a break through by the enemy.

53. Defence posts and defended localities.

1. Defence posts and defended localities are the terms applied to self-contained defence works, whether they are detached defences protecting some isolated point, or whether they are the groups of trenches or defended areas forming the framework of a complete defensive system.

In order to prevent the enemy locating the exact extent of such important defences, it is necessary that they should be extended or connected to neighbouring defences at the earliest possible opportunity, so that their identity may be merged in the general trench system. If this is not done isolated defence posts run grave risks of being obliterated by hostile artillery fire, or are liable to an isolated attack.

2. **Defence posts** consist of a group of trenches or shell-holes suitable for any garrison from a section to one or two platoons. They must be designed and sited to bring fire to bear in any required direction, especially to cover the ground between and in front of neighbouring defence posts. They must be wired, provided with shell-proof cover and observation posts for the garrison, and must be capable of all-round defence should the necessity arise. Storage accommodation must be provided for water, ammunition of all sorts, rations and tools.

Defence posts may be situated in any portion of a defensive system, camouflaged from hostile observation by the intricacies of the trench system in which they are located, or concealed by natural cover such as woods or hedges. They may often consist of short lengths of trench prepared for fire at junctions of fire and communication trenches. Machine gun defence posts will normally be clear of the trench system and will depend on natural cover or artificial camouflage for concealment.

3. **Defended localities** consist of an area of ground organized for defence by a definite unit such as a company, two companies, or a battalion. They will comprise groups of defence posts so sited as to cover all the ground to the front and flanks of the locality with fire and to cover all the ground between it and neighbouring localities. Each defence post within a locality must afford support to its neighbours, and be connected up by communication trenches on flanks and rear.

Accommodation for troops for local counter-attack must be provided in the rear or on the flanks of defended localities, also for ammunition, water, &c., in addition to the shelters within the defence posts.

Obstacles must be provided in the front and flanks and where necessary in rear also, the necessary gaps being left for counter-attack, but these gaps must be concealed from direct hostile observation.

54. *Villages.*

1. Each village will require special treatment according to its situation and extent. Whether a village should be included within the position or not must be decided on the spot, according to the lie of the ground and the advantages which will accrue to the enemy, if he is allowed to occupy it. Villages attract artillery fire and the effect of this is increased by the debris of the buildings ; they also harbour gas. On the other hand, they provide a cover and opportunities for a protracted resistance. Infantry defences should be provided outside the perimeter of the village, and full use must be made of the concealment afforded by hedges, orchards, walls, &c., which generally abound on the outskirts to enfilade and surprise the attack. These defences should be connected by trenches to the cellars within the village. The roofs of cellars should be strengthened, and the cellars should be connected by subways, and provided with at least two entrances. The debris will provide opportunities for concealed Lewis gun positions, and positions for machine guns should be prepared to cover all approaches to and flank the defences (Plates **89** and **90**).

2. A keep should always be prepared within the village, so as to assist in the recapture of the defences by counter-attack.

55. *Woods.*

1. The inclusion of woods within a defensive position depends on their situation and extent. Woods harbour gas and, if heavily shelled with gas shell, they are untenable by attacker or defender. Woods of small extent may be converted into obstacles to break up the attack, by entangling their outer edges with wire, Sec. **16**, para. 5, and siting

trenches and machine gun emplacements to flank them. If they are of large extent their inclusion within the position will be advisable, if possible, because (*a*) they afford cover for reserves, working parties, stores, tramways, &c. ; (*b*) if the enemy is allowed to occupy them, they give cover for large concentrations of his troops.

The system of defence should be designed for (i) the occupation of the wood ; (ii) preventing the enemy from filtering into the wood when occupation of the trenches in it is impossible on account of gas.

2. The position of the trenches should be far enough from the front edge of the wood to prevent the enemy's artillery ranging on the front edge from affecting the garrison. Wide clearings will have to be made in order to obtain a field of fire, and trenches and machine gun positions must be sited to flank these clearings (Plate **91**) ; subsequently the trenches will be connected up into a continuous system. The entry of the enemy into the wood must be prevented by entangling the front edge and providing fire to flank it. In extensive woods, wide rides through the wood should also be cut and trenches and machine guns sited outside to enfilade them.

3. Passages within the wood must be cleared to assist communication and plenty of direction boards provided, so that any part of the defences can be reinforced and counter-attacks can be launched without losing direction.

56. *Shell-proof accommodation.*

1. The construction of shell proof accommodation is described in Chapter XV. The position of all dug-outs, shelters, &c., must be decided, so that work may be begun on them as early as possible.

2. The order of their construction will be decided by the commander under whose orders the defensive system is made, but generally will be :—

    i. Observation posts. Machine gun emplacements. Brigade and battalion headquarters.

    ii. Accommodation for the garrison.

57. *Notice boards and cables.*

1. To facilitate inter-communication throughout the position, there must be plenty of notice boards. These should be used to show :—

    i. The names of the various sectors and trenches.

    ii. The names of the defence posts and defended localities.

    iii. The allotment of troops to the position.

    iv. Lanterns should be provided for use at night.

2. **Buried cable.**—To ensure signal communications throughout a position, cables should be buried in all situations exposed to shell fire. Whenever time permits the trench dug for the cables should be at least 6 feet deep.

58. *Defence of camps and small posts against a badly armed enemy.*

1. When operating in mountainous country against an uncivilized enemy, who is likely to make night attacks, the three principal points to attend to are :—

    i. An outer line of strong self-contained piquets, placed so as to deny to the enemy all ground from which he could fire into the camp.

    ii. A well-defined firing line round the camp itself.

    iii. A good obstacle in connection with it.

When the number of the defenders is insufficient to provide an all-round defence, the perimeter must be defended by flanking fire from works constructed with that object, and especial attention must then be paid to providing as formidable an obstacle as possible.

The positions to be taken up in order to repel a night attack should be marked out as soon as possible after the force has reached camp. If there is only time to do this with a line of stones, it will give the defenders a definite line to occupy and hold on to.

For convenience in camping, troops should generally occupy the same relative positions each night ; but this convenience must be sacrificed to the arrangements necessary for defence, as it is very important that units should camp close to the ground which they would have to hold in case of attack.

In selecting a camp site attention must be paid to the water supply and its protection ; but the first consideration is a good position against possible night attacks.

2. When operating in open country not of a mountainous nature, or in bush warfare, piquets will often be withdrawn by night. In such cases a well-defined firing line combined with an efficient obstacle must be provided round the camp. It may, under certain circumstances, be advisable to apply this principle whenever the force halts in order to provide an asylum for the non-combatants and transport. These improvised defences may take the form of laagers or zarebas. Laagers are enclosures formed with the vehicles accompanying a force, supplemented by breastworks of pack saddles, stores, &c., and strengthened with trenches and abatis. Zarebas are enclosures fenced in by abatis of thorn bushes. It being most important to obtain a clear field of fire, the bush nearest the side of the camp must first be removed and arranged round the perimeter. Subsequently, tracks and hollows by which the enemy might approach may be filled with thorn scrub if time allows.

3. In local operations, hints as to the best design of defensive work may generally be got from the enemy, who will have evolved the types best suited to local materials, as well as to resist the form of attack

and weapons which he will employ against us. Such types, when improved by the light of our own knowledge, modified to suit our weapons, and executed with the aid of good tools and engineering skill, will, as a rule, be suitable for our own use.

Plate **93** gives an example of a defensible post, where the block-houses are arranged to enfilade the lines of obstacles. South Africa produced corrugated iron and shingle blockhouses, surrounded by barbed wire (Plate **61**); on the north-west frontier of India, stone sangars are the rule; in the Soudan, breastworks of sand and thorn zarebas. Where railway stations have to be protected, blockhouses, stockades and splinter-proofs made of rails, and loopholed buildings will predominate.

59. *Field defences of a coast line.*

1. The first object of an enemy landing force is to secure his foot-hold by exploiting his success laterally and inland sufficiently to obtain communication inland, and to interfere with the arrival of reinforcements for the counter-attack.

The enemy has the advantage of the choice of the actual points of attack, and is consequently able to concentrate forces superior in numbers to the defenders at the start.

The attack will probably take the form of a landing, supported by naval guns and the weapons of the infantry.

2. The works intended to resist the attack of a landing force may, therefore, be sited, chiefly with a view to obtaining the best fire effect on the enemy while he is approaching in boats or in the act of landing; concealment of works from direct view of the enemy artillery is of secondary importance.

Generally speaking, the system of defence should be one of mutually supporting defence posts or localities by which troops may be econo-mized and the risks attached to the provision of a number of detached works may be avoided. Any advanced trenches, which may be found necessary, must be definitely affiliated to a defended locality.

3. The organization of obstacles in the defended area should be as already described in Chapter VI, with the exception that, in the case of coast defences, there is less objection to the obstacle round a defended locality being conspicuous.

4. The rules for siting machine gun emplacements (Chapter V) apply equally to this nature of defence works.

5. Special attention should be paid to establishing " Road Blocking Posts " to command the junctions of roads leading from the coast inland, to ensure that cyclists or other fast moving troops may be prevented from penetrating.

6. The development of a defensive system behind the defences on the coast line follows the principles laid down for positions inland.

7. The coast line may be classed under the following heads :—
- i. Shingle beaches.
- ii. Sand hills.
- iii. Marshy shores with sea walls.
- iv. Clay cliffs.
- v. Cliffs of chalk or rock.
- vi. Town fronts.

i. **Shingle beaches**.—The chief difficulties presented by these are :—
- (a) The ever-varying nature of the beach under the action of the sea.
- (b) The risk of casualties caused by the shingle under artillery or machine gun fire.

A shifting beach involves constant changes in the field of fire, and considerable damage to, if not complete obliteration of defences and obstacles.

It must, therefore, be recognized that a position in this class of coast line calls for the highest order of vigilance and industry on the part of its garrison.

The defences should be of the simplest nature ; trenches should be sited on the crest of the shingle bank ; obstacles of a portable nature should be made and held in reserve to supplement the permanent obstacle at such places where it is most liable to damage by the action of the sea. Sandbags filled with sand or earth should be stored at definite places, to rectify small changes in the height of parapets and to give protection against flying pieces of shingle during a bombardment.

ii. **Sand hills**.—Positions among sand hills have disadvantages similar to those described for shingle beaches ; trenches and obstacles, and sand drifts under the action of wind and sea are often totally effaced.

Sand hills by their peculiar irregularities afford valuable cover and concealment for the troops in occupation.

The defences must, therefore, be sited to develop the fire effect on the beach and foreshore to the highest degree possible to prevent the enemy from establishing himself in a position from which it is very difficult to expel him.

Shells even of the heaviest natures do not penetrate sand to any great extent ; the liability to casualties from splinters is, therefore, considerably greater than in alluvial ground.

iii. **Marshy shores with sea walls**.—The properties of this type of coast line are, shallow water on the fore-shore, poor roads inland, good obstacles provided by the drainage dykes. Defended localities may be sited, in such cases, at greater distances apart. The sea wall gives good cover to the defender, but usually provides a parapet for frontal fire only—flanking fire must be developed. Special attention must be paid to the defence of road junctions, bridges, &c.

iv. **Clay cliffs.**—The constant changes in these cliffs due to erosion by the sea and weather render the occupation of the face dangerous and sometimes impossible.

The main system of defence should consist of a continuous entanglement on the top of the cliff with the defended localities or posts close to it.

Sites for hasty trenches or machine gun emplacements on the lower slopes should be selected and constantly checked and revised, and material for obstacles should be stored in the immediate vicinity, to be put in hand and completed on the necessity arising.

Access to these lower positions must be arranged for beforehand. Every opportunity to develop cross fire from the top of the cliffs on the water and beach below must be seized.

v. **Cliffs of chalk or rock.**—However inaccessible a cliff may appear, it should never be assumed to be so.

Where it is impossible to obtain effective fire over the foreshore, defended parts must be established at the heads of all ravines debouching on to the shore, and arrangements made to ensure that the line of cliffs is watched when the necessity arises.

vi. **Town fronts** are usually strong against frontal attacks because of the cover afforded to the defenders by the buildings from which the esplanade can be swept. The sea wall usually provides an obstacle to an advance from the beach. This can be improved as shown on Plate **36.**

Attempts to take these positions would usually be made from the flanks, which should be defended in the ordinary way.

In order that the convenience of the public may not be interfered with unduly, ramps leading to the beach, and streets leading from the beach inland must not be blocked, but portable obstacles must be made and stored at convenient places for use when necessity arises.

The beaches should be swept by machine gun and rifle fire arranged from inconspicuous positions in the esplanade wall and steps. Piers should be prepared for demolition, by explosives and fire.

Groynes.—The shifting nature of beaches constantly affects the degree of cover which groynes may afford to an enemy.

They must be frequently inspected and the defences must be revised from time to time to ensure that the beaches are adequately swept by fire.

CHAPTER IX.

CAMOUFLAGE.

60. *Considerations affecting camouflage.*

1. **Definition.**—Camouflage is the art of concealing that something is concealed. Its keynote is deception. By it, objects are rendered indistinguishable or unrecognizable by means of a covering or disguise.

Concealment in the limited sense of hiding from view, is screening, not camouflage (Sec. **65**).

2. Before deciding how to camouflage an object, the enemy's means of observation must be examined.

This may be by :—

    i. Aeroplane photos.
    ii. Air observation.
    iii. Ground observation.

As the definite position of objects on the map is usually fixed by air photos, efforts to camouflage must be directed in the first place to deceive :—

    (*a*) The camera.
    (*b*) The expert examiner.
    (*c*) The ground observer.

To effect any success in camouflage, the *camoufleur* must be skilled in reading air photos, so that having made an examination of the photo of the area in which it is proposed to construct a work, he is able to camouflage it without **interference with the normal aspect of the locality with which the enemy has become familiar.**

3. All camouflage must be erected before work is commenced, and on a sufficient scale to conceal everything, viz., excavation, spoil, materials, &c. Erection of camouflage after completion of work is simply a waste of time and material, as the work will have already been located.

Time, labour and materials are important factors in this work as in others, and economy in all of them can only be effected by strict attention to the following principles :—

    i. Do not try to camouflage a work which obviously cannot be successfully camouflaged, such as long lengths of trenches, a large work in full view of ground observers, a large camp in the open, &c.

    ii. Do not attempt to economize by using too little. If assimilation to surroundings is being aimed at, the area of camouflage should be considerably greater than that of the object to be concealed.

    iii. Do not employ artificial, if natural material is available. The latter is probably better in every way.

iv. Judicious siting and taking advantage of natural cover or accidents of terrain will reduce the amount of material necessary for camouflage.

The most common fault committed is that of neglecting to conceal, or to deal with, evidences of occupation or construction, such as tracks, dumps of materials, spoil, trampled ground, &c., with the result that the work stands out like a veritable bull's-eye.

61. *Aeroplane photographs.*

Photographically, the effect of colour is not so marked or important as the effect of light and shade. Earth appears light, and grass or vegetation dark, **not** because of their respective colours, but on account of the amount of **contained shadow** or **texture**.

A billiard table or top-hat illustrates this quality. Brush them the wrong way, against the nap, and their tone is lowered to dark green in the one case and dead black in the other ; brushed the right way, they appear very noticeably lighter in tone. The reason is that they gain **texture** when brushed the wrong way, and lose **texture** when brushed the right way. Nap is constituted of millions of slender hairs, each one throwing a shadow when erect, but casting none when " laid." Grass, or vegetation, possesses this same property to a marked degree. The longer it is, the darker it appears on a photograph ; but when it has been pressed down the amount of shadow thrown is lessened, and, consequently, it appears lighter.

For this reason, a slightly worn track in grass, which is quite inconspicuous from the ground, shows up vividly on a photo. Earth, on the contrary, contains little texture, and the longer it has been turned up and exposed to rain and sun the less it has. A beaten track is, however, noticeable, as it contains no texture at all, and will, therefore, reflect more light.

The reason for the mottled effect of a patchy mixture of grass and earth, which blend imperfectly into each other in a photograph is, therefore, evident.

It is essential, when judging the colours of a locality, to view it vertically, and not obliquely, as one is accustomed to see a flower bed. A field of young corn viewed from the ground appears green, but, from above, the earth only is seen, and this is darker in tone than the normal, owing to the shadows cast by the young blades of corn. Similarly, with a field of ripe corn, the actual light tone of the straw and ear will be somewhat darkened by the shadows.

62. *Concealment.*

1. The clues by which any new work is recognized are :—

    i. Disturbance of soil (especially among vegetation).

    ii. Tracks.

    iii. Shadows.

    iv. Blast marks (guns only).

    v. Regularity.

2. **Disturbance of soil.**—Among vegetation it is important to conceal **all trace** of the spoil from an excavation. In a locality which is bare of vegetation, it will be sufficient to scatter it.

3. **Tracks.**—The success of camouflage of a work depends upon the discipline observed as regards tracks. It is essential to decide from the first what is to be done and to rigidly enforce this decision.

It is almost impossible to camouflage a track in the open, but observance of the following precautions will render them less conspicuous :—

    i. Traffic must be confined to one track.

    ii. Tracks must be kept as narrow as possible by wiring in on both sides.

    iii. They may be run under shelter of natural cover, such as trees, hedges, &c., or in the limits of the shadows cast by them, taking care to thicken the cover where necessary. This will occur where tracks go through woods or under hedges, which become bare in winter. The thickening must be done before the leaves fall.

    iv. They may be run along strong lines of demarcation shown in air photos, e.g., between two plots of dissimilar cultivation. In this case no short cuts or cutting corners must be permitted.

An example of this is given on Plate **94.** The real communication trench follows the ditch along the edges of the fields, while an easily seen dummy is provided to draw the photograph interpreters' attention from the real one.

Earth from the real communication trench was disposed of as follows :—

    (a) A certain amount was used along intermediate or dummy parts of the rectangular trenches between fields, as a thin wall 12 inches high, so as to increase the shadow effect.

    (b) The remainder was scattered amongst the beet plants in the surrounding fields.

The dummy trench is 5 feet wide and 6 inches deep. The sides are vertical or undercut so as to throw as much shadow as possible and the excavated earth well spread out on each side.

    v. If no natural cover exists, the track must be continued beyond the work it serves to join another track existing or made for the purpose, avoiding obvious loops. This extension must be used occasionally to keep up its appearance as a real track. The track should pass as close as possible to the actual work, and great care should be taken in the case of battery positions to avoid any apparent connection between each gun pit and the track.

vi. A track in front of the gun pits can be led right along the edge
of the pits themselves, thus avoiding any connecting links
and making the blast mark problem easier; the back of
the pits can be closed to all traffic. In this case, traffic
on the track may be held up while the battery is in action,
and a source of dust is formed. Plate 95 shows the different
aspects of strict track discipline, and the reverse with
reference to the identification of a possible headquarters.

4. Shadows.—A study of an aeroplane photo will show how the
shadows of objects in the area fall. The shape of any object, and,
therefore, its nature, can be deduced from its cast shadow.

The methods by which the shadows of camouflage may be disguised
are :—

i. To take advantage of existing definite shadows, such as provided by trees, hedges or houses.

ii. To erect a cover, the shadow of which will be so fantastic
that its appearance will not be attributed to human agency,
in fact, it will appear " natural."

iii. To endeavour to eliminate shadow altogether, or failing this,
to cast one that is vague and indefinite. This can be attained
by adopting a **flat-topped** cover with **no sides** (as
opposed to a mound-shaped cover). An area of material,
sufficiently opaque to conceal excavation and spoil, forms
the centre ; from this the opacity diminishes gradually
towards the edges.

The opacity of the thinned portion will be sufficient to mask or blur
the shadow cast by the opaque centre when the sun is low, but at
the same time is not sufficient to cast a definite shadow itself. Further,
the increasing transparency towards the edges will allow the camouflage
to blend gradually with the surrounding ground, which shows through
it.

Such a cover must be considerably larger than the work to be concealed ; the higher above the ground it is erected, the further will be
the shadow cast, and, consequently, the larger it must be. For
instance, if erected at a height of 6 feet above the ground, the total
cover should be nearly three times the area of the actual work which
has to be concealed. Further, it is most important that its contours
should correspond to those of the ground it represents. Sagging
introduces new-cast shadow, and must be obviated.

5. Blast marks are always looked for in a photograph. On grass
they appear light because the grass is destroyed. On earth they are
hardly apparent In snow they appear dark. The shape, resembling
a fan, is definite, and corresponds to the amount of switch. The size
increases the nearer the muzzle of the gun is to the ground level.

The blast marks cannot be detected if the gun can be sited so as
to fire over a road, bare earth or water.

To conceal the marks either of the following can be used :—

    i. Removable camouflage, which must be opaque.

    ii. Expanded metal. This must be very firmly fixed to withstand the blast. One thickness photographs like short grass, two like medium, and three like long grass. It is not necessary to paint it.

It is always advisable to distort the shape deliberately, so that it may bear no resemblance whatever to the characteristic form of a blast mark.

Blast marks can be prevented by placing a strong sloping platform under the muzzle to deflect the blast from the ground.

6. Regularity.—Anything of a rectangular or regular shape arrests the attention when a photograph is examined. Nothing in Nature appears regular, and, consequently, anything that **is** regular must be the work of human hands, and is, therefore, suspicious.

Regularity is manifested in two forms :—

    i. In the shape of the camouflage and the shadow.

    ii. In regular spacing or dressing of emplacements.

It is **most important** to distort any straight line or rectangular shape, **unless** they are deliberately designed to conform to the existing ground pattern, or to simulate normal objects. This distortion must be made on such a scale as will show on an air photo taken at 10,000 feet.

In broken ground the photographic pattern is intricate ; all that is necessary is to conceal the straight lines of excavation and other work.

In open country the principle of irregular shape is equally applicable to the **flat-topped** camouflage which is the best suited to it.

Dressing and spacing.—A peculiar mark on a photograph which might escape notice, if unique, must be regarded with suspicion if it appears four or six times at regular intervals and in a straight line, as in the case of a battery.

The following methods of camouflage may be applied :—

    i. The whole battery to be included under one cover, especially if the alignment is determined by some artificial feature such as a road or hedge, or line of demarcation between two contrasting tones, e.g., adjacent plots of roots and plough.

    ii. Irregular spacing and grouping if they are consistent with tactical requirements and fire control.

    iii. A different disguise for each gun.

63. *Materials supplied and their employment.*

The materials supplied (Appendix IV, Table 8) must be modified in colour and opacity to suit the area in which they are required.

Fish netting or scrim which is too green can be toned down by the application of dead grass or mud. Light-coloured materials, such as sandbags, timber and spoil which would show through netting, must be covered with dark-coloured scrim or brushwood. The shadow thrown by a covering can be judged by its outline on the ground when the sun is low.

Camouflage is usually inflammable ; it must be, therefore, protected against ignition by the flash of discharge.

All camouflage must be maintained in good order, or it will deteriorate and betray itself.

64. *Examples of how camouflage may be effected.*

1. Area camouflage.—In open agricultural country, regular chessboard pattern presented by cultivated ground in an air photo may be reproduced in camouflage material. Large areas have been treated successfully in this way, affording cover from view for troops, guns and aeroplanes.

2. Heavy guns may be camouflaged :—

    i. By painting the gun to suit a definite site, and confining attention to concealing evidences of activity, ammunition, and light railways.

    ii. By disguising the gun as a railway truck, dump of material, haystack, &c.

3. Observation posts.—The indications which betray the site of observing posts are :—

    i. The shadow cast by the loophole.

    ii. Tracks converging to one spot combined with the knowledge that it is suitable for an observation post.

    iii. Light showing through from the back of the loophole.

    iv. Reflection of sunlight on lenses of observer's field-glass.

    v. Visibility of the observer.

The shadows may be disguised by the methods already described in Sec. **38**, para. 3 (loopholes), and Sec. **62**, para. 4 (shadows), or by covering the loophole with wire gauze (Plates **96** and **97**) behind which field-glasses can be used. Periscopes obviate the necessity for loopholes.

Dummy loopholes attract the enemy's fire from real loopholes.

The use of veils and coats of a colour to match the background is useful (Plate **97**). In a trench in which sandbags are used, an empty sandbag worn over the head is a good disguise. Men wearing veils of brown or green gauze are difficult to detect against backgrounds of earth and grass respectively. Grass, weeds, wood or branches may give concealment.

Tree observation posts are imitations of existing trees : these are cut down and replaced by the imitations in one night.

Periscopes can be used in any object which can accommodate a 4-inch pipe, *e.g.*, tree, wire entanglement post, telegraph pole, and brick wall.

4. **Dummy gun positions.**—These should supply, not too conspicuously, all the clues of an active battery, *e.g.*, blast marks and tracks.

5. **Snow camouflage.**—Blast marks will give away the presence of an active battery, unless steps are taken to conceal them. They show as a dark blur, black near the muzzle of the gun and gradually fading away, and very much larger than usual.

Remedies for blast marks are :—

    i. Covering them up with fresh snow after firing.
    ii. Covering them up with flattened chalk. It is best to supplement this with snow.

Footpaths do not show immediately unless the snow is thin, or very soft, but the usual track precautions must be maintained.

Artificial camouflage will not hold snow, and consequently each pit will show up as a dark square unless precautions are taken. The only remedy is to fix large islands of white calico on it.

The comparative warmth of an emplacement will often result in the melting of the snow all round the outside ; fresh snow should be used to correct this.

When the snow is of sufficient depth to cover the landscape completely, and the whole countryside is one unbroken sheet of white, the blast marks and the nets are most conspicuous unless considerable attention is paid to covering them. Dummy positions afford the best protection if realistic blast marks are made with soft, irregular edges, and the real blast marks are carefully concealed.

6. **Trenches.**—It is not practicable to conceal long trenches. Short lengths of trench, such as saps or connections between shell-holes, may be concealed, but the camouflage **must be prevented from sagging**, and arranged to thoroughly conceal the spoil. The minimum width of material necessary is 50 per cent. more than the maximum ultimate width of spoil banks and trench. The best material is painted scrim.

7. **Spoil from dug-outs, &c.,** is treated with painted scrim, if it cannot be carried away and dumped in some unfrequented place or old trench. If open to direct observation, the scrim must be supplemented with netting and canvas knots.

8. **Mining work and forward dumps.**—These can be concealed in the same way as any defensive work or battery position.

9. **Machine gun emplacements.**—The area to be camouflaged must be restricted as much as possible, and therefore, among vegetation, all spoil should be removed and traces of work concealed so that the cover over the gun itself may be almost flush with the ground and very slightly

larger than the actual trace of the emplacement. **To render the gun** in action invisible to a low flying aeroplane the emplacement should be covered by camouflage supported on a light wooden frame which is made to move up and down in a vertical plane. The cover must be concealed from ground observation at comparatively short range. All evidence of activity and occupation (tracks, &c.) must be suppressed.

The embrasure of a machine gun emplacement in a brick wall may be concealed by a combination of dummy brickwork and gauze on a hinged flap which will enable the embrasure to be completely hidden when the gun is not firing, whilst at the same time admitting of observation (Plate **98**, Fig. 2). A similar method, adapted for firing through a door, is shown on Plate **99**.

On grass, coloured raffia (or gardener's bass) or natural grass, suitably coloured is best (Plate **98**, Fig. 1).

On earth or among shell-holes painted scrim is sufficient.

10. **False work.**—It is possible to paint dummy trenches on canvas, but the painted shadow remains fixed and does not alter with the position of the sun.

A new trench is best concealed by making it look like an abandoned one, by covering it with painted canvas of an appropriate width and pegging it out on the ground with real spoil laid along both sides of it.

11. **Camps.**—Nothing short of a comprehensive overhead camouflage will obliterate roofs, shadows and paths. Practically all that can be done is to paint the sides and roofs of huts a monotone approximating to the tone of the ground, with a matt surface to minimize reflection in moonlight. The matt surface can be produced by sprinkling sand, dust, hay, &c., on the paint or tar while it is wet.

12. **Tanks.**—May be treated like heavy guns ; the best treatment being a flat-topped cover of fish netting.

65. *Screening.*

1. The primary aim of screening is concealment from view in order to permit free and unobserved circulation of traffic.

Screening may be carried out by means of :—

    i. Artificial screens.

    ii. Natural screens.

2. Artificial screens are made of :—

    i. **Wire netting** woven with grass, brushwood, or canvas strips (Plate **100**, Fig. 1).

    ii. Brushwood and tree branches interwoven on horizontal wires stretched tightly between two uprights.

    iii. Canvas, or coir, suspended on strongly braced poles.

Of these, canvas and coir do not stand the weather well and require more maintenance on this account.

3. **Natural screens** are made by supplementing the height or thick-ness of existing hedges, copses or fringe of trees to render them more effective without making the fact that they are screens conspicuous.

A screen should be sufficiently opaque to hide movement from any but very close and continuous scrutiny. The efficiency of any screen, except one absolutely opaque, is influenced by the background. A com-paratively transparent screen may be successfully used in combination with a background of hedges and trees, or if it is set obliquely to the enemy's angle of view.

In order to hide movement at ranges between 2,000 and 4,000 yards three-quarters of the surface of the screen should be opaque. Screens may be either plain or camouflaged.

4. **Plain screens** are those which are put up without any idea of dis-guising the fact that they are screens. Newly erected screens always draw fire but if the damage is regularly repaired, the attention paid to them rapidly diminishes.

These screens have been used with effect as follows :—

    i. The act of screening an area or battery position before they were required, drew fire from the enemy for a period during which the screens were regularly repaired.

        When the screens were no longer shelled, they fulfilled the functions for which they were erected without further interference.

    ii. To conceal a party working behind them.

    iii. To draw fire while work was being carried out at a distance to the flank.

Although plain screening affords protection from view after the enemy has ceased to notice it, a careful reconnaissance of the area to be screened should always be made in order that full advantage may be taken of the natural features.

5. **Camouflaged screens** are made of canvas or wire netting com-bined with canvas, brushwood, grass, &c., painted to reproduce a definite locality such as a brickwall ruin, hedge, &c., or a general landscape. These should only be used in places where required for a short time because the paint does not stand the weather long and they require constant watching and careful maintenance. These screens have been used with effect as follows :—

    i. Imitation brick walls painted on canvas backed with wire netting to screen a much used thoroughfare (Plate **100**, Figs. 3 and 4).

    ii. Imitation hedges of a combination of raffia, canvas strips, and brushwood, on wire netting were made to conceal a battery position which would otherwise have been under direct observation. In this case an existing hedge which was too far back to be used was removed and the imitation hedge substituted for it **in front** of the guns.

6. If an **area** occupied during the summer is likely to be occupied during the winter months as well, the problem of screening should be considered early so that the loss of cover due to the leaves falling from the trees may be made good with brushwood beforehand, and no noticeable change in the landscape when the trees and hedges are bare.

7. Road screens.—Roads running perpendicular to the front line are best screened by hanging vertical screens between trees or houses, or poles, across the road (Plate **101,** Fig. 1).

For roads running parallel to the front short lengths, of about 30 yards, placed in echelon and overlapping each other are better than long continuous lengths. This method permits of plenty of passage ways, and limits damage by shell fire (Plate **101,** Fig. 2).

Roads running obliquely to the front can be concealed by screens facing the front, arranged in echelon (Plate **101,** Fig. 3).

8. Flash screens.—Screens have been successfully used to hide gun flashes at night from the front and from a flank.

In one case where the flashes of a battery were visible from a flank, six small screens were erected, one about 4 yards to the right of the muzzle of each gun, and running out about 8 yards to the front. They were about 8 feet high. They were dismantled during the day and re-erected each night, in socketed holes.

66. *Manufacture, erection and maintenance of screens.*

1. Manufacture.—Screens should be made up in bays of 30 feet, with supports 10 feet apart.

In order to localize the effect of shell fire, each longitudinal width of wire netting should be suspended independently on a longitudinal wire between the uprights.

The screens are made of strips of canvas interlaced in wire netting. The strips should be $2\frac{1}{2}$ inches wide, and threaded through every third or fourth mesh vertically, leaving no horizontal interval. Opacity can be considerably increased by the use of alternate vertical bands of plain and dark-coloured canvas, each band being about a foot wide. There should be a strong contrast between the plain and coloured canvas. Such a screen is effective at ranges of a mile and upwards (Plate **100,** Fig. 1).

Still better results can be obtained if the colour is arranged on the vertical bands so as to produce a chequered effect.

Irregularity of outline, if necessary, can be given by not commencing the threading of every strip at the top of the wire netting, or by inserting a brushwood crown.

2. Erection.—Whenever possible, screens should be attached to existing objects such as trees, hedges, houses, &c. If poles have to be used instead they should be sunk well into the ground and well guyed. The screen is fastened with staples to the poles. The uprights are joined by longitudinal wires 3 feet apart, and cross diagonal bracing between each pair, which should be guyed in the normal way (Plate **100**, Fig. 1). These wires also form horizontal braces to the uprights which are cross braced and guyed as shown on Plate **100**, Figs. 1 and 2. Uprights should be of at least 3-inch timber. Guys should consist of at least four strands of No. 14 gauge wire, or the equivalent.

Screens should not be erected piecemeal, or unnecessary casualties from shelling will be incurred. The material should be laid out on the ground beforehand, so that the whole screen can be erected in one night.

CHAPTER X.

THE ORGANIZATION OF WORKING PARTIES AND THEIR TASKS.

67. *Responsibility for the execution of work.*

1. All arms are responsible for the construction of ordinary field works without any technical assistance from the engineers. Such work will include the construction of defensive systems, the provision of cover from fire and the construction of obstacles. Field works should be regarded as a military duty and should be executed as a military operation.

Commanders of formations and infantry units will be responsible for siting, organizing and constructing field defences. Officers of all arms must study the most suitable types of defences and the details of siting, depending on armament, ground, concealment, &c.

The engineers will normally be employed only on works requiring technical skill, special tools, or specially elaborate organization such as the construction of battle position or rear position defences. There are not enough engineer units available with formations to enable the engineers to undertake any but this technical work.

2. Field works may, therefore, be divided into two classes :—

    i. Work for which units or formations other than the engineers are responsible. This will be carried out under the orders of infantry commanders with materials supplied by the engineers, but without engineer assistance or supervision, other than technical advice or minor assistance in technical details, such as fixing of timbers in complicated shelters, &c. The provision of this technical advice or minor assistance is the duty of engineer liaison officers with infantry formations.

ii. Work for which the engineers are responsible. This will be carried out solely by engineer units, or by engineer units with the assistance of working parties from infantry or other units, or civil labour.

In the case of work under class ii there will be two principal officers involved in the work.

(a) The engineer officer in charge of the work.

(b) The officer in command of the working party.

3. The responsibilities of each are as follows :—

The officer in charge of the work is responsible for—

i. Making the preliminary reconnaissance.

ii. Tracing out the work.

iii. Demanding the working party.

iv. Supplying materials and extra tools, if necessary.

v. Supplying guides to ensure that the working party actually arrives on the site of the work.

vi. Seeing that the work is completed as designed.

He must also see that arrangements are made for provision of any covering party, that is required, **in addition** to the working party.

The officer in command of the working party is responsible for the disposal of the men on the work, for issuing all orders as regards smoking, lights, talking, and that the work is completed to the satisfaction of the officer in charge of the work, but the officer in command of the working party is responsible for quantity of work done.

4. The senior military officer must obtain definite orders from the authority ordering the work as to whether, in the event of suffering serious casualties, the men should be withdrawn temporarily, or an attempt should be made to carry out his task at all costs.

68. *Alternative methods of organizing working parties*

For straightforward work, where daylight reconnaissance out of enemy observation is possible and when there is sufficient time to make the necessary reconnaissance and to send the " demands " back to brigade or battalion headquarters, the procedure in Sec. **69** should be followed.

It must be recognised, however, that enemy observation and urgency (which may limit time for reconnaissance) make modifications necessary.

The **time** taken to carry out all the steps and duties detailed in the notes, including the sending back of the " demands " and subsequent detailing of, and the approach march of, the working party, may be eight hours or more.

If the time available is less than eight hours, a working party must be detailed and the rendezvous and time fixed, **before** the reconnaissance is made.

Details of this procedure for hasty work are given in Sec. **70.**

69. *Procedure when reconnaissance is possible before sending in demands.*

1. Immediately on receipt of orders to put any work in hand, the officer who is appointed to take charge of the work will visit the site, by daylight if possible, to make a reconnaissance of the work. This reconnaissance enables him to decide on the approximate numbers of the working party, what sort of tools and materials are required and whether there are any particular difficulties in carrying out his orders, e.g., hard ground, roots, &c., which call for special measures to meet them.

Whenever the conditions of the operations permit, in order to save time, he should take with him a tracing party of suitable strength to mark out the work, and the guides which will be required to direct the working party from the rendezvous to the work. If possible, the officer who will be in command of the working party and some of his non-commissioned officers should take part in this reconnaissance.

Before starting they should all be informed where the work is, what it is, and its purpose. Note should be taken of the following points :—

    i. The route to the work which involves the least fatigue and delay and the time required to march to it ; plenty of time must be allowed, remembering that a large party moves very slowly, especially across country or in trenches in single file. If the enemy's harassing fire is heavy it may be necessary to tape out routes across country which will involve special measures being taken to cross trenches.

    ii. Landmarks which assist in locating the site of the work and the route to be followed ; if none exist, artificial marks should be erected.

    iii. The guides should be shown the rendezvous at which their parties will be ordered to report, if it has been decided upon.

The work should next be marked out ; the officer commanding the working party, if present, will decide in consultation with the officer in charge of the work, how best to distribute his men ; this should be done by platoons and companies, so that each commander shall interest himself in the work of his men, for he is responsible that it is completed.

The limits of each platoon and company should be clearly marked, and each guide should be shown the point to which he is to bring his party and the extent of its task. The formations in which the party are marched to the work depends upon the amount of interference the enemy will bring to bear, but there should be no bunching on the march up, and the arrival of the parties must be so timed that one party does not have to wait while another is being put on to the work. In normal times, the rate of arrival should be 50 men (actual workers) at 15 minutes interval, and when harassing fire is heavy 25 men at 5 minutes interval.

With carrying parties the loads should be so distributed that it is possible to carry on if a portion of the party fails to arrive.

If it has not been possible for the guides to go over the ground, they should be given their instructions in writing, but in any case they must be at the rendezvous punctually.

2. The officer in charge of the work must now make a rough estimate of the amount of each sort of work, such as clearing, entanglements, digging, carrying, &c., and by the help of the tables in this book (Appendices II and III) or his own experience of the abilities of the men who will form the working party, he will be able to make an estimate of the number of men and tools and of the materials he will require.

It must not be supposed that under active service conditions that this estimate will be accurate, but it will be near enough to enable him to form a fair idea of the size of the working party and of the time required to complete the work, or if the time is definitely laid down, how many men will be required to complete the work in a given time.

If too many men are detailed the working party is crowded together, and the risk of noise and consequent discovery by the enemy is increased, and while the work is not speeded up a greater number of men are deprived of their needed night's rest, and the efficiency of the unit as a whole is impaired.

If too few men are detailed they become disheartened, and in consequence, turn out less work, the completion of the work is unnecessarily delayed and the risk of interference by the enemy is increased.

On active service the actual strengths of formations vary so quickly from day to day that it is generally impossible for the officer in charge of the work to demand his working party in the form of so many companies, platoons, &c., however desirable this may be. His original demand for a working party submitted to the proper authority, must be based on the estimated number of tasks or men required to finish the work in a given time. It is the duty of the officer who orders the working party to be provided to detail the numbers of actual workers required in the approximate numbers of formations or parts of formations, e.g., battalions, companies or platoons. This ensures that the men work under their own officers and that a proper proportion of noncommissioned officers, stretcher bearers, &c., are also detailed. It may not be possible to adjust these exactly, but the importance of detailing complete formations as against detachments of certain numbers is paramount.

3. A detail of the tools required involves a knowledge of what tools are available and where they are carried. This information may be obtained from the Army Form G. 1098 of the unit providing the working party. For the simplest forms of work the proportion of tools in Appendix II will give a guide as to what is required in ordinary circumstances.

In position warfare when tool dumps are formed all tools will be drawn from and returned to the dumps.

4. Having arrived at the numbers of men, the tools and the materials required, the officer in charge of the work has these additional duties :—

i. To fix exact location of the rendezvous where the working party are to assemble, the stores and extra tools are to be dumped or drawn, and where his guides will meet the working party to conduct them to their work, and to notify this to all concerned. The rendezvous should be on the way up to the work, but not necessarily near it. It must be distinctly marked (a map reference is not enough) especially if the party arrives at the rendezvous after dark, and should be near some sort of cover.

It must not be conspicuous from the enemy's point of view, e.g., a cross road, a lone tree, or an isolated building. In a trench system, it should not be a trench junction except when the working party is detailed from the troops holding the trenches.

ii. To fix the hour of arrival at the rendezvous. This must be carefully worked out so that the men are marched from thence to their work without delay. If men are kept hanging about in the dark they fall asleep and are difficult to rouse up.

The enemy's observation and the state of the light must also be considered. For night work, work should commence just after dusk or before moon-rise.

iii. To arrange for the necessary materials, tools, &c., and for transport of the same ; and to ensure delivery of these at the rendezvous before the arrival of the working party.

iv. To see that arrangements are being made for the provision of any covering party that may be required, so that there is no risk of the working party being reduced by having to find the covering party out of the numbers detailed for work.

He is now in a position to submit his demand for :—

(a) The necessary working party for work which will occupy a certain number of hours.
(b) Stating the hour at which they must be at
(c) the rendezvous
(d) with a certain proportion of tools from their own equipment, and haversack rations, if necessary, and where

(e) they will meet guides provided by him and draw additional tools and material as may be required for the work in hand, which

(f) they will carry or escort in the transport he provides

(g) to the site of the work,

(h) which the tracing party has already marked out on the ground,

(i) and on which the working party will be extended, while they are protected by

(j) the covering party, which will be in position before work is commenced.

The demand for working party should be made out in quadruplicate. A typical form is shown in Appendix I.

On receipt of this demand the headquarters of the formation will issue orders to the unit supplying the working party, which will ensure that the party meets the guides and brings the necessary tools, &c. A typical form for these orders is shown in Appendix IA.

When the work is continuous over long periods, much clerical labour can be saved by the headquarters of the formation responsible for the work issuing, without demand, orders giving full particulars of all daily working parties and the unit responsible for providing each.

Fictitious names, which should be obviously of an engineer nature, are allotted to the guides, and these names should be permanent always, i.e., the guide for No. 1 Party will always be Sapper Shovel, whatever his real name may be, or however often he may be changed.

5. If the **officer in command of the working party** has not been present at the tracing out of the work he will in consultation with the officer in charge of the work decide upon the best method for distributing his men on the work.

There are various methods of filing on to the work; two in general use are given below.

First method.—The leading man halts as he arrives at his place and the others pass behind him, each halting as the beginning of his task is marked; this method is the quickest, but it causes the men to bunch and is unsuitable for front line work in trenches.

Second method.—The leading man goes right through to the far end of the work and the remainder space themselves out behind him along the line of the work. The commander of the party then checks and corrects their intervals, starting from the leading man and working backwards. In this method the men are always well extended, and as a result, although it takes slightly longer than the first method, it is usually necessary to adopt it in front line work. The tendency to close up when the front man halts for fear of losing touch in the dark must be checked.

Working parties of second or subsequent reliefs on trench work, should not be allowed to move along the partially dug trench, unless the tactical situation demands that they should do so.

6. When working in close contact with the enemy, the commander of the party may decide to work with arms slung; this greatly hinders the work and should not be done unless there is danger of attack. In the forward area it is generally sufficient if each man lays out his arms and equipment close at hand for use in an emergency. When the work is below fire step level and there is danger of an attack, arms should be laid on the parapet and all the earth should be thrown on to the parados.

In rear areas arms and equipment may be left under guard in a convenient spot.

Gas helmets must always be in alert position in front line work.

7. Work can be carried out either by :—

    Task work, *i.e.*, a definite amount of work is given to each individual or preferably each formation, *i.e.*, section, platoon or company ; or

    Time work, *i.e.*, the working party is required to work for a certain number of hours.

Task work should be given whenever possible.

In most work, especially that of a straight-forward nature like digging or wiring, better and quicker work is obtained by **task work.** The allotment should be by small units, *e.g.*, sections, platoons, &c., and each party must be allowed to withdraw on the completion of its task ; this is the whole essence of task work.

Great care and considerable experience is required in setting tasks and, once fixed, they must not be altered even if they have been wrongly estimated. Some figures showing the amount of work to be expected are given in Appendix II, but the figures may require modification according to the distance the men have to march to and from their work.

When working with reliefs on task work, care must be taken that all parties of one relief have finished their tasks before the arrival of the next, so that the latter are not kept waiting. An interval of 30 minutes should be allowed between the estimated time of the first relief finishing and the arrival of the second relief.

Supervision over task work must be strict, and the tasks which are usually on the easy side must be rigidly enforced. In deliberate trench work the dimensions of the trench should be checked by a templet, *i.e.*, a skeleton pattern of the section of the task, to ensure that the trench has been dug to full dimensions.

A templet or a 6-foot rod, should be carried by each platoon commander, by which he can explain the task.

If time work is decided on, regular periods for rest, say, 10 minutes every 2 hours, should be arranged for.

8. It has been proved that the best is got out of a working party in 4 hours—after that period, the men tire rapidly, and the amount of work they do will seldom justify their being kept out.

Work in the face of the enemy is carried out during hours of darkness and in single reliefs, but it may happen that work must be done in continuous reliefs, that is, a fresh working party is provided every 4 hours.

Continuous reliefs are difficult to manage in the case of large numbers.

Relieving parties should not pass each other on their way to or from their work. There must be an up-and-down route in a trench system : communication trenches must be definitely allotted as up-and-down routes by the general staff.

If task work is being used, it is essential that all tasks shall be thoroughly cleaned up before the men leave work, or there is abundant occasion for grumbling.

Each relief should arrive complete with all tools required for work and should return them to the dump from which they were drawn. It is not possible to hand over tools from one relief to the next in the dark.

If the work is being carried out in single reliefs at night, a clearing-up party should be detailed for work in daylight in order to square up the trenches and correct faults, and leave the trench fit for straightforward task work for the next relief.

This party must not be so numerous as to attract the notice of the enemy.

70. *Procedure when demands have to be sent in without previous reconnaissance.*

The following procedure should be followed when there is not sufficient time to send back " demands " after making a reconnaissance
of the work :—

    i. Orders are issued by higher authority to engineers and infantry
       to carry out the work.

    ii. **The officer in charge of the work** at once sends in the
       following information, based on previous knowledge of
       the ground and large scale maps :—

    (*a*) An estimate of the numbers of the working party.

    (*b*) Details of tools to be carried, or tools or stores to be picked
       up at the rendezvous.

    (*c*) Rendezvous for the working party.

    (*d*) Time for the working party to be at the rendezvous.

    (*e*) Duration of work.

    (*f*) Probable time of arrival back in billets.

He will also :—

    (*g*) Detail guides to lead the working party from the rendezvous
       to the site of the work.

    (*h*) Request that **the officer in command of the working
       party** with officers or non-commissioned officers re-
       presenting companies or platoons, be at the rendezvous
       1 hour before the remainder of the working party arrives
       there. This party will be guided immediately on arrival
       at rendezvous to the site of the work where the officer
       in command of the working party will meet the officer
       in charge of the work.

iii. **The officer in charge of the work**, with tracing party and guides for working party proceeds from rendezvous to site of work, where a reconnaissance is made and work marked out. If under enemy observation this marking out may not take place until dusk.

iv. **The officer in command of the working party** with party enumerated in ii (*h*) above, on arrival at rendezvous 1 hour before the remainder of the working party, is guided by the guide detailed by the O. i/c work to site of work, where he meets O. i/c work, who explains the task in hand, and hands over necessary templets.

v. **The officer in command of the working party** decides how to distribute his companies and platoons.

vi. Main body of working party, on arrival at rendezvous at appointed time, is taken by guides to site of work, where under the instructions of the officer in command of the working party, the men are marched at once straight to their tasks.

71. *Tracing and setting out work.*

1. In order to avoid delay, any work that is to be executed should be marked out beforehand by tracing tapes (each tape is 50 yards long), or Hambro' lines, pickets, spitlocking, &c. The following instructions refer to tracing trenches, but they apply in principle to setting out wire entanglements, roads, tracks, tramways, &c.

Tracing has to be done frequently by night, in which case preliminary reconnaissance is important; the direction should be checked at frequent intervals by compass or by measuring the distance at right angles to a known line, such as the front line trench in the case of a tape laid for wire in front of forward positions. The steel helmet and box respirator must be taken off before the compass is used; both these articles of equipment affect it.

All officers and non-commissioned officers should be practised in tracing by day and by night.

2. In the case of deliberate defences, if the line is marked by flags (Sec. **12**), the trench should not be laid out in a series of straight lines from flag to flag; but each fire-bay should be carefully sited and marked with pegs by the tracing officer; the traverses should be fitted in by a non-commissioned officer following behind.

Similarly, if the line is marked by a ploughed furrow, this must not be slavishly followed in tracing the fire-bays.

3. Types of trace are illustrated on Plates **37, 38, 39, 40** and **42**; they may be applied to the ground by :—

i. Varying the angle of the traverse. This angle, must, however, never exceed 135°, or be less than 90°.

ii. Varying the length of the fire-bays, or of the legs of the traverses.

iii. Changing direction in a fire-bay.

4. With the bastion trace (Plate **37**, Fig. 2), until experience in laying out has been acquired, a tape templet should be used (Plate **102**, Fig. 1). This consists of a quadrilateral of tape of the dimensions of a normal traverse with two diagonals of tape to act as braces. It must be regarded as a guide, and not as a standard pattern for a traverse. After some practice, it is quicker to dispense with the templet, the sides of the traverses may be paced and the angles judged by eye.

5. The front limit or " cutting line " of a fire trench is the line which should be traced, but it is a great advantage, if sufficient tape is available, to trace both sides.

6. **Organization of tracing parties.**—The party should be divided into groups as follows :—

    i. An officer and 1 O.R. with extra men as carriers. The officer traces out the fire-bays, driving pegs in at the ends of them ; he must select the position of each fire-bay with regard to the shape of the ground.

    ii. An experienced non-commissioned officer and 2 men with extra carriers if necessary. If a templet is used, it is laid out with its two front corners near the inside pegs of two adjacent fire-bays, and the back pegs of the traverse are driven in near the back corners of the templet, under the direction of the non-commissioned officer.

    iii. A number of men, varying with the nature of the ground, to clear crops, bushes, &c., from the line of the tape.

    iv. One man running out the tape and fixing it to the pegs, with carriers for extra tape.

The duties of the various parties are tabulated below :—

Party.	Composition.	Tools and stores.	Duties.
No. 1	1 officer, 2 or 3 men.	1 mallet, bundles of pegs.	Peg out fire-bays.
No. 2	1 N.C.O., 2 to 4 men.	(Tape templet) 1 mallet, bundles of pegs.	Peg out traverses.
No. 3	As required by nature of ground.	Reaping hooks, &c....	Clear line for tape.
No. 4	2 or 3 men.	Tape	Carrying tape and fixing to pegs.

Total : 1 officer, 1 non-commissioned officer, and 6 or more men, depending on the amount of material to be carried, and the amount of clearing to be done.

French wire entanglement staples or wire " hairpins," are good substitutes for pegs ; the tape is laid very loosely at first, and then pulled back taut round the traverses (Plate **102**, Fig. 2).

PART II.—BRIDGING.

CHAPTER XI.

KNOTTING AND LASHINGS.

72. *Knots.*

The following are the most useful knots and their principal uses :—

Thumb knot and Figure of 8 knot (Plate **103**, Figs. 1 and 2).—To make a stop on a rope, or to prevent the end from fraying, or to prevent its slipping through a block.

Reef knot (Plate **103**, Fig. 3).—To *bend* or join two dry ropes the same size.

Single-sheet bend (Plate **103**, Fig. 4).—To join two dry ropes of different sizes.

Double-sheet bend (Plate **103**, Fig. 5).—To join two ropes with great security, or for wet ropes of different sizes.

Hawser bend (Plate **103**, Fig. 6).—To join large cables.

Bowline and running bowline (Plate **103**, Figs. 7 and 8).—To form a loop or **bight** on a rope which will not slip. The loop formed by passing a bight through a bowline loop at the end of a rope is called a **running bowline.**

To secure the ends of ropes to spars, pickets, &c., or to other ropes, the following hitches are used :—

Clove hitch (two half-hitches) (Plate **104**, Figs. 1 and 2).—Generally used for the commencement and finish of lashings.

Timber hitch (Plate **103**, Fig. 9).—For holding timber, &c., where the weight will keep the hitch taut.

Two half-hitches (Plate **104**, Fig. 3).—For making fast the running end of a rope on to its standing part.

Round turn and two half-hitches (Plate **104**, Fig. 4).—For belaying (or making fast) a rope so that the strain on the rope shall not jam the hitches. This will be used for making fast a rope to a bollard or anchorage. Should the running end be inconveniently long, a bight of it should be used to form the half-hitches.

Fisherman's bend (Plate **104**, Fig. 5).—For making fast when there is a give-and-take motion, *e.g.*, for bending a cable to an anchor.

Lever hitch (Plate **104**, Fig. 7).—For drawing pickets by a lever and fulcrum fixing the rounds of a rope ladder, fixing bars to dragropes, &c.

Man harness hitch (Plate **104**, Fig. 8).—To form a loop on a drag-rope which will not draw tight ; the loop being of a size to pass over a man's shoulder.

Running knot (Plate **104**, Fig. 6).—To form a loop which will draw taut round an object.

Catspaw (Plate **105**, Fig. 1), or,

Single Blackwall hitch (Plate **105**, Fig. 2), or,

Double Blackwall hitch (Plate 105, Fig. 3).—To secure a rope to a hook.

Draw hitch (Plate **105**, Figs. 4, 5, 6).—To secure boat's painter, &c., to a post, ring, or rope, so that it can be instantly released. This knot will stand a give-and-take motion, and can be instantly released by a jerk on the running end.

Magnus hitch (Plate **106**, Figs. 1 and 2).—To make fast to round spars when much friction is necessary to prevent slipping.

Rolling hitch (Plate **106**, Fig. 3).—To make fast a rope end round an object or secure a rope to a hook.

Stopper hitch (Plate **105**, Fig. 7).—To transfer the strain of one rope to another for use on occasions when it is necessary to shift the strain off a rope temporarily.

73. *Lashings.*

1. A **rack lashing** consists of an 8-foot length of 1½-inch rope, with a pointed stick at one end. It is used for fastening down ribands at the edge of the roadway of bridges.

2. **Square lashing.**—To lash one spar square across another, commence by a clove hitch on spar (*a*) below (*b*) (Plate **106**, Fig. 4), and twist the ends together, carry at least four turns round the spars, as shown in figure, keeping outside previous turns on one spar and inside on the other ; two or more **frapping** or cross turns are then taken, the corners of the lashings being well " beaten in " during the process ; finish off with two half-hitches round the most convenient spar.

When the spars are the leg and transom of a trestle or frame, the clove hitches should be on the leg below the transom, and the lashings should be finished off on the transom outside the leg. When the spars are leg and ledger, the clove hitch should be on the leg above the ledger.

3. **Diagonal lashing.**—To lash two spars together that tend to spring apart. Begin with a timber hitch or running bowline round both spars and draw them together, then take three or four turns across each fork and finish with frapping turns and two half-hitches (Plate **106**, Fig. 5). When the spars are not horizontal, the lashing should be finished off above the junction.

4. **Wooden wedges** with well-blunted points are often useful for tightening lashings. They are generally used by builders in scaffolding, and should be driven in at the top of the lashings.

5. **Hemp-rope lashings** soon become loose, and require frequent re-making. Wire lashings should be used in their place when possible. These can be made in a similar way to hemp-rope lashings ; but, unless staples are available, the wire should be finished off round a set of returns, and jammed between them and the timber. It is of little use attempting to finish off on a round spar. No. 14 gauge steel wire may be taken to have a quarter the strength of 2-inch cordage.

6. **To lash a block to a spar.**—The back of the hook is laid against the spar, a clove hitch is taken round the spar above the hook, then several turns round the hook and spar, and finished off with two half-hitches round the spar below the hook (Plate **106**, Fig. 6).

7. **To mouse** the hook of a block take some turns round it with spun yarn or very light lashing, commencing with a clove hitch on the back of the hook and finishing off with one or two frapping turns and a reef knot (Plate **106**, Fig. 6).

8. **To seize** the end of a rope to the standing part with spun yarn or string, make a clove hitch round one of the ropes with the spun yarn near its centre, taking each part round both ropes in opposite directions, leaving one end long enough to make two frapping turns between the ropes, and tie the two ends with a reef knot (Plate **105**, Fig. 7).

9. **To whip** a rope is to tie a piece of twine round the end to prevent it from untwisting and unfraying.

10. **Fishing spars.**—To fish spars is to strengthen them, by lashing other spars parallel to them.

The fishing spars should be against the spar to be fished, so that they may take off as much strain and be in as close contact as possible, and the lashings must be tightened with wedges. The lashing rope or ropes having been made fast by one end being laid along the spars, so as to be covered by the returns, are then passed round and round all the spars until, as it were, one spar is formed, the end being made fast by taking the last four or five returns rather slack, the end then passed backwards under them, the returns tightened up and the slack hauled through.

Several separate lashings may be applied instead of one continuous one.

74. *Drag-ropes, slings, &c.*

1. **Drag-ropes** for general service are of two natures, heavy, which are of 3-inch white rope, 5 fathoms long, and light, of 2-inch white rope, 15 feet 6 inches long. A hook is spliced into one end.

2. **Slings** are made of rope, wire or chain. Rope slings are made by splicing the ends of a rope together.

Rope slings are designated by their circumference in inches and by their length in feet when laid out straight and stretched.

In describing chain slings the size of link given is the diameter of the iron of which it is made.

Chain slings are much heavier than rope slings of similar strength, but they are more durable and free from stretch.

3. **Selvagees** are issued for the purpose of slinging projectiles, but, when passed round an object, form a convenient means of attachment for the hooks of tackles : they should always be used on the double, and are only suitable when the strain to be borne is small (Plate **106**, Fig. 7).

75. *Holdfasts.*

1. The artificial holdfasts most frequently used are picket, baulk, anchor, or buried holdfasts.

2. A **picket holdfast** consists of one or more pickets driven into the ground (Plate **107**, Fig. 1).

The usual combinations of pickets are 1.1, 2.1, or 3.2.1.

3. The safe stresses with 5-foot pickets driven 3 feet into good ground are :—

Single picket	...	...	...	...	... 7 cwts.
1.1 picket holdfast	...	...	...	... 14 cwts.
2.1 „ „	...	...	...	... 1 ton.
3.2.1. „ „	...	...	...	... 2 tons.

In the case of the 1.1 pickets used with a baulk holdfast, if spaced 18 inches apart along the baulk (Plate **107**, Fig. 2), each of the 1.1 pickets may be considered able to take 12 cwts. ; thus, with 6 pickets front and rear, the safe stress would be about 3½ tons.

4. Before driving pickets, it should be ascertained, by plan or otherwise, that cables or pipes will not be interfered with. In driving pickets of different sizes, the strongest should be nearest the weight, and at right angles to the direction of strain ; they are grouped thus :— 3 in front, then 2, then 1.

The eyes of the pickets should be to the front. Pickets should be driven at such a distance apart that the angle between the lashing and the pickets is approximately a right angle. The stress should always be applied close to the surface of the ground. In lashing pickets together, make fast the lashing rope to the head of the front picket, and lash to the foot of that behind, and frap next to the larger, so as to get the returns of the lashing as close together as possible. To prevent the lashing slipping up, a collar of spun yarn may be applied round the rear pickets above the lashing.

5. In drawing all pickets care should be taken that they are drawn out in the same line as that in which they have been driven, otherwise they may be broken.

6. A good method of drawing 5-foot pickets by using a stout 12-foot or 14-foot pole as a lever is as follows :—Place a wooden block in rear of and adjacent to the pickets to act as the fulcrum (*see* Sec. **76**) and the pole against the pickets, so that one rests on the block and the other on the ground in front of the pickets.

Pass a clove hitch on a soft drag-rope round the lever and pickets, take in the slack and pass the ends round the pickets in opposite directions two or three times, twist them together, and hold on while the lever is lifted.

7. A limber can be employed in drawing pickets as follows :—It is run up with its hook immediately in rear of the pickets to be drawn and the pole raised. A clove hitch, made in the centre of a drag-rope, is passed round the picket or pickets low down ; the ends are passed round the hook in opposite directions two or three times, twisted together and held on to. A collar of spun yarn round the picket above the clove hitch keeps it from slipping up. On the pole being pulled down the picket is drawn.

8. A **baulk holdfast** (Plate **107**, Fig. 2) may be made by driving a row of pickets, placing a heavy baulk behind them in such a way that it bears evenly against each picket, driving a second row of pickets immediately in rear of the first row, and lashing the pickets as shown : the pickets should be at equal distances on either side of the centre of the baulk. The baulk is inclined by sinking the rear edge, in order to give a fair bearing against the pickets. Sufficient soil is removed to allow of straps or ropes passing round the baulk.

9. When a **buried log** (Plate **107,** Fig. 3) is used for large strains, a trench is dug long enough to hold the log, and the cable is given one complete turn round it and passed up through a narrow incline constructed at right angles to the trench. The running end is then seized to the standing part in two or three places. The method of calculating the depth of the trench and the precautions to be taken when using this anchorage are given in Military Engineering, Vol. III.

76. *Levers.*

1. A lever is a rigid bar capable of motion about a fixed point. The bar may be straight, as a handspike, or bent, as a claw hammer used to extract nails.

2. The point about which the lever turns is called the FULCRUM. The distances from the fulcrum to the points of application of the power and weight are called the **lever** and **counterlever** respectively.

3. There are three orders of the lever.

First order.—When the fulcrum is between the weight and the power.

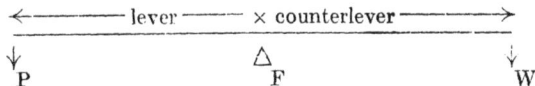

$$\xleftarrow{\hspace{1.5cm}} \text{lever} \xrightarrow{\hspace{0.5cm}} \times \text{counterlever} \xrightarrow{\hspace{2cm}}$$

$$\downarrow \qquad\qquad\qquad \triangle \qquad\qquad\qquad \downarrow$$
$$\text{P} \qquad\qquad\qquad \text{F} \qquad\qquad\qquad \text{W}$$

The handle of a lift pump is an example of a lever of the first order.

Second order.—When the weight is between the fulcrum and the power.

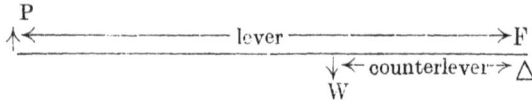

P
△ ←———————— lever ————————→ F
  ↓ ←— counterlever —→ △
  W

A wheelbarrow is an example of a lever of the second order.

Third order.—When the power is between the fulcrum and the weight.

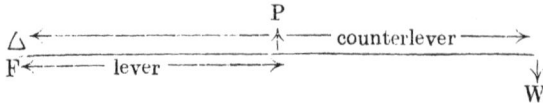

                          P
△ ←————————————————△—— counterlever ———→
F ←——— lever ————→
                                          ↓
                                          W

The elevators of a H.P. mounting form an example of a lever of the third order. In this case the gun is the weight, the power is applied by the ram and the pins of the elevator are the fulcrum.

4. The power, P, required to raise a weight by lever is—

$$P = \frac{W \times CL}{L}$$

where L is the length of the lever, CL is the length of the counterlever.

77. *Strength and size of cordage, &c.*

1. The chief requisites of rope are strength, suppleness and durability: service ropes are made of hemp, manilla fibre, and coir: of these hemp is the strongest and most reliable: it is issued white or tarred.

Manilla stands exposure to weather and damp better than hemp, and does not kink so much, but cuts, where knotted.

Coir is very weak but light. It will float in water and does not rot: it is suitable for towing purposes.

2. Under a heavy stress a new rope exhibits a tendency to unlay itself, and so to twist: new rope must therefore be stretched before use.

To uncoil a new coil of hemp rope, pass the end which is at the core through the core to the opposite side and draw it out; the turns will then run out without kinking.

A new hemp rope, when subjected to its working stress, will stretch about one-twentieth of its length.

Cordage must be returned to store dry, and should not be coiled down in a wet condition.

3. The following precautions should be observed when working with cordage :—

    i. Sudden or snapping stresses should not be allowed to come on a rope.

ii. A rope exposed to chafe should be " parcelled " by wrapping it in canvas well daubed with tar and bound with spun yarn.

iii. Ropes should as far as possible be kept from chafing against sharp edges by interposing rollers or handspikes.

iv. Rope which has been thoroughly saturated should not be exposed to a greater stress than two-thirds of what would otherwise have been its safe working stress.

4. The size of a rope is denoted by its circumference in inches, and its length is given in fathoms (a fathom is 6 feet). Cordage is usually issued in coils of 113 fathoms, and steel wire ropes in coils of 100 fathoms.

5. White rope new is 30 per cent. stronger than tarred hawser rope and 20 per cent. stronger than tarred bolt rope. White rope is also more reliable, since it is made of the best Italian hemp.

Rope when wet shrinks, and when thoroughly saturated is weaker than when dry. White rope may lose a third of its strength if soaked for two or three days.

6. **Working stress.**—If C be the circumference of a rope in inches—

i. For unselected cordage, C^2 cwts.

ii. For white service rope in good condition, $2C^2$ cwts.

iii. For steel wire rope, $9C^2$ cwts.

7. The **strength of wire** varies greatly; as a very rough rule it may be taken that the breaking weight in pounds equals three times the weight per mile in pounds (Appendix IV., Table XII). This rule holds good for iron and hard-drawn copper wire, while steel wire may be taken as about twice as strong.

8. **Strength of lashings.**—The safe holding power of cordage lashings is $\frac{4}{5}$ of the working stress of cordage multiplied by the number of returns; with wire lashings $\frac{2}{3}$ of safe stress of the wires. These fractions are to allow for the unequal straining of the cordage or wires.

Thus a square lashing gives four returns for each complete turn of the lashing, so that a square lashing of four complete turns of $1\frac{1}{2}$-inch hemp cordage has a strength of

$$\frac{3}{5} \times 16 \times \left(\frac{3}{2}\right)^2 = 21 \cdot 6 \text{ cwts.}$$

To find the number of turns required to lash a block to a spar, multiply the stress on the block by $1\frac{1}{2}$, and apply the above rule. The factor, $1\frac{1}{2}$, is to allow for the angle the turns of the lashing make with the direction of pull. Thus the number of turns of 2-inch cordage required to lash a block having a pull of 1 ton to a spar is $1 \times 1\frac{1}{2} \times 20$ divided by $4 \times \dfrac{3}{5} \times 2$ or 7 complete turns.

CHAPTER XII.

BLOCKS, TACKLES AND USE OF SPARS.

78. *Blocks.*

1. **Blocks** are used for the purpose of changing the direction of ropes or of gaining power.

They are called single, double, treble, &c., according to the number of sheaves which they contain. The sheaves revolve on a pin, which should be kept well lubricated.

Snatch blocks (Plate **108**, Fig. 1) are single blocks with an opening in one side of the shell, to admit a rope without passing its end through. This opening is closed by a hinged strap.

2. A **fall** is the name given to the rope with which the tackles are rove.

The **standing end** of the fall is the fixed end, the other the **running end**.

A **return** is any part of the fall between two blocks, or between the ends of the rope and any block.

To **overhaul** the tackle is to separate the blocks.

To **round in** is to bring them closer together.

The blocks are said to be **chock** when they are brought together.

To **rack** a tackle is to fasten two opposite returns together, so that the blocks may retain their relative position when the running end is let go.

3. A tackle is rove by two men, back to back, 6 feet apart ; the blocks should be on their sides between the men's feet, hooks to the front and the coil of the rope to the right of the block at which there are to be the greater number of returns. Beginning with the lowest sheaf of this block, the end of the fall which is to be the standing end is passed successively through the sheaves from right to left and then made fast.

4. To **overhaul** a heavy tackle :—

> The block from which the running end comes off is made fast to some holdfast, a handspike is placed through the hook or shackle of the other block, and a drag rope made fast to it.

> To overhaul the tackle the handspike and drag rope are manned, whilst one or two men overhaul the returns from the standing block.

> **Rounding in** is the converse : the numbers on the drag rope and handspike hang back, whilst a few numbers heave on the running end.

> The handspike, both in overhauling and rounding in, is to keep the tackle out of the dirt, which would clog the sheaves.

5. In using tackle great care must be taken to prevent it from twisting. The best method is to place a handspike between the returns, close to the movable block, with a rope to each end, by means of which it can be steadied.

79. *Tackles.*

1. Various **tackles** are shown on Plate **108**. The simplest form of tackle is a **whip**, that is, a single movable block, rove with a rope one end of which is attached to a holdfast, while the other is hauled on (Fig. 3).

When one tackle is bent to another, the total mechanical power is the product of the powers of the two tackles, *e.g.*, if a tackle giving a power of four to one is bent to a tackle giving a power of two to one, the total power gained is eight to one, *e.g.*, whip upon whip tackle (Fig. 2).

Theoretically in any system of two blocks the power required to raise a weight, W, is W divided by the number of returns at the movable block. An addition has to be made to overcome friction and resistance of the ropes to bending. Suitable tackles for most operations required in the field can be selected from Appendix V., Table III.

2. The fall, in lifting heavy weights, can rarely be worked by hand, but has to be " **led** " to either a capstan or winch by which power is gained and a steady pull ensured.

3. In using tackles with sheers, gyns or derricks, the running end of the fall should always be led through a " **leading block** " lashed, as a rule, to one of the spars a few feet above the ground ; a snatch block is most convenient for the purpose.

4. Before using a tackle it should be seen that :—

   i. The straps, blocks and fall are in good condition.

   ii. The blocks are well lubricated and free from grit and dirt. No block in good working order should " complain," that is, make a noise.

   iii. The pins securing the hooks or shackles, and sheaves of the block are made fast.

   iv. The standing end of the fall is properly made fast.

   v. The fall is free from kinks, runs freely over the sheaves and has a fair lead. Whenever possible the pull should be down hill.

5. During use it should be seen that :—

   i. When the fall is taut it is not jarred by being struck or by men treading on it.

   ii. The returns near the blocks are not touched when moving unless absolutely necessary, and then only those moving away from the block.

   iii. The stoppering of the fall, when necessary, is correctly carried out, and that the stopper is equal to the stress it will have to bear.

   iv. The position of the men is such that they will not be injured in the event of an accident.

v. When going round a curve the running end is led off so that the resultant strain is in the direction of travel.

vi. When a suspended weight is eased off it is done uniformly and not by jerks.

vii. In all cases when working near the safe limit of a tackle, before leaving a weight suspended, the tackle should be eased off slightly after raising or raised slightly after lowering.

6. Tackles should be carried and not dragged along the ground.

7. Two tackles should not be hooked into the same sling if it can be avoided.

8. Men should be taught to pull together silently. The fall should be **double banked** as this will keep the pull in a straight line. At the caution **taut** the slack is hove in, and at the word **heave**, they pull together and keep what they get.

80. *Single derrick.*

1. A **derrick** is a single spar set up with four guys at right angles to one another, secured to the tip with clove hitches (Plate **106**, Fig. 6). A block for the tackle is lashed to the head and the derrick can be used for raising and swinging a weight into any position within its reach, which is about one-fifth of its height. The anchorages for the guys should be at a distance from the foot of the derrick equal to twice its height. The foot should be let into a hole in the ground to prevent it slipping and should rest on a bearing plate.

2. To raise a derrick, the spar is first laid on the line joining the footing and one of the guy anchorages, with the butt nearly over the footing. A foot-rope is secured to the butt and to a holdfast on the same side of the footing as the spar is on, and close to it. The four guys having been made fast to the tip and passed to their holdfasts, the tip is lifted as high as possible by hand. The back guy is then hauled on and the fore guy let out until the derrick is in the desired position.

3. In carrying a spar, the party should be equally divided on either side of it, and sized from one end. The spar should then be lifted in two motions on to the inner shoulders of the men. In lowering a spar the party should slowly face inwards, and lower first the butt end and afterwards the tip. One man should always give the word for lifting and lowering.

81. *Swinging derrick.*

1. A **swinging derrick** consists of a standing derrick with a swinging arm attached to it near its foot. The tip of this swinging arm is connected to the upright spar by a connecting tackle, and the main or lifting tackle is attached to the tip of the swinging arm or jib (Plate **110**). The upright spar is practically the same as a standing derrick, with the exception that as it will frequently be erected at the

edge of a wharf or other place where it is not convenient to use a fore guy, a strut or struts must be used instead. A good method is to use two struts each about half as long again as the upright spar, lashing the three together as in a gyn (Plate **109**, Fig. 3), and then to erect it so that the standing derrick is vertical and the two struts are at right-angles in plan, and situated symmetrically with respect to the edge of the wharf. Two guys can also be used, one over each strut in plan, but secured to holdfasts at the customary distance ; or three guys can be used, one a back guy and the other two side guys, but set back about 20° from the edge of the wharf, to allow room for the loads to be landed. It is important that these guys should not stretch too much, so that the upright spar may remain vertical. It is, therefore, better to make them of wire rope, and in any case a tackle should be included in them to take up any slack. The stress in these guys is much greater than that in guys for a standing derrick to deal with the same weight.

2. The jib is most conveniently formed of two spars lashed together at their tip, and separated at their butts to a distance about equal to the diameter of the upright spar. They are lashed about 12 or 18 inches from their butts to a stout cross-piece, and the end thus formed encircles the butt of the upright (Plate **110**). The jib is supported by a length of chain secured at its centre by a clove hitch round the upright, and prevented from descending by a collar of rope, and each end secured to one of the arms of the jib. The latter is enabled to swing under the control of two side guys or reins attached to its tip. The length of the jib may be of the same length as the upright. The inclination of the jib can be altered by the connecting tackle, and the radius of its circle of operation is thus determined. The weight can be lifted or lowered by the main tackle and the jib swung by the reins. The jib can only be swung right over the edge of the wharf on that side of the upright on which are the leading blocks of the main and connecting tackle. If it is necessary to swing both ways, duplicate leading blocks for the falls of these tackles must be provided.

3. It must be noted that the weight must be allowed to hang vertically from the tip of the jib. If it is hauled towards the butt of the standing spar, the thrust on the jib is greatly increased ; if it is hauled away from this butt, the stresses in the back guy and connecting tackle are largely increased. If the weight has to be brought in otherwise than by swinging to a flank, it should be done by raising the head of the jib by means of the connecting tackle.

4. The number of men required to erect derricks to lift weights up to 4 tons is from 20 to 25.

82. Sheers.

1. **Sheers** require only two guys—" fore " and " back." They should be fastened to the legs above the crutch by clove hitches, the

back guy to the fore spar, and *vice versa*, so that their action may tend to draw the spars closer together and not strain the lashing (Plate **109,** Fig. 2). The minimum distance of the anchorages from the legs should be double the height of the sheers. The upper block of the tackle is hooked to a sling of rope or chain passed over the crutch. Sheers can, as a rule, be used for heavier weights than derricks, but can only move them in a vertical plane passing between the legs. The feet of sheers must be secured or let into holes in the ground. The distance apart of the legs should not be more than one-third the length of the leg up to the crutch, and the sheers should not be heeled over more than one-fifth of their height.

2. In order to lash the legs they are laid side by side on a skid, and kept 2 inches apart by a wedge. The lashing is commenced with a clove hitch on one spar, carried six or more times upwards round both spars without riding, then two frapping turns, and finished off with two half hitches round the other spar (Plate **109,** Fig. 1).

83. *Gyns.*

1. A **gyn** consists of three spars lashed together at the tips, the butts forming an equilateral triangle on the ground. It requires no guys but can only be used for a vertical lift.

2. In order to lash the legs, the spars are laid on a skid, a clove hitch is made on one of the outside spars, and the lashing taken loosely over and under the three spars six or eight times : frapping turns are taken round the lashing between each pair of spars, finishing off with two half hitches on the other outside spar (Plate **109,** Fig. 3).

3. The two outside spars are then crossed until the distance between the butts equals half the length of the leg : a ledger is lashed across these two spars about 1 foot from the ground. The gyn is then raised by using the centre spar as a " prypole," and the two remaining ledgers lashed on.

4. Many varieties of " service " gyns are available : of these one of the most useful is known as the gyn, triangle, light, to raise 2¾ tons ; this gyn weighs 2 cwts. is made of steel and consists of two cheeks, 13 feet 3 inches long, a prypole, and a shackle connected by a headbolt and a 3-ton differential tackle. The cheeks have loops at their lower ends for a bracing rope. When erected for use, the cheeks should be 8 feet apart, and the foot of the prypole 12 feet from the cheeks.

5. **Points to be observed when working with gyns :—**

i. The prypole is always considered to be the front.

ii. The foot of the prypole should be equi-distant from the feet of the cheeks.

iii. All three feet should be on the same level and properly supported. If, as sometimes happens, the prypole has to be lower than the feet of

the cheeks the maximum load put on the gyn must be considerably reduced.

iv. The height of lift can be increased by placing the feet on planks or skidding.

v. The gyn should be placed with its head over the centre of gravity of the weight to be raised, except where a swing lift is required when the head of the gyn should be over a point midway between the points at which the weight has to be picked up and lowered.

vi. A suspended weight can be hauled straight towards the centre of the cheeks or towards the prypole without risk of upsetting the gyn, but hauling it to either flank or taking a swinging lift is liable to cause the gyn to capsize and should only be done with the greatest caution. In no case should the lower block be allowed to come outside the line joining the prypole and cheek. If a considerable swing is required light side guys should be attached to the head of the gyn, or the cheeks lashed down to prevent them rising.

vii. Before taking the weight, the foot of the prypole should be lashed to the cheeks.

viii. A gyn should not be left standing on pavement, concrete, or any smooth surface, though there may be no weight suspended, without securing the feet from slipping.

ix. Any slip of the fall round the windlass, called " surging," should be carefully guarded against, as it may increase the stress on the tackle and the gyn as much as 50 per cent. If the windlass is slippery, or if working to a close margin, extra turns should be taken. Sand or grit should never be applied to a windlass.

x. Steadying ropes or tackles should always be attached to the weight to keep it from swinging when raised.

84. *Size of timber, tackles and ropes for derricks, &c.*

1. The design of derricks, sheers and gyns depends on the weight, the height to which it is to be lifted and the stores available. The length of the spars must be sufficient to allow for the length of slings and for blocks becoming chock in addition to the height which the weight has to be lifted. A tackle when chock-a-block will occupy 4 or 5 feet.

In the case of derricks and sheers the distance from the base at which the weight has to be picked up may determine the height since this distance must not exceed $\frac{1}{5}$ of the height.

2. The stresses in individual members are best determined graphically, but as this is not always possible, Tables I, II, III, and IV are given in Appendix V., from which suitable spars, ropes and tackles can be selected.

3. Example :—*A weight of 3 tons has to be lifted 10 feet and moved 4 feet horizontally. Compare the principal stores required for doing this with sheers and with a standing derrick.*

The minimum height of either sheers or derrick to the point of attachment of the main tackle is 4 by 5 feet = 20 feet. This will be sufficient to raise the weight 10 feet allowing for tackles and slings. In both cases the main lifting tackle is the same, viz., two treble blocks with 1½-inch wire rope (Table III).

For sheers the stresses are (Table I) :—

In leg with leading block	$3 \times$ ·9 ton = 2·7 tons ∴ use 10½-inch spar (Table II).
In other leg	$3 \times$ ·7 ton = 2·1 tons ∴ use 10-inch spar (Table II).
In back guy	$3 \times 1·3$ tons = 3·9 tons ∴ use two treble blocks and 2-inch wire rope (Table III).

For a derrick :—

Stress in spar	$3 \times 1·5$ tons = 4·5 tons ∴ 12-inch spar is required (Table II).
In running guys ...	$3 \times$ ·5 ton = 1·5 tons ∴ two treble blocks with 3-inch cordage are suitable (Table III).
In other guys	$3 \times$ ·3 ton = ·9 ton ∴ 3-inch cordage used double will do (Sec. 77, para. 6).

A reference to Table IV shows that, while with the sheers a capstan or winch will be necessary for lifting tackle and back guy with the derrick, the guys can be manipulated, if necessary, by hand power.

Note.—From the tables square baulks and other systems of tackles can readily be selected to suit the stores available.

CHAPTER XIII.

ROAD BRIDGES AND THE PASSAGE OF GAPS.

85. *Classification of bridges and of the materials of which they are built.*

1. Road bridges may be required in a theatre of war :—

  i. To supplement existing bridges so as to enable the roads to carry the increased military traffic.
  ii. To enable an army to force a crossing over an obstacle in the face of opposition.
  iii. To replace bridges destroyed.

2. **Classification of bridges.**—Bridges are classified according to the loads which they are designed to carry :—

  i. *Light.*—To carry cavalry, infantry, field artillery and first line transport.

ii. *Medium.*—To carry medium artillery, lorries, motor buses, &c.

iii. *Heavy.*—To carry heavy artillery and special heavy vehicles.

iv. *Tank.*—To carry heavy tanks.

Light bridges are of three types :—

(a) Infantry footbridges.

(b) Pack bridges for pack animals.

(c) Artillery and transport bridges.

3. **Material.**—The material for making road bridges consists of :- -

i. The mobile (pontoon) bridging equipment of the army.

ii. Certain stock span bridges.

iii. Such materials as may be obtained locally or from engineer parks.

4. **Use of materials. --**

i. The pontoon equipment is primarily intended for the rapid construction of light bridges for the tactical crossings of rivers, canals, &c. It may be used also for medium and heavy bridges. This bridging material should not be allowed to remain in bridge longer than the tactical situation demands but must be replaced by other types of bridges, so that the mobile equipment may be always available for tactical crossings. Instructions for using this equipment are given in Military Engineering, Vol. III.

ii. Stock span bridges, together with material for the piers, are held in engineer parks. They are steel bridges designed to meet the requirements of speed in erection and lightness of individual parts for transport to the site. They are medium or heavy bridges used to repair and complete the communication of an advancing army, or to supplement existing bridges in case of a possible retirement. Only a limited number of these bridges is likely to be available, so that they will be used only when a bridge is required in the shortest possible time. The erection of these bridges is described in Military Engineering, Vol. III.

iii. Bridges can be made of many types out of local materials, supplemented, if necessary, by timber and rolled steel joists from engineer parks. Except in the case of infantry footbridges, which will be prepared beforehand for a particular operation, these bridges require more time for erection than the other types, but will be built when time and local resources admit, so that the stock span bridges may be available when required for urgent work.

86. *Considerations affecting the position of bridges and the arrangement of bridges and approaches for assaulting across a gap.*

1. **Locality of bridges.**—Light bridges will be sited according to tactical requirements ; medium and heavy bridges according to the

requirements for traffic control. Approaches generally determine the locality of bridges, as long diversions from existing roads require more time, labour and material for their construction than the bridges themselves. A bridge site, however, in the direct line of existing approaches must be reserved for the heaviest type of bridge, which will eventually be built in any locality ; the sites for medium and light bridges being chosen clear of the site for the heavy bridge. Also when building a single traffic bridge on existing abutments, the bridge should be placed, if possible, clear of the centre line of the abutments, so that a second bridge can be built alongside with the minimum of labour.

2. In the case of an assault across a gap, the number of tactical crossings will be limited only by the amount of bridging material which can be delivered at the sites. Vehicles carrying this material must be given priority, but if too many crossings are attempted other traffic may be unduly delayed. The infantry crossings should be numerous and arranged in pairs, about 5 yards apart, and spare material, equivalent to that required for one crossing, should be available at each pair of crossings.

The crossings within a divisional front must be clearly numbered from the right (*i.e.*, No. 1 infantry, No. 2 ford infantry only, No. 3 field guns, No. 4 infantry, and so on), both at the crossings and on the forming up line ; tapes being laid from the forming up line to the crossings either beforehand if the gap can be got at by the bridging parties, or unrolled as the bridging parties advance at the time of assault. Illuminated signs must be provided on the forming up line by night ; lights in petrol tins, pierced with small holes, cannot be seen by aeroplanes.

3. During a daylight operation arrangements must be made to show troops, &c., coming up, which crossings are " in action." Flags are the best method, one colour for infantry, one for pack animals, and one for guns and vehicles : they should be fixed on the line of approach to, and 200 yards short of, the crossing. For pack animals, guns and vehicles, a " control " line should be established well back from the gap and out of direct observation by the enemy, the flags for pack animals, &c., being repeated along the line of approaches back to this " control " line. Animals, vehicles, &c., except guns going into action on the near side of the gap and vehicles with bridging material, both of which should be provided with passes, may only pass the " control " line when the flag of the bridge, which they are detailed to use, is flying. The junctions of the approaches and the " control " line must be very clearly marked with notice boards.

4. All military bridges are, in the first instance, constructed for single traffic only, and arrangements must be made to establish traffic controls directly the bridge is open for traffic. The officer in charge of construction is responsible for erecting the necessary notice boards

(*see* Sec. 89, para. 10), the traffic controls for seeing that the notice boards are strictly obeyed. Traffic, too heavy for the bridge, must be stopped sufficiently far away from the crossing to allow of it being directed elsewhere. By night when, owing to the presence of enemy aeroplanes, the notice boards cannot be illuminated, especial vigilance is required of the traffic controls. The arrangements for illumination are made by the officer in charge of construction ; their maintenance is the duty of the traffic control.

87. *Selection of the exact site for a bridge and measurement of the gap.*

1. The selection of the exact site for a bridge depends on the nature of the banks and approaches, the nature of the bed, the width to be bridged, the depth of the gap, and, when there is water, its depth, the strength of the current, the probability and extent of floods, and, if a tidal river, the rise and fall of the tide. The headway to be allowed over a road or railway is 12 feet 6 inches ; that over a canal varies in different countries, but is generally about 12 feet 6 inches. In navigable waterways provision must be made for a " cut " of 20 feet in floating bridges, or alternately for swinging bridge.

2. The **approaches** at both ends of the bridge are of paramount importance, to obviate any crowding on the bridge. Approaches, other than those consisting of properly metalled road, cut up very quickly, and, if neglected, become almost impassable in wet weather : some form of roadway, sleeper or corduroy (*see* Sec. **111**), must be laid down. Timber approaches have the advantage that animals get accustomed to the sound before coming on to the bridge. Approaches must be made up so that there is no impact or bump when vehicles pass from the approaches on to the bridge.

When ramped approaches are necessary, the slope for animals and wheeled vehicles should not be steeper than 1/10 ; in ramping down on to a bridge, the approach should be level for 3 or 4 yards immediately before the bridge in order to lessen the impact.

Except in close proximity to the enemy the approaches must be clearly outlined by stout pickets painted white ; tracing tapes must be used where this is impossible.

3. **Section of the gap.**—As soon as the site has been fixed, a section will be made to determine the type of bridge which best suits the gap and the gap will be divided into spans according to the material available for bridging.

A section is taken as follows :—

A string or rope, marked into these spans with knots of white tape, &c., is stretched as taut as possible across the gap at " shore transom " level, and a section of the gap obtained by using a pole marked in feet and inches. If the rope has much sag the method shown on Plate **111**, Fig. 1, must be adopted.

When taking the section of rivers, canals, &c., if no boat is available, something must be improvised to support the man holding the pole. The single barrel raft (Plate **120**, Fig. 2) and the bivouac sheet boat (Plate **125**) are examples.

For bridges with floating piers nothing more is necessary ; the section will show any place where there is insufficient depth of water to float the piers when the bridge is under its maximum load. In such places fixed piers must be built.

For bridges with fixed piers the mean vertical height of each pier from the foundation or " footing " to the level of the shore transom is found from the section ; to this height must be added :—

   i. An allowance for " camber " (*see* Sec. **88**, para. 3).

   ii. An allowance for " splay " when the piers are trestles with splayed legs ; this splay equals 1 inch in 6 feet along the length of the trestle leg.

   iii. An allowance for soft bed, if the bottom is muddy ; a trestle, in soft mud, providing " shoes " (*see* Sec. **91**, para. 4) have been fixed, will not sink more than 12 inches. If the bottom is very uneven a section for each side of a trestle bridge will be required.

88. *General description of bridges.*

1. **Types of bridges.**—Bridges, whether light, medium or heavy may be divided into three types :—

   i. Single span bridges.

   ii. Bridges with intermediate fixed piers, such as trestles, crib piers, piles, &c.

   iii. Bridges with intermediate floating piers, such as barrel piers, log piers, boats, &c.

   iv. A combination of ii and iii.

2. **Parts of a bridge.**—A bridge consists of a roadway carried on two or more piers over an obstacle, or, as it is called a gap. The gap may be wet or dry. The distance between any two piers is called a " span " or a " bay " ; the length of a bridge from shore to shore is called the " total span."

The parts of a bridge described in the following paragraphs are those of a " light " bridge. Medium and heavy bridges are dealt with in Military Engineering, Vol. III.

3. The **roadway** (Plate **111**, Fig. 2) consists of decking carried on roadbearers or baulks, which rest on transoms (the top beams of the piers) : the decking is held down by ribands or wheel-guides.

The normal **width of roadway** in the clear between ribands is 9 feet.

The roadway is generally constructed with a slight rise towards the centre of the bridge to get loads on to the bridge quietly and easily and to assist traffic off the bridge; this is technically called the **camber**, and is obtained by giving a rise of 1 in 30 for about 30 feet from each end of a bridge.

4. **Decking** should consist of planks 3 inches thick, but 2-inch hard wood (beech) road slabs may be used if available. In all decking there should be $\frac{1}{2}$-inch spaces between the planks to allow for drainage; the planks are nailed to the outer roadbearers. The decking should not project more than a few inches beyond the outer roadbearers. Straw, rushes, &c., must be laid on the decking if it becomes slippery. Battens or earth must not be used; the former get knocked off and leave nails, which lame horses, while the latter becomes very slippery and blocks drainage.

5. **Roadbearers** may consist of timber in scantling, round logs, rolled steel joists, rails, &c. The number and size of the roadbearers for any span may be obtained from Appendix V, Table 5.

Roadbearers are spaced evenly over the width of the roadway; if they differ in strength, the strongest should be placed beneath the wheel tracks. The outer roadbearers should be spaced 9 feet apart in the clear. In many bridges the roadbearers of adjoining bays overlap on the transom; they should therefore be from 2 to 3 feet longer than the span. When round spars are used, the ends on any transom must be all butts or all tips. Roadbearers should be spiked to the transoms, when tactical conditions permit; otherwise lashings of wire or cordage may be used; chocks should be fixed on the transom between the roadbearers to keep them in position.

6. **Transoms** carry the roadbearers and must be adequately supported. For bridges with spans up to 15 feet, round logs of 8 inches average diameter, or 6 inches by 6 inches squared timber, form suitable transoms if supported as shown on Plates **117** and **118**; if carried on two supports only they must be considerably heavier (Plate **119**). A transom should project sufficiently far on either side of the roadway to allow of the fixing and strutting of handrails (*see* para. 8 below).

A **bankseat** or **shore transom** must be provided on either side of the gap to support the shore-ends of the roadbearers. A 9-inch by 3-inch plank, laid flat, makes a suitable shore transom (Plate **111**, Fig. 3); care must be taken that the plank "bears" throughout its length, and that it is securely anchored. It must be placed at a sufficient distance from the edge of the gap to prevent the soil breaking away under the pressure—this distance should never be less than 1 foot (in chalk, &c.) and may be 3 or 4 feet; if the sides of the gap are liable to collapse, they must be revetted.

7. **Ribands** or **wheel-guides** hold the decking together, and prevent the traffic going off the bridge; they should consist of 6-inch round logs, or 6-inch by 6-inch squared timber spiked to the decking immediately above the outer roadbearers. The ends of adjoining wheel-guides should either butt or be halved into one another. Wheel-

guides should be painted white, and the shore ends splayed to facilitate the approach on to the bridge. A heavy bumping post, well let into the ground, is useful at the extreme end of the wheelguide.

8. **Handrails** (4-inch round timber or 6-inch by 3-inch scantlings) should be fixed to all bridges, except when they would disclose the position of the bridge to enemy ground observers ; they should be 3 feet above the decking, and at least 9 inches clear of the inner edge of the wheel-guides, the posts being fixed on the transoms and strutted (Plate **117**). These posts must not be less than 4 inches in diameter and should be painted white. Screens of canvas or branches, 6 feet high from the decking, should be securely fastened to the handrails to prevent animals seeing the water or the depth of the gap below them. Flapping screens are worse than none.

9. The best **fastenings** for bridge work are iron fastenings, but wire or rope lashings may have to be used ; the latter are difficult to obtain, and are the least satisfactory, owing to the difficulty of keeping them taut.

Iron fastenings may consist of dogs, spikes, drift-bolts, nails, bolts, chains, &c. With dogs the position of each must be chosen with the definite object of preventing a possible distortion of the frame. They should be on both sides of the frame. Dogs should not be driven within 3 inches of the edge or 4 inches of the end of a piece of timber.

Spikes, when driven in pairs, should incline towards each other. They run from 5 inches to 10 inches in length (Appendix IV, Table 13). Spikes with chisel points should be driven so that the edge is across the grain.

Drift-bolts are made of round iron, pointed at one end and with a small head at the other. They may be of any length, and are especially useful for fastening horizontal timbers to the top and bottom of upright ones. Holes slightly smaller than the bolts should be bored to receive them.

10. On the completion of the bridge the officer in charge of the construction is responsible that :—

    i. The bridge is strong enough for the loads which it is intended to carry.

    ii. **Conspicuous notice boards** are erected at either end of the bridge stating what loads the bridge will take ; these notice boards must be supplemented by similar notice boards sufficiently far back on the roads leading to the bridge to allow of traffic, too heavy for the crossing, being directed elsewhere. Arrangements must be made to illuminate the boards at night. " Break step " notice boards must also be erected at either end of the bridge.

    iii. A maintenance party of adequate size is left to maintain the bridge. This party will improve the bridge, maintain the approaches, and police the traffic pending the arrival of

military police. He will ensure that this party knows how
to communicate with him, and that it is not withdrawn without definite orders from the higher command.

.iv. Material is available on the spot for repairs, that arrangements
have been made to bring up such material, or that the
maintenance party knows where to obtain it.

11. **Gauges.**—Whenever there is water in the gap, a clearly
marked gauge must be established, so that any rise or fall of water may
be detected.

89. *Light bridges—infantry footbridges.*

1. Infantry footbridges may be extemporised from planks with or
without intermediate supports, but it is best to provide light portable
bridges made up either as single spans or, if more than one span is
required, in two parts—(a) the roadway section ; (b) the pier section.

If there are " n " piers, there will be " n + 1 " roadway sections.
Such a bridge must be so designed that it or any one of its component
parts can be carried by not more than four men over 1,000 yards of
rough country in the dark without undue exertion, and that it is
simple to assemble, or put together.

2. **Single-span infantry bridges.**—The simplest form is a plank :
a 9-inch by 2-inch plank, 10 feet long, will take two armed men at a
time over a gap 8 feet wide. Two planks should be placed side by
side to give a roadway of 18 inches. For gaps 8 to 14 feet in
the clear some form of trussed beam must be used (Plate **112**).
The roadway should be covered with wire netting or grillage. Great
care must be taken to get a level bearing surface for the shore ends,
or the bridge may take a tilt, and quickly become useless.

3. **Infantry bridges with fixed piers.**—The piers may be light
two-legged or four-legged trestles or light piles. These bridges have
the following disadvantages :—the trestles are difficult to handle and
place in the dark, and, when previous reconnaissance is impossible, the
adjustment of the transoms takes time ; they are very liable to distortion when the weight comes on them. The construction of light-pile
piers involves considerable time and noise.

4. **Infantry bridges with floating piers** are the best method of
crossing streams over 14 feet in breadth ; if properly constructed
they are stable, and a level roadway is ensured ; they can be placed
in position in a very few minutes by trained personnel. The materials
most generally available for piers are petrol tins, metal cylinders or
cork slabs. The method of making these up into floats is shown on
Plates **113** and **114**.

The roadway consists of light " duckboards " with bearers arranged
so as to interlock on the saddle of the float, and fitted with clips to
grip the saddle.

To give lateral rigidity so that the bridge can be boomed out rapidly
across the river tie baulks are slipped on over the ends of the handles

and secured by hooks. As the bridge reaches the far bank two men double across, lay the shore bay and secure the mooring rope. Where there is little current one mooring rope passed through the rings on the ends of the piers, and made fast to pickets on each bank, may suffice to hold the bridge in place ; but additional mooring ropes, or anchors, will be needed where there is much current.

Cask footbridges (Plate 115) may often be useful but take rather longer to construct.

90. *Light bridges—single span.*

1. Small gaps may be bridged by filling in (Sec. 110, para. 7) or by laying roadbearers, selected according to the span from Appendix V, Table 5, across the gap, resting on bankseats and supporting 3-inch decking.

2. **" Artillery "** **bridges** are portable bridges, consisting of two longitudinal sections of roadway, designed to enable a battery to cross a small gap (Plate 116). They are not intended for continuous traffic.

3. **Cantilever and suspension bridges.**—When the gap cannot be spanned by the above methods, and neither a trestle bridge nor a floating bridge can be used (as in crossing a ravine with deep precipitous sides), recourse must be had to a cantilever or suspension bridge. It will only be on rare occasions that such bridges will be required. Both forms of bridge take a considerable time to construct.

Details of suspension bridges are given in Military Engineering, Vol. III, Chapter VII.

91. *Light bridges—fixed piers.*

1. **Fixed piers** may be made of an infinite variety of materials, *e.g.,* brushwood cylinders, or wooden crates, filled with stones, carts, &c., but the material most likely to be available is timber ; this may be used in the form of crib piers or trestles.

2. **Crib piers.**—When timber is plentiful, crib work (Plate 116) is useful and speedy up to a height of 4 feet ; if sleepers are available and close at hand, piers up to 7 or 8 feet are economical in time. In water, a tray should be formed in the bottom of the cribs and filled with stones. These piers, if necessary, can be floated into position and sunk by loading the trays, but if the bottom proves to be very uneven it will be difficult to keep the piers vertical. Figs. 4 and 5 on Plate 116 show a crib causeway, which was constructed in the war of 1914–18, to enable tanks to cross a small river. As all the structure was below water level, the existence of the bridge was concealed from the enemy before it was used.

3. **Trestles** are the most useful form of fixed piers. The best are those which are framed together with the transom resting on the head of the legs. When using squared timber all joints should be flush ; tenons, notches, &c., are unnecessary and detract from the strength of the timber.

A framed trestle is shown on Plate **117** suitable for bays of light bridge up to 15 feet and for a height of trestle up to 12 feet. It weighs about 12 cwts. and is easily handled. It can be used for height of trestle up to 15 feet by using four 6-inch by 6-inch legs evenly spaced or three legs 7-inch by 7-inch. A similar trestle can be made out of round spars, but in this case the timber must be notched to get a bearing surface and 8-inch round spars must take the place of the 6-inch by 6-inch scantlings with 4-inch braces, and the trestle will be somewhat heavier.

A plank trestle can be made of similar type as shown on Plate **118**. In this case all the joints are nailed and the legs are hollow, but are packed solid where the braces are nailed to them. The planks forming the legs are nailed together and should be bound with hoop iron every 8 feet. A trestle of the dimensions shown is suitable for bays of light bridge up to 15-foot span and for a height of trestle up to 12 feet.

Framed trestles require a level foundation and the height of transom cannot be readily adjusted when the trestle is in place.

When the bottom is uneven and cannot be levelled a type of trestle with two legs only must be used, the ledger being placed high enough to clear any obstruction. In this type the joints are usually made with lashings of wire or rope, which allow of a certain adjustment of the height of transom when the trestle has been launched, and the legs extend above the transom to take a hand rail. Lashings, however, require constant attention and are only suitable for temporary bridges.

A trestle of the dimensions shown on Plate **119** is suitable for bays of light bridge up to 15 feet, and height of trestle up to 15 feet.

The transom, the ledger and the butt of one of the braces are on the same side of the legs, the other butt and the tips of the braces on the opposite side. All lashings are "square" except when the braces cross, when a diagonal lashing is used. The trestle must be squared by making the diagonals equal before the braces are finally lashed.

4. Ledgers and shoes.—When the bottom is soft, the ledger is placed near the bottom of the trestle legs. When the bottom is very soft, arrangements must be made to prevent the trestle sinking by placing horizontal timbers at the feet of the legs, or by furnishing them with flat shoes about 15 inches square of 1½-inch planking, spiked to the feet of the trestle legs.

5. Placing trestles.—If the gap is wet, trestles can be carried out and placed by men working in the water. When the water is too deep for this, they may be carried on to the bridge and lowered feet first down inclined spars (Plate **120**), or taken out on a raft and tipped into position by means of guys.

6. Trestles are kept vertical by fastening the roadbearers to the transoms, and by cross-bracing from each trestle to its neighbours. When using lashed trestles, the nearest trestles to the bank on either side should be rigidly connected to bollards on the banks by light spars fitted to the legs about 3 feet above the transom. These light spars are put on before the trestle is launched, and help to get it into position, and must be secured before the first bay is used for placing the second trestle.

7. **Scouring.**—The presence of a pier in running water previously unobstructed causes an underscouring action by the water to commence on the upstream side of the pier, which may eventually capsize the pier.

This can be temporarily guarded against by surrounding the upstream side of the pier with boulders or sacks of small stones, but the waterway (Sec. **92**, para. 1) must not be obstructed.

92. *Light bridges—floating piers.*

(*For buoyancy see* Sec. **95**).

1. **Floating piers,** in the absence of pontoon equipment, may be constructed of casks, boats, logs, &c. Each pier must have enough available buoyancy to support the heaviest load that can be brought on to one span of the bridge. The length of each pier should be twice the breadth of the roadway for the sake of steadiness, and with the same object, they must be connected together at their ends by tie baulks (Plate **122**). The waterway between the piers should never be less, and should, if possible, be more than the width of the piers.

2. A bridge can be put into position in the following ways:—

i. By **booming out,** *i.e.,* when the head of the bridge already constructed is continually pushed out into the stream, fresh materials being added at the tail. This method cannot be used with steep banks and deep water close in shore.

ii. By **forming up,** *i.e.,* when material is continually added to the head of the bridge, the tail being stationary. This method is uninfluenced by the nature of the banks, no men being required to work in the water. Its only drawback is the distance the roadway materials have to be carried.

iii. By **rafting,** *i.e.,* when the bridge is put together in different portions, or rafts, along the shore, each raft consisting of two or more piers, and these rafts are successively warped, rowed or towed into their proper positions in the bridge.

This method has the advantage that a large number of men can be employed simultaneously, and, if secrecy be an object, the various portions can be constructed at some distance from the eventual site of the bridge, and a favourable opportunity seized for its construction.

iv. By **swinging,** *i.e.,* when the entire bridge is constructed alongshore, and then swung across with the stream.

A long bridge can be constructed by a combination of two or more of the above methods.

3. The **bridge ends** to floating bridges require careful consideration. In tidal rivers, and in non-tidal rivers, when the effects of droughts and floods may cause considerable variations in level, much ingenuity, time and labour are required in order that the bridge may be available at all times for traffic (Military Engineering, Vol. III).

A floating pier should not be allowed to ground, as it will be liable to be crushed when a load comes upon it. If it **has** to ground, it must be built of large barrels and the bottom levelled, or a cradle made of reeds, brushwood or sacks of earth.

4. **Anchors.**—As a rule, there should be an up-stream and down-stream anchor to every second pier of a floating bridge : in tactical operations under fire there must be an up-stream anchor to every pier.

For ordinary bridge work 56-lb. anchors, with a reserve of 112-lb. anchors, will generally suffice for moderate streams. The following substitutes may be employed :—Two or more pick-**axes** lashed together ; heavy weights, such as large stones or rails (the latter are best when bent).

The cables are generally of 3-inch cordage. The length of cable " out " should be ten times the depth of the stream, and rarely less than 30 yards. The cable is attached to the ring of the anchor by a fisherman's bend ; a buoy should be attached to the anchor by a buoy line of 1-inch rope to mark its position, and to serve as a means of tripping it. One end of the buoy line is fastened to a ring of the buoy by a fisherman's bend, and the other round the crown of the anchor with a clove hitch split by the shank, and two half-hitches round the shank.

In the absence of anchors, or in a very rapid current, when the anchors are liable to drag or to pull the piers down by the head, a hawser (preferably wire rope) buoyed with floats can be stretched across the river (provided its width does not exceed 100 yards), and its ends secured to anchorages on each bank at a distance up-stream of about one-fourth the span : cables from the bridge piers are then secured to it. The danger of using this method under fire is that one shot may destroy the bridge.

5. **Passage of traffic.**—If a bridge has to remain down for some time, arrangements may have to be made for the passage of river traffic. This can be done by having two or more rafts, at the centre of the bridge, arranged for forming " cut " ; or the two halves of the bridge may be swung, to afford the requisite passage.

6. Arrangements must always be made, up-stream, for the protection of a bridge from damage by floating substances, either by a boat patrol or by posting men at each pier to pole off such floating objects into the fairway. Down-stream, rafts with buoyed floating ropes should be anchored, to save men who fall off the bridge.

7. **Boats piers.**—Few boats, with the exception of heavy barges, are strong enough to allow of the baulks resting direct on their gunwales. A central transom should be improvised, by resting a beam on the thwarts, and blocking it up from underneath, so that the weight is brought directly on the kelson (Plate **121**, Fig. 1).

In a non-tidal river, the boats should be placed " bow on " to the current and slightly down at the stern : in tidal rivers they must be placed pointing up and down-stream alternately.

8. **Cask piers.**—Piers of casks (Plate **121**, Fig. 2) are made as follows, the number of men required being $2n + 2$, when n is the number of casks :—

The casks are first laid bung uppermost and aligned ; then two baulks, technically known as gunwales (GG), are placed over the ends of the casks by four men, while the remainder of the men stand opposite the intervals between the casks on either side.

The gunwale men at one end place the eyes of the slings (SS) over the gunwales ; the gunwale men at the other end secure the slings to their ends of the gunwales with a round turn and two half-hitches. The brace men keep the slings under the casks with their feet, and, as soon as they are secured, adjust the braces as follows, working simultaneously by word of command :—

The eye of the brace is passed under the sling in the centre of the interval between two casks, the end passed through the eye and hauled taut, the sling being kept steady with the left foot. The brace is then brought up outside the gunwale immediately over the eye, and a turn round the gunwale taken to the left, the foot is removed from the sling and each man then hauls up the standing part of his brace with the left hand, holding on to the turn with the right ; as soon as the brace is taut, the turn is held fast with the heel of the left hand, and the remainder of the brace, in a coil, is placed on the cask to the left. Each man then takes his opposite neighbour's brace from the cask on the right, and passes it between the standing part of his brace and the cask on his left, then back between his brace and the cask on his right, keeping the bight so formed below the figure of 8 knot on his own brace, and placing the end on the cask to his right. Each man then takes back his own brace from the cask on his left, passes it under the gunwale to the left, of the standing part, places his foot against the gunwale, and hauls taut. The pier is then rocked backwards and forwards, all the brace men taking in the slack of their braces and hauling taut until the word **steady** is given, when they take a round turn round the gunwale to the left of the previous turns, and make fast with two half-hitches round the two parts of their own brace close to the gunwale, drawing the two parts close together and placing the spare ends of their brace between the casks. The pier is then turned up on one side, and the sling adjusted below the third hoop of the casks, and a breast line attached to a sling at each end ; it is then lowered and turned up on the other side, the other sling adjusted, a sledge, technically called the **ways,** brought up into position, and the pier lowered on to it ready for launching.

For a pier of the size shown in the figure the following are needed :—
Gunwales, 21 feet by 4 inches by 5 inches ; slings of $2\frac{1}{2}$-inch rope, 6 fathoms long, with an eye splice 1 foot long at one end ; braces of $1\frac{1}{2}$-inch rope, 3 fathoms long, a small eye splice at one end, and a figure of 8 knot 1 foot 5 inches from the eye.

9. When **log piers** are used, it is usually necessary to place the logs in two layers in order to give sufficient waterway between the piers ; the logs are laid side by side, thick and thin ends alternating ; they should then be strongly secured with ropes or dogs, and, if possible, by cross and diagonal pieces of timber fastened by spikes ; a central raised transom must be used. The up-stream end of the pier may, with advantage, be slightly convex. Such piers are most easily put together and manipulated in the water.

93. Rafts.
(For buoyancy see Sec. **95**).

1. When there is insufficient material for the construction of a bridge, or when it is only necessary to establish a ferry, rafts may be used, constructed of bridging equipment, of barrels, boats, logs, or, as a very temporary measure, of waterproof material, such as tarpaulins or ground sheets, stuffed with straw, heather, ferns, &c. Such rafts may be towed, rowed, poled, hauled backwards and forwards across the gap by means of ropes, or, when there is a good current, utilized as "flying bridges" (see para. 8 below). Free rafts are generally most easily moved by towing ; rafts with horses and vehicles cannot be rowed or poled, except under most favourable circumstances.

2. **Rafts** consist normally of two piers connected by baulks on which the decking is laid ; the length of each pier must be twice the width of the platform of the raft. Three-pier rafts, when loaded, are unmanageable in a stream and are not recommended. When loading a raft with infantry, the men should sit down on the edge of the raft as close as possible, and then the central part of the raft should be loaded.

If the raft consists of one pier only (such rafts may be constructed of barrels or logs) the central quarter only of the platform should be loaded.

3. Rafts of two or more piers are merely sections of a floating bridge ; for the construction of barrel or boat piers, see Sec. **92**, paras. 7, 8 and 9 ; the decking is constructed in exactly the same way as a roadway ; the lashings must be constantly watched.

4. **Log rafts** consist of one pier only ; and are made in the same way as log piers in a bridge.

5. The **deck space** required for rafts may be estimated from the following dimensions :—

	Length. ft. ins.	Width. ft. ins.
Armed man, sitting	3 6	2 3
Horse, harnessed	8 0	4 0
18-pr. gun	14 6	6 3
18-pr. limber	5 9*	6 3
G.S. wagon	13 9*	6 1

* Exclusive of pole.

6. When animals are carried, hand rails must be provided, and screens are always desirable, but may have to be dispensed with in a wind.

7. **Landing stages.**—Planks should be carried to enable men to get ashore from the raft ; when animals and vehicles are carried, landing piers and landing gangways are necessary ; landing piers are usually constructed of trestles, but in tidal waters, elaborate structures are necessary (Military Engineering, Vol. III), if the piers are to be available at all times. Landing gangways consist of five or six baulks lashed into a frame by cross baulks 2 feet from their ends. This frame forms a sliding span, sufficient decking being kept at each landing pier to complete the span.

8. **Permanent ferries.**—In the simplest form of permanent ferry, boats or rafts are hauled backwards and forwards from bank to bank by means of ropes stretched across the river. Such a rope should, if attached to the boat, &c., be made fast at the stem or stern, not amidships. If it is not convenient to stretch a rope across the stream on account of traffic or other reasons, or if the current is rapid and regular, a flying bridge may be used. This is a form of bridge in which the action of the current is made to move a boat or raft across the stream by acting obliquely against its side, which should be kept at an angle of about 55° with the current (Plate **123**, Fig. 1). Long, narrow, deep boats with vertical sides are the best for the purpose, and straight reaches the most suitable places, as they are generally free from irregularities of current. It is necessary to have a vertical surface for the current to act against. If, therefore, the boat is a shallow one, or if the raft is made of casks or other material with a curved surface, vertical boards, called lee boards, must be lashed to its side. These lee boards consist of two or three planks held together by battens nailed to them. The depth of the lee board must be kept to a minimum, as the action of the current on the lee board causes the raft to tilt ; a deep lee board may swamp the pier.

The cable, which should, if possible, float, can either be anchored in midstream or on shore at a bend of the river, and the raft can swing between two landing piers. The length of a swinging cable should be from one and a-half times to twice the breadth of the river, and it will work better if supported on intermediate buoys or floats to prevent it from dragging in the water. A number of telegraph wires, buoyed as above, make a good swinging cable.

Another way is to stretch a wire cable across the river, and arrange for the raft to travel along it by means of a block with a large sheave.

A spare anchor and cable must always be carried on the raft.

9. An **extemporised raft** to carry an 18-pr. field gun without limber, or a weight not exceeding 24 cwts., may be constructed of four 18-foot by 15-foot tarpaulins stuffed with straw as follows :—

Make a light framework of poles, 6 feet square by 2 feet 6 inches high, on the ground (a hole of similar dimensions will do almost as well). Then place two lashings about 24 feet long across the framework each way, and over these the tarpaulin, well soaked. Fill the tarpaulin with straw or similar material and trample it well down. The ends and sides of the tarpaulin are then folded over the straw, and the whole made into a compact bundle by securing the lashings across the top (Plate **124**, Figs. 1 and 2).

Two of these floats are then lashed together by means of two 14-foot spars. This forms half the raft. The other half is made in a similar manner. The two halves are then lashed to one another, 3 feet apart, by means of four 16-foot roadbearers. The raft will then measure 15 feet by 12 feet by 3 feet 6 inches. In such a raft the buoyancy is greatly in excess of that actually required to carry the load, but this is necessary owing to the kind of material employed and the short length of the piers. With good tarpaulins the buoyancy will remain good for at least eight hours.

The stores required are :—

Tarpaulins	4
Straw (tons)	1½
Planks	16
Spars (average 4 inches diameter), four 16 feet, four 14 feet, and two 12 feet	10
Lashings, 1-inch, about 3 fms. long	40
Do. 1½-inch, about 6 fms. long	16
Ropes, 2-inch, length according to width of river...	2
Punting poles...	2

10. Smaller rafts can be made in a similar manner by stuffing ground sheets with straw and placing them in a frame ; 24 of these made into a raft will support a load of 1,800 pounds (Plate **124**, Figs. 3 and 4).

11. An extemporised boat made of a bivouac sheet stretched over timber framing is shown on Plate **125**.

94. *Fords.*

1. A ford, to be passable, should not exceed the following depths :—

For cavalry	4 feet.
For infantry	3 feet.
For artillery	2 feet 4 inches.
For lorries	2 feet.

2. The positions of fords are usually indicated on maps. They are often found just below weirs. The local inhabitant is the best source of information. A river which is not fordable straight across may sometimes be found passable between two bends as at A, B, Plate **123**, Fig. 2.

3. The **approaches** to a ford break down rapidly under continuous traffic, owing to the drip from men, horses and vehicles. They should be " corduroyed," the " pull out " side being done first, and ditched for 100 yards on either bank ; the corduroy must be carried well into the water.

4. A ford with a sandy bottom is likely to become heavy. The bottom of a ford must be carefully examined before use, all holes being filled with stones or other hard material. Large stones and any obstacles to traffic must be removed.

5. For dry or fordable gaps, no provision is necessary for infantry, except in the case where the sides of the gap are precipitous, when provision must be made for ropes, knotted every 3 feet, with screw pickets to act as holdfasts, and ladders. Every endeavour should, however, be made to bridge a fordable stream on the immediate front of the attacking troops, in order to start the men dry.

6. A shallow (not exceeding 3 feet of water) muddy river may be made temporarily passable for infantry by laying mats made of canvas and wire netting on the top of the mud. A rope must be firmly fixed from bank to bank on either side of the mat.

7. **Marking fords.**—All fords must be clearly marked by strong pickets (4 inches and over), driven into the river bed above and below the ford ; these pickets should project at least 2 feet above water level ; their heads should be painted white, and should be connected by a strong rope securely anchored to holdfasts at each shore end. Four of the pickets in the deepest parts must be clearly marked in feet and inches so that the depth of the water may be easily seen.

8. Notice boards should be erected at each end of the ford, giving fordable depths for each arm.

95. *Buoyancy.*

1. **Bridges.**—When calculating the buoyancy required for a pier of a light floating bridge, the weight to be supported by any pier may be taken as the weight of the greatest load on one bay of the bridge plus the weight of one bay of superstructure. A pier that has sufficient buoyancy to carry infantry in fours will also carry light field artillery and first line transport. The maximum weight that can be brought on the bridge by infantry in fours is 5 cwts. per foot run of bridge, and the weight of superstructure may be taken at $1\frac{1}{2}$ cwts. per foot run up to 15 feet span, so that the total weight will be $6\frac{1}{2}$ cwts. per foot run.

2. Rafts.—In the case of rafts, each pier must be capable of supporting the greatest load which may come on to it. On a raft designed to carry vehicles, the greatest load will be brought on a pier when the raft is being loaded or unloaded.

The superstructure will be the same as for a bridge and its weight must be allowed for.

The weight of an armed man may be taken as 200 lbs., and he occupies 8 square feet of deck space when sitting. A light draught horse weighs about 1,400 lbs. and requires 8 feet by 4 feet of deck space.

The loads brought on a raft by light artillery and G.S. wagons and the space occupied by them are given in the table below.

Loads on rafts due to horse and light field artillery and G.S. wagons.

Vehicle.	Weight in tons fully loaded.		Distance, axle to axle.	Overall width.	Overall height.	Wheel track.	Width of tyres.
	Front or limber wheels.	Rear or gun wheels.					
			ft. in.	ft. in.	ft. in.	ft. in.	in.
Q.F. 13-pr.	·65	1·00	9 11½	6 3	4 8¾	5 3	3
Ammunition wagon and limber for above	1·62	1·62	7 3½	6 3	5 0	5 3	3
Q.F. 18-pr.	·71	1·40	9 11½	6 3	5 3	5 6	2·6
Ammunition wagon and limber for above	1·94	1·94	7 4¾	6 3	5 2	5 3	2·625
Q.F. 4·5-in. howitzer ...	·74	1·34	10 0½	6 3½	5 9	5 3	3
Ammunition wagon and limber for above	2·00	2·00	8 7½	6 3½	4 11½	5 3	3
Wagon, G.S.	1·35	1·00	7 1	6 4	6 11	5 3	2¼

3. The available buoyancy of a boat may be determined by loading it with unarmed men to within 12 inches of the gunwale and multiplying this number by 160; this gives the available buoyancy in pounds. In rough water or in a violent current a margin for safety must be allowed.

4. When using closed vessels such as casks, petrol tins, oil drums, &c., for floating piers, the **safe** buoyancy for bridging purposes must be taken at nine-tenths the **actual** buoyancy, so that the roadway will be clear of the water.

The actual buoyancy of a closed vessel is the weight of water it will displace when fully immersed, less its own weight. Thus, to find the actual buoyancy, the volume and weight of the vessel must be ascertained. In this connection, it is useful to note that a gallon of fresh water weighs 10 lbs. and that 1 cubic foot equals 6·25 gallons.

PART III.—ACCOMMODATION.

CHAPTER XIV.

CAMPING ARRANGEMENTS.

96. *Bivouacs, billets, and hutting.*

1. **Sites.**—The choice of sites for bivouacs, camps or hutments is determined by the tactical situation.

The nearer the enemy the more important are the tactical considerations ; the further the enemy, the more attention can be paid to the comfort and health of the men.

Under all circumstances strict attention to the sanitary conditions of men's accommodation has a direct bearing on the efficiency of the force.

2. The site for a camp or bivouac should be dry, on grass, and on a gentle slope. Steep slopes, woods with undergrowth, low lying meadows, the bottoms of valleys and newly-turned soil must be avoided. The water supply should if possible be within one mile.

3. **Bivouacs.**—Simple shelters may be formed in many ways. One method is to drive two forked sticks into the ground with a pole resting on them ; branches are then laid resting on the pole, thick end uppermost at an angle of about 45°, and the screen made good with smaller branches, ferns, &c. A hurdle may be supported and treated in a similar way.

A shelter tent for four men may be formed with two blankets or waterproof sheets laced together at the ridge, the remaining two blankets being available for cover inside.

When materials are available extemporised huts can be made by erecting a rough lean-to roof of light poles and brushwood rods, roughly thatched with branches, brushwood, pine brush, or covered with a tarpaulin (Plate **126**).

4. Men sleeping on the ground in tents or bivouacs suffer discomfort from two main causes :—

i. Rain water.

ii. Draughts.

i. *Disposal of rain water.*

    (*a*) See that the roof does not leak, and that joints between W.P. sheets, blankets, canvas, &c., are well overlapped.

(b) See that the water from the roof is conducted to the ground and that it cannot run back under the sides over the floor of the bivouac or tent. Plate **127** shows the right and wrong ways for disposing of rain water from tents and bivouacs. By the right method the water is conducted from the roof to the ground and is led away in the trench to the lowest ground. The earth from the trench protects the floor of the tent from any rush of water over the area of the camp caused by a sudden downpour. By the wrong method the rain running down the sides of the tent or bivouac soaks the earth which is placed on the canvas and the water leaks through the canvas and over the floor.

ii. **Draughts.**—The proper arrangement for the disposal of water also prevents draughts along the level of the floor.

5. A method of protecting men sleeping in tents from splinters of shells and bombs is shown on Plate **128**, but it is open to the objection that the earth placed against the canvas walls of the tent will interfere with the efficient disposal of rain water as explained above.

6. **Billets.**—When villages are used as rest billets or staging areas, the accommodation may be greatly increased by erecting bunks in barns and farm buildings.

The walls of barns may be repaired by using wattle and daub, i.e., trimmed brushwood rods daubed over on both sides with clay in which is a proportion of any fibrous substance, such as straw, grass, &c., chopped into short lengths.

Roofs may be repaired with tarred felt on boarding, or with corrugated iron.

Bunks should be limited to two tiers, however much headroom there is : the bunking space required is 6 feet 6 inches by 2 feet per man. The bed consists of grillage of hoop iron or plain wire : the mesh must not exceed 4 inches by $2\frac{1}{2}$ inches, otherwise the occupants can insert the heels of their boots between the wires and the bunks are rapidly damaged. Rabbit wire netting, covered with canvas, may be used, but sags badly under a load and does not last long.

Windows of oiled canvas should be provided ; these add to cleanliness and comfort.

7. **Hutting.**—When hutted camps are necessary, the design of the huts depends upon the theatre of war. This subject is dealt with in Military Engineering, Vol. VII.

If huts are constructed of material drawn from the engineer parks, the most convenient form is rectangular in plan, 17 or 18 feet wide, to allow of bunks on either side. When attack by aeroplanes is expected only one tier of bunks can be provided, and that must be as close to the ground as possible. The floor should be of hard material or wood, the sides and roof of corrugated iron or of wood covered with tarred paper or felt, and the windows of oiled canvas or glass.

Splinter-proof walls of earth, shingle, &c., must be constructed round each hut, as a protection against splinters from shells or bombs from aeroplanes. To economize material and labour and to increase the protection afforded, the floors of all huts should be as close to the ground surface as possible ; on sloping ground the site of each hut should be levelled by cutting into the slope of the hill.

8. **Stoves.**—If stoves are provided, the floors, walls, and roofs of huts and billets must be specially protected with sheet iron or tin where the stoves stand, or the stove piping passes through.

9. **Stables** must be sited near to an existing road ; Plate **129** shows a typical site plan. A properly made approach road (Chapter XVI), is necessary, and to save labour and materials this should be as short as possible. The section on Plate **129** shows a type of shelter which affords accommodation for two rows of horses ; the central passage includes the road and serves as a harness room, until a separate shed can be provided. The sides and ends must give protection from wind, and the roof should be weather proof.

Standings must be made of the hardest material obtainable, e.g., concrete, brick or corduroy of logs, and they must be well drained. Stables are particularly vulnerable to attack from aeroplanes using bombs. Earth or mud walls should be built at the sides and ends to stop splinters and in long stables traverses must divide the stable into compartments holding not more than 20 horses (Plate **129**).

97. *Water supply.*

1. **The quantity of water** required per diem is as follows :—

In bivouacs :

1 gallon per man.
8 gallons per animal.

In standing camps, rest billets, &c. :

5 gallons per man.
10 gallons per animal.

Hot weather and hard work will nearly double ordinary requirements, and, in making any calculation of the amount required, these factors must be considered.

2. **The sources of supply** usually available are streams, ponds and wells.

Small ponds and shallow wells should be avoided for drinking purposes. Water in shell-holes must never be used for drinking, as it may be poisoned.

3. **The rough average yield of a stream** may be measured as follows :- -

Select some 15 yards of the stream where the channel is fairly
uniform and there are no eddies. Take the breadth and
average depth in feet in three or four places. Drop in a
chip of wood and find the time it takes to travel, say, 30 feet.
Thus obtain the surface velocity in feet per second. Four-
fifths of this will give the mean velocity, and this multiplied
by the average sectional area in square feet will give the
yield per second in cubic feet of water (one cubic foot equals
six and a quarter gallons).

The yield from a well may be gauged by pumping to lower the
level one foot, and then, noting how long it takes to fill to the original
level. The contents of circular wells per foot deep is as follows :—
3 feet diameter, 44 gallons ; 4 feet diameter, 78 gallons ; 5 feet diameter,
122 gallons ; 6 feet diameter, 176 gallons.

Water can be obtained, if necessary, from a marsh by the method
shown on Plate **130**, Fig. 1.

4. **Quality of water.**—The source of supply must be carefully in-
vestigated, and measures taken to prevent the pollution of the water
en route to the drinking supply. Wells must be tested at the earliest
opportunity, and each well clearly marked as fit for drinking or for
washing purposes only.

5. **Methods of purifying water.**—Water may be purified—

i. **By boiling.**—This is the best method, but it is not always
possible to provide means of boiling water on a large scale,
or for cooling it. The water should be kept on the boil
for at least 5 minutes, and should be aerated before use by
passing it through an empty biscuit tin, pierced with small
holes and suspended over a storage tank. This process
must be done with great care or the boiled water will take
up fresh impurities.

ii. **By filtration.**—This may be effected by passing the water
through—

(a) Charcoal, sand or gravel. This method does not necessarily
render the water pure. The chief use of such filters
is to remove visible impurities, prior to treatment with
a sterilizing filter, as described below.

(b) A sterilizing filter.—The filtering medium consists of cylinders or candles of porous earthenware. Water is pumped
through the candles outside to inside, leaving its impurities on the outside. The best known patterns of sterilizing
filter are the " Berkefield " and the " Pasteur." Army
pattern water carts are fitted with such filters, and can
deliver 3 to 4 gallons of filtered water per minute. The
capacity of a water cart is 108 gallons.

Dirty water must be strained before filtering as described in (a) above, or by passing the water through a sheet tacked on a wooden frame so as to form a bag, containing wood ashes as a filtering medium.

iii. **By the addition of chemicals**—

(a) Muddy water may be cleared by adding *alum* at the rate of one teaspoonful to ten gallons. This must be done some hours before the water is required.

(b) Chloride of lime destroys the micro-organisms in water that has been cleared and is frequently employed for this purpose. A teaspoonful of this substance is dissolved in a quart of water. This forms the stock solution. One teaspoonful of this solution per 2 gallons of water will generally be sufficient, but for further details *see* Millitary Engineering, Vol. VI, Chapter XVII.

(c) **Permanganate of potash** (Condy's fluid) removes offensive smell from water and, to some extent, oxidises dissolved organic matter. It should be added until a faint tint remains permanent.

6. **Raising water.**—The pump in general use in the service is the "lift and force pump," weighing 84 lbs. complete. It is worked by two men. When in good order it can lift water from a maximum depth of 28 feet and force it 32 feet (*i.e.*, 60 feet in all at a rate of 12 gallons a minute). Four are carried in the field by each field squadron, and four by each field company of engineers and each field park. To obtain the best results the height of lift or suction should be reduced to a minimum, and can rarely exceed 20 feet. The end of the suction pipe must never be allowed to rest on the silt or mud at the bottom of a well or stream.

From deep wells, unless power driven pumps are available, water must be raised by buckets and ropes, windlasses being improvised as soon as possible.

7. **Control of water supply.**—Immediately on the arrival of the troops (before if possible) the available sources of supply must be distributed under the following headings :—

i. Watering places for men (white flags).
ii. Watering places for animals (blue flags).
iii. Places where water for washing purposes may be drawn (red flags).

In a stream men should draw drinking water above the place for animals, while water for washing, &c., must be drawn below it.

Every precaution must be taken not to foul the stream : there may be other troops down-stream. Sentries or patrols must be established to see that the sources of supply are used as detailed.

8. **Watering arrangements for men.**

i. Cooking utensils must not be dipped in ponds or streams : they must be filled by dippers.

ii. **Arrangements** must be made as soon as possible to enable water carts to be filled quickly by installing storage tanks with hose deliveries.

iii. Special provision for filling water-bottles and petrol tins is most useful.

At all water cart filling points, besides the hose required for the carts, a water-bottle filler should be provided. This is a 2-inch iron pipe, fitted with ¾-inch bibcocks at intervals to which nozzles small enough to fit into water-bottles are fitted. If funnels are provided they should be made so as to allow air to escape from the bottle or tin.

9. **Watering arrangements for animals.**

i. Animals must not enter streams or ponds : watering must be by bucket or from a system of trenches as shown on Plate **130**, Fig. 2, the source of supply being fenced off.

ii. As soon as possible pumps and troughs must be installed : there should be sufficient troughing to enable all animals to be watered in one hour ; an animal should be allowed five minutes to drink and four feet of lateral space.

Troughs may be of corrugated iron or wood : canvas troughs do not last long : if used, they must be slightly raised off the ground and it is essential to protect them from damage by a stout guard rail all round, one foot higher than and one foot clear of the troughs.

Troughs must be placed, if possible, so that horses proceeding to and from water shall not use roads required for general traffic. They must never be placed at the side of a traffic road.

iii. Watering places should be surrounded by stout barricades with narrow IN and OUT openings to regulate traffic.

iv. The ground at watering places cuts up rapidly ; corduroy or brick standings should be laid for a width of 10 feet on either side of the troughs.

v. A special railed-in trough must be provided for animals under suspicion of infectious disease, and marked with conspicuous notice boards.

10. **Washing arrangements.**—No washing should be allowed within 30 yards of the water supply ; empty biscuit tins, &c., should be used to draw water for this purpose ; ablution benches (Plate **131**) should be made as soon as possible and drains provided.

11. **Miscellaneous.**—All watering points must be carefully drained.

Grease traps.—Waste water from ablution benches, kitchens, &c., must be cleared of grease before being discharged into soak pits. Designs of grease traps for this purpose are shown on Plate **139**.

Surface water must be prevented from running into wells by brick or concrete copings.

Direction boards to the watering places must be provided liberally.

12. The subject of water supply is dealt with in detail in Military Engineering, Vol. VI.

98. *Cooking arrangements.*

1. **In the open.**—The simplest arrangement for cooking in the field for any party over 20, if the halts are not of long duration, is to place a proportion of the kettles on the ground in two parallel rows about 9 inches apart, handles outwards, block the leeward end of the trench so formed with another kettle, lay the fire between the kettles and place one or two rows of kettles on those already in position.

Mess tins can be arranged similarly, but in their case not more than eight should be used together.

2. **Trenches.**—The most economical method when time is available, is to dig or raise a narrow sloping trench for the fire on which the kettles are placed. The interstices are then filled up with clay so that the fire, fed from the windward end, may draw right through. A chimney may be built at the other end to increase the draught.

A type of raised kettle trench is shown on Plate 132. This can be built of tins filled with earth or mud, bricks or sods. All joints should be made airtight with clay, &c. The chimney consists of empty tins, with the sides or ends knocked out to form flues, resting on a foundation of tins filled with earth. The end and chimney should be well built round with clay, sods, &c., and round tins or stove piping can be added to give additional height.

3. Types of ovens and cookers are given (Plates 133, 134, and 135). Those shown on Plates 133 and 134 are best built in brick, but with care can be built with sods, if good sods are obtainable, or with biscuit tins.

The type shown on Plate 135 being portable, is suitable for small detached parties.

4. Weatherproof cover should be provided for cooks to enable them to prepare food properly and to provide for the storage of rations. A simple timber framework with end, sides and roof of corrugated iron will suffice. The roof should have a good fall.

5. For storing rations a fly-proof safe is essential in warm weather. A safe of light timber framing provided with hooks from which to hang meat and covered with fly-proof gauze can be made to any size required. A portable pattern is shown on Plate 136.

99. *Latrines.*

1. **Sites** for latrines must be very carefully selected. They should be situated as far as possible from the water supply, and when practicable, to leeward of kitchens. Flies are greatly responsible for spreading infection. No filtration must reach the water supply.

2. Latrines and urinal pits must be dug immediately on arrival of troops in their bivouac or camp. These should be replaced as soon as possible by urinals of the type shown on Plate **137**, and by deep trench latrines (Plate **138**). This type without the trench is also suitable for the bucket systems, the back being made removable to allow of the buckets being emptied.

3. **Accommodation.**—Shallow trenches should be provided at the rate of five trenches per 100 men. Five yards run of deep trench is required per 100 men. Seating accommodation should be provided, if possible, for 10 per cent. of the men.

4. **Screens** of canvas or bushes should be provided, and latrines must be clearly marked " Officers," " N.C.Os.," or " Men."

100. *Refuse.*

1. **Camp refuse** must be collected and burned, the residue being buried.

A type of temporary incinerator is shown on Plate **140**. An oil drum with the bottom knocked out supported on a grid of iron bars resting on bricks or stones is equally good. Spaces must be left below the grid to form air holes and for raking out ashes. A more permanent type of destructor may be built in brick.

In camps and trench systems numerous small receptacles (sandbags, X.P.M. gabions, &c.) must be provided for paper, cigarette tins, &c.

2. **Disposal of manure.**—Definite places must be allocated where manure may be dumped ; these should be at a considerable distance from camps and horse lines ; the heaps must be properly built up, and covered with 1 foot of earth.

3. **Disposal of dead animals.**—Dead animals must be skinned, cut open, and buried, the place being clearly marked as " foul ground."

CHAPTER XV.
SHELTERS AND DUG-OUTS.

101. *Degrees of protection.*

1. **Shelters** for the protection of troops, armament and ammunition against the effect of enemy projectiles are provided, either in specially-constructed buildings on the surface, or in underground chambers and passages called dug-outs.

2. **Protection** is given in three degrees.

    i. **Splinter-proof :** against penetration by splinters of shells of all calibres.

    ii. **Medium shell-proof :** against direct hits by shells of all calibres up to 6-inch.

    iii. **Shell-proof :** against direct hits by shells of all calibres and aeroplane bombs.

The construction of shell-proof buildings above ground involves a large amount of steel, concrete and highly-skilled labour ; this subject is dealt with in Military Engineering, Vol. II.

102. *Types of shelters.*

1, To assist in the construction of protected accommodation, both above ground and under ground, the following special materials are provided (Appendix IV, Tables 5 and 6).

    i. Small elephant shelters.
    ii. Large elephant shelters.
    iii. Curved corrugated-iron sheets.
    iv. Troughing plates in 6 and 9-foot lengths.
    v. Timber, rails, steel joists, logs, &c.

2. **The small corrugated steel shelter** (Plate 141).—The complete shelter consists of five segments or arches, each composed of two sheets, 2 feet 9 inches in width, which overlap 12 inches, and are fastened together by six half-inch bolts, 1½ inches long, through holes drilled in the sheets for this purpose ; each segment overlaps the next by half a corrugation (3 inches).

To erect, lay out the 4-inch by 3-inch timber bearers, put the curved segments into position, drive in the clasp nails provided, and nail on the 2-inch fillet (Plate 141).

Five segments of a small elephant shelter make a shelter 12 feet 9 inches long, as shown in perspective on Plate 141. If more head room is required than 3 feet 9 inches, it can be raised on timber frames (Plate 142, Fig. 1) ; it should not be supported on a sandbag wall.

3. **The large corrugated steel shelter** consists of seven segments or arches, each composed of three sheets of corrugated steel 2 feet 9 inches wide, which overlap 18 inches, and are fastened together and erected in the same way as a small elephant shelter.

An example of the use of a large elephant shelter is given on Plate 143.

The sheets of steel shelters may be used singly, in which case wall plates must be provided. Single sheets should not be used to support big weights.

4. **Curved corrugated iron** can be used resting on duckboards on edge. This makes a quickly constructed trench shelter as shown on Plate 144. They can also be used on timber frames as shown on Plate 142, Figs. 2 and 3.

5. **Troughing plates**, 6 feet or 9 feet long and 3 feet 3 inches wide, are considerably stronger than curved corrugated iron (Plate 145).

A centre support is required.

103. *General instructions.*

1. Before commencing any shelter or dug-out, decide what degree of protection is to be provided, and what number of men are to be accommodated.

2. Dug-outs intended to be splinter-proof, and medium shell-proof, are made on the " cut and cover " principle, that is to say, an excavation is made in which a shelter is built, and then covered up.

When corrugated steel shelters are used for this purpose, the end not used as an entrance must be closed and firmly strutted. The framework on which the shelter rests must be braced to prevent collapse, and the arch must be supported centrally throughout its length.

Steel shelters erected inside houses should be placed well back from the walls and covered with sandbags or concrete. The interval between the shelter and wall should not be filled in with loose earth or rubble, as these materials will only serve to transmit the shock of a bursting shell.

3. All shelters and dug-outs must be constructed to resist the effect of the explosion of a shell near them and consequent collapse, even if they are not designed to resist a direct hit.

The framework must, therefore, be in the form of a box braced in every direction. The essential points in construction are :—

i. Sides must be prevented from collapsing inwards, by being strutted top and bottom. When square timber is used, the heads and feet should be kept apart by a spreader nailed on ; cleats are useless (Plate **146**, Fig. 3). Notches must not be used.

ii. The whole box must be prevented from distortion by diagonal bracing on sides and ends.

iii. Except in hard chalk or rock, sills or bearing plates must be placed under the uprights supporting the roof.

iv. Joists must be laid on edge to obtain the full strength (Plate **148**).

v. They must be placed so that the spans are as short as possible.

vi. Timber joists having large knots should be placed so that the knots are in the upper and not in the lower surface (Plate **146**, Fig. 1).

vii. The ends of timber joists must have a good bearing on reliable supports (Plates **146**, Figs. 2 and 4, and **147**).

viii. Uprights should be round or as nearly square as possible.

ix. When fastening heavy timbers together, dogs and spikes must not be driven within 3 inches of the edge or 4 inches of the end of the timbers ; dogs must be placed on both sides of the frame. Auger holes must be bored for spikes, or the latter will split the timbers.

104. *Roofs.*

1. The roof must be weatherproof. Corrugated iron or similar material, used for this purpose, must be graded to throw off water, and this and other surface water must be prevented from entering the shelter or dug-out.

The grading of the roof is done by having one side or end slightly higher than the other.

When laying corrugated iron on a slope, the lower layers are laid first, the upper layers overlapping the lower ones. Nails should be driven through the ridges, not through the valleys of corrugated iron.

2. Where two girders cross each other they must be firmly clamped together to prevent lateral movement; distance pieces must be fitted between parallel girders.

3. Appendix V, Table 6, gives the safe load which can be carried by certain girders and rails. For timber joists and round poles Table 5 can be used thus—to find the safe load per joist or pole for any span of roof divide 16,000 lbs. by the number given in Table 5 for the joist or pole at the required span. Example: 9-inch by 3-inch joists will safely carry $\dfrac{16,000}{5}$ lbs. per joist in a roof of 7-foot span, 5 being the figure given in Table 5 for 9-inch by 3-inch joists over a span of 7 feet.

4. To find the weight of earth in lbs. which may be supported by one girder or joist, multiply the span in feet by the distance apart of the girders in feet by the depth of earth in feet by 100 (a cubic foot of earth weighs roughly 100 lbs.).

Thus, to support a roof of 2 feet 6 inches of earth, if the span is 8 feet, 40-lb. rails may be used, spaced 2 feet apart, or 5-inch by 3-inch girders 3 feet apart (Table 6).

105. *Splinter-proofs.*

1. Splinter-proof protection is given by 12 inches to 2 feet 6 inches of earth. Earth is not shell-proof until some 20 to 30 feet is used, and anything more than 2 feet 6 inches only serves to bury the occupants of a shelter. A bursting course of broken bricks, stones, &c., is always a useful addition to splinter-proof cover, but the depth of the whole roof covering should not exceed 2 feet 6 inches.

2. The earth and bursting course is usually supported on one of the shelters described in Sec. 102 and illustrated in Plates 141 to 145, but a revetted trench can be roofed with corrugated iron sheets, hurdles, planks, &c., supported by joists, poles, &c., laid across the trench. Fire and communication trenches must not be treated in this way as they quickly become blocked under shell fire, but recesses should be dug off them preferably behind a traverse as shown on Plate 144;

slit trenches (Sec. **33**, para. 2) provide good splinter-proof protection. Protection of huts and tents from splinters is dealt with under those headings in Chapter XIV, but in large camps this protection should be supplemented by slit trenches.

106. *Medium shell-proofs.*

1. The cover required to give full protection from shells up to 6-inch is as follows (Plate **149**).

    i. Burster. This turns the nose of the shell and causes it to burst before it has penetrated too far.

    ii. A cushion to absorb shock.

    iii. Distributing course. This spreads the stresses caused by the explosion over a large area of the roof.

    iv. A second cushion. This acts as a buffer between the distributing course and the roof.

    v. A thin layer of hard material immediately above the roof, to stop splinters.

2. A burster of non-rigid material—broken bricks, stone sets, or hard chalk—about 2 feet thick, has been found superior to slabs, rails, concrete, &c., for, although it must be thicker, it is less susceptible to permanent damage by shell, and is more easily replaced and repaired.

The burster must be carried well over the front and sides, so as to protect them as well as the roof.

3. The cushion should be made of the spoil obtained from the excavation of the dug-out. It should be about 3 feet thick.

4. The distributing course should consist of logs, rails, &c., tied together with stout wire, so as to form a mat. The material should be laid touching and, if possible, in two layers.

5. The second cushion may be similar to the first.

6. The inner layer may consist of 6 inches of bricks, stone or concrete, laid on boards or corrugated iron.

7. Plate **149** shows some details of construction for a dug-out proof against a 6-inch shell. The timber construction can be dispensed with by using one of the steel shelters described in Sec. **102.**

107. *Tunnelled dug-outs.*

1. Shell-proof protection under ground is obtained by tunnelled dug-cuts. Detailed information on this subject will be found in Military Engineering, Vol. IV.

The thickness of cover which gives shell-proof protection is as follows :—

Hard chalk	...	...	...	...	...	20 feet.
Weathered chalk	...	...	...	...	...	25 feet.
Clay	...	...	...	...	...	30 feet.

In chalk, economy in timber may often be effected by sinking slightly deeper than necessary for cover to reach hard sound chalk in which it is not necessary to timber galleries and chambers.

2. **Tunnelled dug-outs** have the following advantages over concrete and cut-and-cover dug-outs :—

    i. Their construction involves less labour in proportion to the accommodation given and affords more immediate results.

    ii. They give complete protection, both from actual penetration and serious concussion effects.

    iii. Their exact position can, as a rule, be better concealed, although the spoil removed from them may indicate their existence in a particular locality.

The disadvantages of tunnelled dug-outs are :—

    i. Difficulty of exit owing to the depth at which they have to be made.

    ii. Ventilation, lighting and drainage are often very difficult problems.

    iii. The nature of the strata may prevent a deep dug-out from being made in the best tactical position.

3. **Disposal of soil and camouflage.**—Concealment of the spoil excavated is of the utmost importance (Chapter IX), and the total amount to be disposed of must be considered before work is begun.

4. **Material.**—The material generally available for lining galleries and chambers is :—

For inclines and galleries :—

    9-inch by 3-inch timber sets, 6 feet 6 inches by 2 feet 9 inches internal dimensions (Plate 150, Fig. 1).

    9-inch by 3-inch timber sets, 5 feet by 2 feet 9 inches internal dimensions (Plate 150, Fig. 2).

For chambers :—

    Pit props, 4½ to 6 inches in diameter. Steel girders (R.S. joists), 5 inches by 3 inches and 9 feet long.

    Lagging (1½ to 2-inch boarding), in various lengths.

5. The **accommodation** which may have to be provided includes the following :—

    i. Command headquarters for company, battalion and higher fighting formations (Plates 151, 152, 153).

    ii. Accommodation for machine gun personnel (Plate 154).

    iii. Accommodation for personnel working in observation posts.

    iv. Living dug-outs for infantry and artillery.

    v. Subways.

    vi. Dressing stations (Plate 155).

6. Design.—Two main types of dug-outs are given, viz. :—

Type " A " (Plate **156**, Fig. 1), which is more suitable for offices and officers' quarters.

Type " B " (Plate **156**, Fig. 2), which is more suitable for men's quarters. This type involves less excavation for the accommodation provided and gives better ventilation.

7. Entrances (Plate **157**).—Every dug-out must have at least two entrances not less than 40 feet apart.

No attempt should be made to strengthen the head of an incline. No practical means will make it proof against a direct hit, and the use of concrete and girders only render clearing and repair more difficult.

It is essential to prevent entrances becoming sumps for the drainage of the trench. For this reason they should never start from the bottom of the trench. Flooding is best prevented by commencing the incline at the end of a short return of such a length as to allow 5 feet between the side of the trench and the step at trench board level (Plate **157**, Fig. 1). This space allows of extra steps being added as necessary, without blocking the entrance. The return should be made weatherproof (not splinter-proof) and camouflaged.

8. Inclines.—Inclines should be driven at a slope of 45° and should be close cased with sets 3 inches thick. The minimum width admissible is 2 feet 9 inches, as in the standard set. There are two methods of timbering, viz., vertical and normal.

Vertical timbering (Plate **157**, Fig. 1) is not recommended for unskilled men, as they have difficulty in cutting the steps properly, with the result that they crumble and the frames slip under the shock of a shell. Legs shorter than 6 feet give insufficient headroom. Steps should have 9 inches tread and 9 inches rise. Steps narrower than 8 inches are dangerous and should not be used. Vertical timbering is safer when rising through bad ground.

Normal timbering (Plate **157**, Fig. 2) is stronger, requires legs of shorter length to give the necessary headroom, and irregularity in the width of the setts used does not affect the stairway as steps are put in afterwards.

9. Excavation and timbering.—For either type first excavate the bottom and place the ground sill truly. Then put in the legs, excavating only enough ground to place them. Then excavate for and place the top sill. Dig out the remaining ground.

Never attempt to excavate for several sets to be timbered later ; this endangers the lives of those working, and, if the face or sides begin to " run," involves an immense amount of labour and leaves a weak entrance.

Each set should be " laced " back to the previous one, immediately it is put in, at top and bottom, by means of a short length of wood about 1 feet 6 inches by 4 inches by 1 inch nailed to both sets. In the case of an incline timbered normal to the slope, the side pieces of the steps form the bottom lacing.

Afterwards all sets should be strung together with 4-inch by 1-inch wooden runners, spiked about 6 inches from the top and bottom of each side or flat bar iron specially slotted.

Solid chalk is most easily broken to a " cut " (*i.e.*, a narrow excavation, usually down on one side). If timber is put in as above there is a cut, top, bottom and both sides. Picks should be systematically worked to break to one or other of these cuts and not used indiscriminately over the face.

For normal timbering a templet in the shape of a 45° triangle with 2-foot sides and a plummet should be provided.

Tool recesses should be provided at the bottom of each incline.

10. Galleries.—In bad and doubtful ground galleries should be close timbered, but in ordinary chalk and dry clay the sets may be spaced up to a maximum of 4 feet, thus economising timber. Top lagging (1½ to 2-inch boarding) must always be used and, except in solid chalk, side lagging of 1 to 1½-inch boarding, corrugated iron or expanded metal, &c., is also necessary.

When there is any doubt as to the stability of the roof, work should be carried out by " spiling " (*i.e.*, supporting the roof ahead of the excavation). In loose, heavy ground this is the only method to adopt.

Plate **158** illustrates spiling. The whaling board, of 4-inch by 2-inch timber, rests on distance pieces placed on top of the top sill of the ordinary frame. The distance pieces are large enough to allow the spiling boards (*i.e.*, overhead lagging) to be hammered forward between the whaling board and top sill.

The spiling boards are maintained at the original angle by using a spare top sill as a distance piece, bearing on the spiling boards of the set behind.

In very heavy ground the spiling boards may bend with the weight before they can be driven home ; in this case intermediate temporary sets are used. The forward set supports the end of spiling boards, while the back one serves as a distance piece to maintain the angle of drive. The boards are driven from underneath the cap. This method necessitates the excavation of the ground between sets in two distinct operations ; first, to place the intermediate set and then to place the permanent set. If it is necessary to pick out the ground ahead of the spiling boards to facilitate driving, only pick out enough to allow one board to be driven at a time.

Spiling boards should be at least 1 foot longer than the span between sets.

Excavation.—As for entrances, always break to a " cut."

11. Chambers.—Pillars of solid ground of a minimum thickness of 12 feet in chalk and 20 feet in clay, must always be left between chambers.

In ordinary soil chambers should be excavated 9 feet wide and timbered and the roof supported with standard R.S. joists on pit props.

In clay or soft chalk the joists require intermediate props, which should be inserted afterwards. They need not be placed in the centre of a girder but should be arranged to facilitate bunking, &c.

The general rule for spacing girders is at 2-foot centres in clay or sand, and at 2-foot 6-inch to 3-foot centres in chalk.

To prevent the side props from being pushed inwards they are connected to the girders by clips or brackets. The usual patterns are shown on Plate **150,** Figs. **4, 5** and **6.** The clip shown in Fig. 6 must be fixed when the girder has been put in place. Props should not be weakened by notching them.

Girders must be side-strutted to prevent them rolling over and buckling. Four struts of 4-inch by 3-inch timber are wedged between one girder and the next, spaced at intervals of 3 feet along the girder. They must be wedged extremely tight.

In clay, loam or sand, foot blocks 12 inches by 12 inches, or in heavy ground, groundsills are necessary. In good chalk it is sufficient to let the prop 3 inches into the floor.

Excavation and timbering. Except in very solid ground spiling should be employed (Plate **158**). Apart from the danger to men working if any falls occur, they not only delay progress but seriously weaken the overhead cover.

The excavation of the whole face in one piece should not be attempted. Two methods are suggested below :—

No. 1 Method.—First drive a pilot gallery. The most economical size is about 5 feet by 2 feet. This should be timbered and be approximately in centre line of chamber. Such a gallery serves as a useful check on levels and direction.

In connecting between two entrances it is as well to push this right through and secure through ventilation prior to starting the chamber proper. Men work much faster in a well-ventilated gallery. Where speed is very important, however, the face of the chamber can be worked at when this gallery is 10 feet in advance of it. The gallery then forms a " cut " to which the sides of the chamber are broken. Two men can work on each side.

No. 2 Method (without a pilot gallery). Cut out the sides of the chamber 2 feet 6 inches wide and the full height, driving forward the spiling boards over this area first. When both sides are removed to a sufficient depth to set the next props, cut out the top of the centre buttresses driving spiling boards forward as before. Then catch up these boards by a beam underneath, supported on either side of the buttress by pit-props. Set forward girder, distance pieces, whaling board and wedges and then pick out the centre buttress.

12. **Drainage.**—In wet ground, *i.e.,* where pumping is necessary, the water in the dug-out should be collected into one or more definite sumps. Correct levels are, therefore, of the greatest importance, and skilled assistance will probably be necessary. The chambers should be 1 foot

higher than the gallery and should drain towards it. Galleries should have a fall of 1/50 towards the sump to counteract small errors in setting the frames.

Trench boards should never be allowed in galleries. They collect filth and obstruct the drainage. To ensure cleanliness, when the difficulties of transport are not too great, the floors of chambers and galleries may be covered with a layer of fine concrete, 1 inch thick, laid on expanded metal which is nailed to the ground sills.

To prevent the roof of a dug-out leaking in shattered chalk, all shell-holes above it should be filled in so as to develop a slight mound.

13. Ventilation.—Ventilation is best assured by adopting type "B" for men's quarters (Plate **156**, Fig. 2), and, in big systems, by siting the kitchens so as to assist ventilation.

Vertical shafts, when used, should be utilized as kitchen flues and sited accordingly. In chalk, sound earth, and clay, holes should be bored to the surface to take kitchen flues and to assist ventilation. Special precautions must be taken to prevent gas entering by the bore-holes.

14. Bunking.—A method for a 9-foot chamber is shown on Plate **156**, Fig. 2 ; it provides a seat and a blank wall for hanging kit.

Bunks should be 6 feet 6 inches by 2 feet. This governs the length of the chamber. The cubic air space available allows for three tiers of bunks in chambers 8 feet wide, but only two tiers in chambers 6 feet wide.

15. **Gas curtains.**—Gas protection is dealt with in Sec. **45.**

16. **Working parties and footage.**—The usual working parties, exclusive of those employed in disposal of the spoil on the surface, are given below :—

Inclines.

 1 man picking.
 1 man filling sandbags.
 1 man carrying for each 10 feet of entrance.

Progress per shift of 8 hours should be about 2 feet 6 inches.

Galleries.

 1 man picking. ⎫
 1 man bagging. ⎬ Reliefs for men at face.
 1 man assisting. ⎭

Note.—Two reliefs are necessary if a fair advance is expected.

Progress per shift of 8 hours should be about 3 feet and approximately 300 bags per shift will be produced.

Carrying party.

 1 man can carry 100 bags along 100 feet of gallery per 8 hour shift.
 1 man can carry 100 bags up 10 feet of entrance per 8 hour shift.

The best method is to work in relays every 30 to 40 feet of gallery and every 5 to 7 feet of entrance.

Chambers.

No. 1 Method.

    1 man picking in pilot gallery.
    1 man filling in pilot gallery.

Unless speed is of first importance this gallery is driven in advance. Progress should be 4 feet per shift of 8 hours.

Provided the pilot gallery has been or is being driven :—

    Men employed on chamber face :—

        2 men picking.
        4 men filling bags, setting and supporting timber. These men provide reliefs for pick men.

Progress per shift of 8 hours should be about 2 feet 6 inches, and approximately 560 bags will be produced.

Carrying party calculated as above.

No. 2 Method.

    2 men picking.
    4 men filling and timbering, who relieve picking men as they tire.

Progress per shift of 8 hours should be about 2 feet and approximately 450 bags will be produced.

Carrying parties calculated as above.

Surface party.—The number of men dumping depends on nature of dumps and distance of carry.

1 man can carry 100 bags 200 feet in 8 hours on surface under ordinary trench conditions.

Labour underground can be economized by the use of trolleys and windlasses.

PART IV.—COMMUNICATIONS.

CHAPTER XVI.

CROSS-COUNTRY TRACKS, ROADS AND TRAMWAYS.

108. *Cross-country tracks.*

1. **Cross-country tracks** are made for the following reasons :—

    i. To relieve congestion on main roads by taking all foot and horsed traffic off them, at any rate in dry weather, and so leave them for those vehicles which cannot go across country.

    ii. To avoid villages and other shelled areas ; these are sometimes called " avoiding " tracks.

    iii. To improve and shorten communications generally.

All tracks must be reconnoitred, pegged out, roughly levelled, drained, and provided with signposts at frequent intervals. Batteries and conspicuous points which draw fire must be avoided.

2. **Marking of tracks.**—All tracks must be well marked, so as to be easily followed, both by day and by night.

Tracks can be marked in the following ways :—

    i. By posts.
    ii. By notice boards.
    iii. By a tape line.

Posts should be painted white. Halved pickets painted white on the sawn face or screw pickets to each of which a tag of canvas has been tied may be used.

Posts should be spaced at 20 yards interval ; they should be closer together at corners and difficult places. If both sides of the track are marked, the posts should be placed opposite each other in pairs, not chequerwise. A horizontal wire or tape should be fixed between the tops of the posts ; the wire should have short lengths of tape tied to it at 4 feet or 5 feet intervals otherwise men will not see it.

Notice boards.—Notice boards may be substituted for posts. They have the advantage that each one can be marked with the name, letter, or number of the track, and map references at important points ; black letters on a white ground are better than white letters on a black ground. For infantry tracks they should stand 3 feet out of the ground ; for mule tracks they should not be higher than 18 inches or they will be knocked over by the passing loads.

" Up " and " Down " tracks must be clearly marked and the names of any places near which the track passes should be marked on notice boards clearly visible from the track.

Tapes.—Tapes are only a very temporary expedient; they are soon obliterated by mud, and should not be laid earlier than the afternoon before they are required; they cannot be relied on for more than 12 hours, unless they are raised from the ground. This can be done by running them through the eyes of screw pickets. They are of use to troops on the night after an advance for marking the way from the company H.Q. to battalion H.Q., and from battalion H.Q. to tracks leading to the rear.

Lanterns.—Screened lanterns are useful at junctions and important points. They can be made with candles or small oil lamps in perforated biscuit tins with calico shades.

Maintenance of posts and notice boards should be done by the track wardens, detailed for the purpose.

3. Cross-country tracks are of three kinds :—

    i. For men.
    ii. For pack animals.
    iii. For horsed transport.

4. **Tracks for men.**—The most satisfactory track is one made of trench boards. Trench board tracks should avoid mule tracks or the temptation to lead mules along the trench boards will be irresistible.

A trench board track should be 3 feet wide to enable men to move along it rapidly on a dark night without risk of falling off. A one-way track should first be completed; as soon as possible, this track should be duplicated to give an " up " and a " down " route; the tracks should not be within 200 yards of each other; direction boards must be erected at the terminals and at all places where the tracks cross lateral routes. The number of tracks required depends on the tactical situation, but two pairs (*i.e.*, two " up " and two " down ") per brigade front should suffice.

Lateral communication between tracks should be provided, especially in heavily-shelled areas.

In crossing ridges, the track should be laid in a trench of a sufficient depth that men do not show against the sky line. Trench boards should be laid on 3-inch by 1-inch transoms bedded in the ground; if laid on trestles, they are much more liable to damage by shell fire, and men fall off the track at night. Trestles, however, are necessary in swampy ground, in which case they should be raised 6 to 12 inches from the ground.

To prevent slipping, stout wire netting (trench weaving) should be carefully fixed. No. 8 or No. 10 S.W.G. wire has also been found satisfactory. It should be well stapled down in a diamond-shaped pattern 6-inch to 8-inch mesh. Expanded metal and " rabbit netting " quickly wear out and then cause men to trip.

In sandy soil, a quickly-made and efficient track may be obtained by spreading out rolls of wire netting ($\frac{3}{4}$-inch to 1-inch mesh) directly on the ground and pegging firmly down.

When brushwood is available in the immediate vicinity, marshy ground may be crossed by means of brushwood mats, made of 1-inch rods.

Permanent track wardens must be appointed to repair damage.

5. **Tracks for pack animals** should be made at the same time as the tracks for men. They consist of an earth formation on the best ground available ; the route which involves the least earthwork should be chosen.

Two single tracks are better than one double one, but they should be clearly visible from each other and connected by switches at frequent intervals in the same way as trench board tracks (*see* para. 4). Loops should be made at all dumps.

Infantry trench board tracks should be avoided (para. 4).

The formation should be 4 to 5 feet wide for single traffic, or 8 to 10 feet for double. If less than 4 feet wide the mules will slip off.

Shell-holes must be cleared of water before being filled in, otherwise the filling will always be a soft place in the track.

Surface drainage must be provided, by means of a ditch on each side of the track discharging into large shell-holes ; box drains should be put in where necessary.

Mules' feet are small, and, in wet weather, readily sink into soft ground, rendering it impassable in a very short time. The following methods have been found suitable for crossing boggy patches of ground :—

    i. Fascines, with a layer of earth on them to prevent shoes being pulled off.

    ii. Hurdles.

    iii. Corduroy of logs.

    iv. Beech slabs laid on longitudinal runners.

6. **Tracks for horsed transport** can be used by vehicles in fine weather only, and it may even then be necessary to follow roads for short distances to avoid boggy places. It is very important to fix notice boards where the track enters and leaves the road. These tracks are similar to tracks for pack animals, but should be 18 feet wide and marked out on both sides. Side drains are particularly important.

When making a track across trenches, bridging should be resorted to only when absolutely necessary, *e.g.*, when the trench must be kept open or is too big to fill in.

When the track crosses a trench obliquely, it may be necessary, for the sake of speed, to make the track, in the first instance, at right angles to the trenches. In such cases the track should subsequently be made straight as soon as possible. These crossings should be made with corduroy or fascines. Wheel guides and handrails should be provided.

If a track crosses a road a length of 30 yards on either side of the road should be laid with fascines to prevent mud being carried on to the road.

109. *Roads.*

General principles.—The object of a road is to present a hard, even surface for traffic. A hard surface can be obtained by using hard stone or hard woods. Soft stone, earth, clay, or anything that is soft or will turn into mud must be avoided. The interstices in the surface of a road should be filled with gravel, stone chippings, or sand.

To present an even surface, the road must have good foundations, otherwise time and traffic will cause settlements and depressions.

The foundations may give way by being too weak, or by the failure of the earth formation below. They should be composed of a layer of large stones, 9 inches thick, which traffic will not hammer into the earth formation.

The earth formation below the foundation may give way by getting waterlogged and soft; it must be kept dry by longitudinal drains, the bottom of which are below the lowest part of the foundation.

The surface of the road is kept dry by making it higher in the centre than at the sides, so that rain is at once thrown off; otherwise water will lodge in ruts and holes and soak through the surface and foundation into the earth beneath.

110. *Metalled roads.*

1. **Metalled roads.**—The operations of constructing a road are—

    i. Peg out centre line.

    ii. Mark out side drains.

    iii. Throw the earth excavated from the drains into the centre of the road, so as to form the camber, getting additional earth if necessary from borrow-pits outside the drains. Ram this earth (Plate **159**, Fig. 1).

    iv. Lay foundations, or soling stones, by hand, carefully packed, not forgetting the outer wall of soling stones laid in a trench to prevent the road spreading (Plate **159**, Fig. 2). Soling to be 6 to 9 inches thick, according to the subsoil.

    v. Lay broken stones or macadam (2-inch to $2\frac{1}{2}$-inch gauge) in $4\frac{1}{2}$-inch layers, and roll well in. If possible, lay a second similar layer and roll well in.

    vi. Finish off surface by rolling in stone chippings, gravel and, at the very end, a little sand.

    vii. Put in 6-inch posts close up to the haunches of the road to prevent traffic leaving the metal.

They must be at a slope of 6/1 and should be whitewashed.

2.—i. **Camber** should be 1/30 to 1/40. Too much camber is very inconvenient to wheeled traffic, causing it to slip off the road.

ii. Templets (Plate **159**, Fig. 1) must be used in making the earth formation and in laying and rolling road metal.

iii. Single-way traffic requires a minimum width of 9 feet of road metal; double-way traffic a minimum width of 18 feet.

iv. On a single-way road, passing places, 50 yards long, must be made at intervals of 400 yards. With a double-traffic road on a hillside, pickets 4 to 6 inches in diameter should be driven in every 6 feet on the outer edge with rough plank revetment to stop the road spreading (Plate **159**, Fig. 3).

v. As the centre of the road takes the most traffic and gets most hammering a greater thickness of metal can be put there than at the sides and the camber thus improved.

vi. Roads on sloping ground must never be graded to drain right across from the higher to the lower side of the slope. They should be cambered in the usual way and, when necessary, box culverts provided under the road to evacuate the drain on the uphill side (Plate **159**, Fig. 3).

Where roads cross drains, catch pits for silt should be made in the drains to prevent them from becoming blocked and flooding of the road. The catch pit should be made well clear of the road and above it and should be protected by fencing or a strong cover.

It should be large enough to enable a man to get into it to clean it out, and its depth should be at least 2 feet below the outlet. Catch pits should be revetted with timber, corrugated iron, or more permanently with brick.

They should be cleaned out periodically and invariably after heavy rain.

3. After a severe frost, **thaw precautions,** *i.e.*, suspension of heavy and fast traffic, are necessary on all metalled roads on which there is much traffic ; these precautions may last several days.

A frost followed by a thaw has a tendency to disintegrate the material of which a roadway and its foundations are made, with the result that the roadway would break up under heavy and fast traffic ; chalk is especially liable to this process of disintegration.

4. **Maintenance.**—Every road requires a small maintenance party or it will soon go to pieces. Small neglected ruts become enormous holes under heavy traffic in a few days. Water and mud left on the surface of roads quickly destroy them.

5. **Repairs and improvements.**

i. Many roads are developments of old tracks, and in consequence are " sunk " roads. They are watercourses instead of watersheds (Plate **160**, Fig. 1).

This type of road requires reconstruction to re-establish the drainage. This can often be effected by one of the methods shown on Plate **160,** Figs. 2 and 3.

ii. In clay country the clay, in wet weather, works up through badly made roads and destroys them.

This can be prevented by a 6 to 9-inch layer of chalk beneath the foundation of the road ; this chalk should be well rammed until it is smooth. Small broken chalk is better than large hard pieces as it consolidates better ; this chalk forms a seal. Sand may be used in place of chalk.

iii. When widening existing country roads in a clay country dig out the earth berms in short lengths, deep enough to allow of (*a*) chalk layer ; (*b*) soling ; (*c*) road metal, but still preserve the camber. If the traffic is not of the heaviest, *e.g.*, heavy artillery, tractors, &c., the chalk (if not less than 9 inches thick) may replace the soling.

The treatment is the same in the case of roads paved with setts, which are always laid on sand (Plate **161,** Fig. 1).

iv. **Ruts.**—Cut the rut out square ; if the foundation of large stone has been destroyed, replace it by hand packing soling stone and then lay and ram the surface layer of macadam. In clay country, be careful to renew the chalk layer beneath the foundation whenever it shows signs of destruction. Shell-holes require similar treatment (Plate **161,** Fig. 2). Never cut away the earth berm of a road even if it is liquid mud, without immediately replacing it with chalk or stone. To leave a void for even 24 hours will cause great damage to the metalled centre by allowing it to spread.

v. In taking over the maintenance of an existing metalled road in poor condition the following is the order of urgency of work :—

  (*a*) Establish longitudinal side drains and cut wide gaps through the banks of earth, mud and rubbish so that the water will drain off the road. Never dig away the earth berms.

  (*b*) Sweep mud and water off the road into the side drains ; use brooms for this, not scrapers or shovels.

  (*c*) Deposit all solid mud, debris and spoil clear of the drains and on the far side.

  (*d*) Repair the worst ruts first by cutting out square and filling in, as explained above, taking care to ram well.

  (*e*) If sufficient stone is available restore shape and camber to the surface. To do this, treat half the width of the road at a time, length by length. Pick up the surfaces with pick-axes, spread macadam to the required thickness and section (using a templet), and roll well in. Unless the old surface is picked up the new layer of stone will not key into the old and will soon break up.

  (*f*) If it is found that the road is worn concave and that the chalk foundation has disappeared, it will be necessary to cut out a fresh camber in the subsoil and reconstruct the road on this (Plate **161,** Fig. 3).

6. Craters.—The deviation made round a shell or mine crater to allow traffic to pass in the first instance, must be made clear of the debris on the lip of the crater. This debris is required for filling in the crater.

To fill in a large crater :—

    i. Remove all sludge or water.

    ii. Fill to within 1 foot 6 inches or 2 feet of the surface with alternative courses of filled sandbags and rammed dry earth ; or fill with rammed dry chalk.

    iii. Then lay a slab roadway, as described in Sec. **111**.

Tightly filled sandbags covered with wire netting or expanded metal on which is placed 3 or 4 inches of road metal will form a practicable road for lorries.

Whatever hard material is available, *e.g.*, broken bricks, chalk, &c., must be reserved for filling the top portion of the crater ; the bottom part should be filled with softer material.

A method of dealing with small shell craters is described on Plate **161**, Fig. 2.

7. Causeways are used for road crossings over small streams, where bridging operations are unnecessary.

A causeway consists of i. a culvert to carry off the water, ii. a filling of earth or other material to bring the surfaces of the road to its correct level.

    i. The culvert may consist of—

        (*a*) Bundles of brushwood, fascines, large stones, &c., where only a temporary crossing is required.

        (*b*) 12-inch wooden box drains made of 2-inch timber, which is generally procurable.

        (*c*) 2-foot or 18-inch corrugated-iron culverts which give the strongest form of drain.

        (*d*) Earthenware or concrete pipes, which require care in bedding and time to lay.

The size of the culvert required depends on the width and velocity of the stream, and the amount of water to be carried off, full allowance being made for floods.

The culvert should be laid on a firm and level bed, slightly above the original level of the bed of the stream

Wing walls are required to prevent a false passage of water.

The inlets of small culverts must be protected by screens of wire netting to prevent them from being choked.

ii. Earth filling.—When the culverts have been laid, soil is thrown on and well rammed until a height of 6 inches above the correct level of the roadway is reached : this will allow for settlement.

The sides of the earth filling should be 1/1, and must be revetted as the work proceeds with timber and poles (Plate **162**)

A sleeper or corduroy road, as detailed in the next section, is laid down, the width of the top of the causeway being 15 feet and the roadway 10 feet in the clear.

Handrails and curbs should be added.

8. Plate **162** shows the details of a causeway capable of carrying a tank. Handrails and curbs are omitted.

111. *Slab, sleeper and corduroy roads.*

1. The road may be :—

    i. A single-way road.

    ii. A single-way road, which is to be doubled when circumstances permit.

    iii. A double-way road.

A single-way road is 10 feet wide and a double-way is 20 feet wide. Slabs, sleepers and logs are usually supplied in 10-foot lengths

Slabs are of hard wood (beech) 2 inches thick ; sleepers are 3 to 4 inches thick, usually of fir ; corduroy of round logs split in two.

2. **A double-way track** must provide—

    i. A hard surface, which is provided by the hardwood used, *e.g.*, beech.

    ii. An even surface, which must be ensured by sufficiently strong and well-drained foundations.

3. To ensure drainage—

    i. *The roadway must be above the general level of the ground.*

    ii. Side-drains must be cut ; and

    iii. Surface water must be thrown off by raising the centre of the road.

4. The **foundations** must be made of sufficient layers of fascines, or timber baulks or sleepers to prevent the traffic from hammering the surface into the ground, or making it wavy. The foundations, in fact, must spread the weight of the traffic. The finished camber should be rather less than with metalled roads, about $\frac{1}{40}$ to $\frac{1}{60}$.

5. A double corduroy or slab road is made as follows (Plate **161**, Fig. 4) :—

    i. Peg out centre line.

    ii. Mark out side drains.

    iii. Excavate the drains deep enough to drain the earth formation, throw the earth into the centre of the road and ram.

    iv. Excavate for the reception of the outer ends of the sleepers.

    v. Lay the sleepers at the proper slope, allowing for an initial camber of 4 to 6 inches. Traffic will reduce this to the final camber of 2 to 3 inches. Fill in between with earth, and ram.

vi. Lay the runners and dog the two centre ones together. Spike all runners to the sleepers. The holes for spikes and dogs must be bored with an auger, otherwise the wood will split. Fill in between runners with earth, and ram.

vii. Lay the corduroy close, round sides uppermost, and spike each log to every runner.

viii. Lay the ribands, with a gap of 1 foot every 30 feet, to allow for drainage, and spike or dog them to the corduroy and pickets.

ix. Pickets should be fixed when the runners are laid, but may be added later.

6. **Earth berms** preserve the side drains and permit of the passage of foot and rider traffic. They should be at least 4 feet wide.

7. In laying a **single-way** track 10 feet wide, the above procedure is followed, except that the corduroy is laid flat.

Corduroyed passing places must be provided every **200 yards.**

On hard ground, or for light traffic, the sleepers may be omitted. Fascines can be sustituted for sleepers.

In swampy ground a layer of fascines close together laid in the same direction as the sleepers, viz., across the road must be put down first. The sleepers will squeeze in between the fascines and ensure the runners having a continuous smooth bearing on the fascines. Ribands should be about 6 inches by 6 inches, or 6 inches in diameter.

8. **Turning places.**—Provision must be made for turning places at all points at which wagons or lorries may have to unload. These turning places must be made just like the roadway.

Where " in " and " out " roads to depots branch off, and at "refilling " points, the opposite side of the main road may require widening, to enable lorries to proceed in either direction. A turning radius of 10 yards should be allowed.

9. **Converting corduroy roads to metalled.**—The corduroy must be taken up except when it has sunk into soft ground, and it is possible to lay a full thickness of foundation stones and macadam on top. The old corduroy in this case improves the foundations of the roads.

112. *Tramways in forward areas.*

1. **Purpose of tramways.**—Tramways are required, often in extension of light railways, for the carriage and distribution of ammunition, engineer stores, rations and personnel.

They are a means of getting ammunition up to battery positions situated away from roads, of supplying engineer dumps with material, of feeding the troops, and of transporting working parties to and from their destination, &c.

The purpose for which a tramway is required should be clearly defined before work is begun.

Tramways should, if possible, be run on grades that can be taken over by the light railways, thus ensuring as little break as possible in communication.

2. **Types of tramways.**—There are two types of tramways, the 9-lb. 60-cm. track with steel sleepers and the 20-lb. 60-cm. track with steel sleepers, or a combination of steel and wood sleepers. The former type is used for push tracks or mule haulage and the latter for power haulage.

9-lb. or 20-lb. track means that the rails weigh 9 lbs. or 20 lbs. respectively per yard length of single rail.

Both types of tramway are supplied by the mile, and include a proportion of curved rails and turnouts.

3. **Description of track.**—9 lb. track is made up in the following proportions per mile (Plates **163** and **164**) :—

Item.	No. of sleepers per section.	Sections per mile.	Length per mile.	Weight per section.
			metres.	kilos.
5·00 m. straight sections ...	5	290	1,450	65
2·50 m. straight sections ...	2	64	160	33
2·50 m. curved sections ...	2	40	100	33
Turnouts—				
Right-hand...	—	6	30	130
Left-hand	—	6	30	130
				Tons.
Total per mile	—	—	1,770	24·2

The rails are secured to the sleepers by means of hook or clutch bolts, which clamp the flanges of the rails to a lug or plate which is riveted to the sleeper (Plate **163**, Fig. 1).

In addition to the sleepers, each track section is provided with one joint plate which serves the double purpose of supporting the rail ends as a sleeper and joining the ends of the two adjacent track sections by means of hook or clutch bolts (Plate **163**, Fig. 2).

The curved sections are all of 15 m. (49 feet 3 inches) radius (Plate **164**, Fig. 3).

The turnouts are issued ready-made up in one piece (Plate **164**, Fig. 4), and all sleepers are riveted to the rails, thus requiring no assembling. The turnouts are 5 m. (16 feet 4⅞ inches) long overall, so that they can be laid in at any point in existing tracks without cutting closing rails.

Turntables are special stores which must be ordered separately. They are 4 feet in diameter for trucks whose wheelbase is not greater than 3 feet 4 inches. Turntables should be laid at the same time as the track to obviate cutting closing rails.

4. 20 lb. track is made up in the following proportions per mile (Plates **165** and **166**) :—

Item.	No. of sleepers per section.	Sections per mile.	Length per mile.	Weight per section.	
				Type A.	Type B.
			metres.	kilos.	kilos.
5·00 m. straight sections	9	210	1,050	265	199
2·50 m. straight sections	9	104	260	141	108
2·50 m. curved sections, 30 m. rad. ...	5	40	100	141	108
5·00 m. curved sections, 30 m. rad. ...	9	10	50	265	199
5·00 m. curved sections, 50 m. rad. ...	9	20	100	265	199
5·00 m. curved sections, 100 m. rad.	9	10	50	265	199
Turnouts--					
Right-hand	-–	5	37·5	546	546
Left-hand	—	5	37·5	546	546
				Tons.	Tons.
Total per mile	—	—	1,685	91·3	71·9

Type A (Plate **165**, Figs. 1 to 4), used only in bad ground, consists of 20-lb. rails secured by bolts and clips to steel sleepers (Plate **166**), 5 per 5 m. length and 3 per 2·5 m. length, and these sections are laid on wooden sleepers and spiked thereto—4 wooden sleepers per 5 m. length and 2 per 2·5 m. length.

Type B (Plate **166**, Fig. 5) consists of 20-lb. rails secured by bolts and clips to steel sleepers, 9 per 5 m. length and 5 per 2·5 m. length.

The sections are joined together by means of fishplates and bolts— 4 bolts to each pair of fishplates, the bolt heads always being placed on the inner side of the rail to prevent the flanges of the wheels striking the nuts.

The 20-lb. turnouts (Plate **167**) are issued in three pieces and are made right and left-handed. A right-hand turnout is one which branches off to the right when viewed from the switch end looking towards the siding or frog end.

In order that mules can haul trucks without damage to the track, " mule walk " grids can be laid on the sleepers (Plate **168**). Mules must not be allowed to walk on the track unless these mule grids are provided.

A somewhat lighter form of grid can be substituted when man haulage only is used.

5. Location.—Lines to be located are marked out as nearly as physical features of ground permit in straight lines from point to point with a maximum gradient of 2 per cent. (1 in 50), or, if quite unavoidable, 3 per cent. (1 in 33), and the sharpest curve for 20-lb. track, 30 m. radius. This sharp radius curve should **never** be used if it is possible to put in a curve of easier radius. In 9-lb. track the only standard curve provided is 15 m. radius, which is suitable for the traffic which it is intended to carry.

The shortest line is not always the quickest to construct or the easiest to operate.

A thorough reconnaissance and staking out of the line should always be done well in advance of the earth work.

The following points should be noted :—

    i. Take fullest advantage of cover from enemy observation, even if it involves a detour or negotiating features mentioned in (iii).

    ii. Bogs, marshes, oblique crossings of streams and roads, long cuttings must be avoided.

    iii. Avoid localities such as cross roads and prominent objects which attract shell fire.

    iv. All grades should be in favour of the load.

    v. Tracks must be sited to suit the standard curves provided, *i.e.*, 30, 50, and 100 m. radius, and with 9-lb. track, 15 m.

    vi. Main lines in forward areas should never go direct to large dumps, batteries or headquarters, but should connect with these delivery points by means of branches or spurs at least 200 yards long.

    vii. Lines should never be constructed along the sides of a road surface. It leads to congestion of traffic and damage to the line.

    viii. Lines should always be located with a view to draining of formation level, *i.e.*, they should not go up to the centre of a valley, but should be constructed a little up the side slopes.

    6. **Construction.**—i. A convenient site must be selected for taking delivery and assembly of the track. From this point the complete sections are transported over the newly-formed track to the track laying party.

In laying 9-lb. track in forward areas the parties should be kept as small as possible, and " bunching " should be avoided.

A typical distribution for a party of 32 men is shown below :—

	N.C.Os.	Men.	Tools and stores.
(a) Preparing ground	1	11	Shovels ... ... 12 Picks ... ... 4 Billhooks ... ... 3* Axes. felling ... 2* Wire-cutters ... 2*
(b) Loading and pushing	1	9	Push trollies ... 4
(c) Carrying and laying	1	9	Spanners .. .. 4 Picks .. .. 2 Hammers, 6-lb. .. 2

    * Required in special cases only.

100 to 150 yards per hour can be laid by this party if materials are delivered at the near end of the line.

The above is for a maximum push of 500 yards. Add 2 men and a trolley per 200 yards to this party for longer pushes than 500 yards.

A typical distribution for a party of 100 men to lay 20-lb. track is shown below.

	N.C.Os. or skilled men.	Men.	Tools and stores.
(a) Forming formation level ..	2 N.C.Os.	60	Picks ... ... 20 Shovels ... ... 30 Wheelbarrows ... 8
(b) Laying track '	2 platelayers	10	Steel crowbars ... 2 Spanners ... ... 2 Beater picks ... 4
(c) Carrying track sections ...	1 N.C.O.	20	Push trucks ... 5
(d) Assembling track	1 Fitter	10	Spanners 8

This party of 100 men should lay 100 yards of 20-lb. track per hour under normal conditions : this does not allow for ballasting.

ii. Passing points should be allowed every $\frac{1}{2}$ mile.

iii. In the first instance, as little earthwork as possible should be undertaken, and the line laid on the natural ground and opened to traffic as soon as possible. This line can then be improved as labour becomes available.

iv. Provided that drainage is properly developed, tramways for light rolling stock should not require ballasting. A useful roadbed can be made of chalk if available, provided it is thoroughly well drained. Shell-holes should be used for drainage purposes.

v. In passing over very soft ground, the steel sleeper can be supplemented by wood sleepers, and in the worst cases raft track, as shown on Plate **169**, can be used.

vi. Three typical cross sections are shown on Plate **170** for the cases of construction on level ground (Fig. 1), embankments (Fig. 2), and in cuttings (Fig. 3), respectively.

vii. Road crossings should be constructed with wood sleepers, to enable guard rails and 2-inch longitudinal planks to be spiked to them (Plate **171**). When crossing roads which carry heavy traffic, 20-lb. rail should always be used, even though the track is otherwise constructed of 9-lb. rail.

7. **Maintenance.**—A tramway track in operation requires frequent inspection ; at least once in every 24 hours, and also after every bombardment.

Repairs must be made immediately they become necessary.

The points to be looked for in ordinary inspections are—
    i. Bad packing, water lying on track, rails not at correct levels or alignment, sleepers buckling.
    ii. Adjustment, cleaning and draining of points and crossings.
    iii. Clutch bolts, in the case of 9-lb. track, and fishplates and clip bolts in the case of 20-lb. track, working loose.

Materials for maintenance and repairs to track should be distributed for use of the maintenance gangs at convenient points for quick access. and should always include a proportion of curves. All salved track material should be collected at the same points and returned on trucks which have conveyed up new material.

The quickest way to repair a line which has been cut by a shell or bomb is to cut out the damaged rail sections complete and replace by new rail. A large shell-hole is best dealt with by diverting the track round it ; a small one, by filling or bridging with timber or steel, as available.

8. **Control of rolling stock.**—Issue and despatch of trucks must always be controlled at loading points. The senior man in charge of the haulage party should sign for the trucks as issued by the traffic control N.C.O., and should also report and explain all breakages and deficiencies when he returns.

The traffic control N.C.O. should be in possession of a copy of his orders and duties. He should at once report to higher authority any non-compliance with traffic standing orders, in order that the efficiency of his section may be maintained.

9. **Operating forward tramways.**—To ensure the efficient working of the system an officer must be appointed as officer in charge tramways ; to this officer will be sent all indents for trucks, giving number of trucks required, time, place where required, and loads to be carried ; when the number of trucks is insufficient to meet all requirements, preference will be given to heavy articles which cannot be split up into one-man loads, such as girders, &c.

The officer in charge tramways must insist upon the following rules being carried out :—
    i. Haulage parties must be detailed in addition to working parties.
    ii. Riding on trucks is forbidden.
    iii. Empty trucks give way to loaded trucks.
    iv. Damaged trucks must be returned to the tramway centre.

An efficient system of reporting breaks in the lines, &c., must be arranged.

PART V.—DEMOLITIONS.

CHAPTER XVII.

EXPLOSIVES AND DEMOLITIONS.

113. *Explosives.*

1. **Low and high explosives.**—Explosives are classified as low or high explosives according to whether they function by the process known technically as " explosion " or by that of " detonation."

Explosion is a rapid form of combustion, normally started by ignition and the action of low explosives is a comparatively slow one producing a more or less steady pressure or lifting effect. Such explosives are used in warfare as propellants.

Detonation on the other hand is a far more sudden change from a solid to a gaseous state, which takes place almost simultaneously throughout the whole bulk of the explosive. As a result high explosives are intensely rapid in action, and cause violent local and shattering effects, properties which render them suitable for demolition purposes. In modern warfare, therefore, high explosives only are used for work of this nature. The more sensitive high explosives can be detonated, either by ignition or percussion, but the stabler forms and, therefore, those which can be used in bulk with safety, can only be detonated satisfactorily by severe shock, their tendency when ignited being to burn instead of detonating.

2. **Bulk explosives.**—The following explosives are typical of those which will normally be available for use in bulk on service :—

    i. Wet gun-cotton.
    ii. Ammonal.
    iii. Dynamite.

i. **Wet gun-cotton** is compressed nitro-cellulose, damped with 15 to 20 per cent. of water. The addition of the latter renders it safer to store and handle, and increases its effect on detonation. It is issued in 15-oz. slabs, 6 by 3 by 1⅜ inches in size, provided with a tapered 1¼-inch hole for the reception of a gun-cotton primer. It is packed in sealed tin cases containing 16 slabs each.

Gun-cotton is a very powerful and rapid explosive producing great shattering effect, while owing to the form in which it is provided, it is easily and quickly fixed in position. It is therefore well suited for demolitions requiring comparatively small quantities of explosive, in which close contact with the object to be destroyed is essential, as for instance, the destruction of rails or girders. Complete detonation of gun-cotton, if used in large mined charges, is not always certain and its action in any case is rather too rapid to produce a good lifting effect.

ii. **Ammonal** is a grey composite powder. It is issued in 35 lb. damp-proof tins, 15 by 12 by 12 inches in size. Ammonal deteriorates rapidly with exposure as it absorbs moisture readily, and it should not be left in a damp place for any length of time, except in the damp-proof tins.

Ammonal is as powerful as gun-cotton, but being rather slower in action produces a better lifting effect in the case of mined charges, for which it is specially suitable. As ammonal is a powder, it must be placed in a container, when used for small charges requiring accurate fixing and close contact. For this reason, and from the fact that its shattering effect is rather less, it is not so well adapted for demolitions of this nature as gun-cotton.

iii. **Dynamite** is a plastic explosive, used extensively commercially for blasting rock in mines and quarries. It is manufactured in several forms, gelignite, gelatine dynamite, blasting gelatine, &c. It is usually issued in 2-oz. cartridges wrapped in parchment paper and packed in boxes weighing 5 lbs. and 50 lbs. Dynamite cannot be used after exposure to wet, which separates the ingredients and makes it dangerous to handle. Dynamite freezes at 40 degs. F., and remains frozen at higher temperatures. Complete detonation is impossible with frozen dynamite. Frozen dynamite can be distinguished from unfrozen, by being harder, more brittle than plastic, and of a slightly lighter colour. It should be thawed before use, but this is a dangerous operation and should be left to experts.

Dynamite and its modifications are the most powerful explosives in practical use and their action is even more local and shattering than gun-cotton. They are not so safe to handle or to store as wet gun-cotton or ammonal, but being plastic, and of such power, they are specially adapted for use in small bore-holes and narrow and irregular spaces where it would be difficult to fit a charge of gun-cotton.

3. **Other explosives.**—Numerous other high explosives used commercially may become available on service. The mode of action and method of use of such will, however, be in general similar to those described above.

4. **Methods of firing.**—Explosives may be fired by :—

    i. *Safety fuze.*—The service safety fuze No. 11 consists of a train of gunpowder in a waterproof covering, and is packed in flat cylinders containing 8 fathoms. It burns at the rate of about 2 feet per minute, but, as this rate is liable to vary, it should be checked before use by burning a measured length and timing it. Safety fuze will burn under water.

ii. *Electricity.*—By this means charges can be fired from a distance, leads from an exploder being connected to the electric detonator or detonators in the charge. Electrical methods of firing are explained in Sec. **116** and more fully in Military Engineering, Vol. IV.

114. *Auxiliary explosives—primers, detonators, &c.*

1. Fulminate of mercury is a highly sensitive and violent explosive, too dangerous to use in any but very small quantities. It detonates with slight friction or percussion and readily on ignition. It is owing to the latter fact that it is used in the form of detonators, as a medium for the detonation of bulk explosives by ignition.

2. Detonators.—Two forms of detonators are used for explosives on service :—

i. The No. 8 detonator for use with safety fuze. It consists of a cylindrical copper tube, 2 inches long and $\frac{7}{32}$-inch in diameter. The lower end is closed and $1\frac{1}{4}$ inches of the tube is filled with fulminate of mercury composition; the rest of the tube is left open to receive the fuze. The fuze placed in the open end of the tube burns down to the fulminate ignites it and causes it to detonate. No. 8 detonators are painted red and are packed in red tin cylinders containing 25.

ii. The No. 13 electric detonator. This is a fulminate of mercury container like the No. 8, to which it is similar in action, but the method of igniting the fulminate is electrical, and explained in Sec. **116**.

Detonators should be stored separately from other explosives, handled with great care and never left lying about. No attempt should be made to tamper with the fulminate of mercury.

3. Commercial caps are the equivalent in civil life of the No. 8 detonator, to which they are similar, but as they contain less fulminate of mercury, their action is weaker.

4. Primers.—The few grains of fulminate composition contained in a detonator are not sufficient to detonate unaided wet gun-cotton and certain other stable high explosives. The detonating shock set up by the detonator, therefore, has to be amplified by a primer of explosive more sensitive than wet gun-cotton which acts as a medium of detonation between the detonator and the bulk explosive.

The service primer is made of dry gun-cotton.

5. Dry gun-cotton.—Dry gun-cotton is compressed nitro-cellulose without any additional water. It is very inflammable, burning with a fierce hot flame. If exposed to the sun's rays for a long period it may detonate spontaneously. Gun-cotton primers are issued in the form of 1-oz. tapered cylinders, provided with a hole in the centre for the reception of the detonator. The whole is coated with paraffin wax to keep it dry. The primers are packed in sealed tin cylinders, containing 10 primers.

6. **Cordeau detonant** consists of a lead tube $\frac{5}{32}$-inch in diameter filled with high explosive (melinite or T.N.T.). To use cordeau detonant it must be detonated ; its action is practically instantaneous. It will not detonate if ignited.

The principal uses of cordeau detonant are :—

>   i. To fire a number of charges simultaneously when firing by fuze. It may sometimes be used in the same way in conjunction with electric firing to avoid complicated connections and circuits.
>   ii. To avoid the use of excessive lengths of safety fuze, which would otherwise be required in certain demolitions when firing by fuze, e.g., mined charges.

The method of detonating cordeau detonant is shown on Plate **172**, Fig. 1 ; two primers and a No. 8 detonator with safety fuze are required.

The explosive in cordeau detonant deteriorates on exposure to air ; before using, therefore, about 6 inches of open end should be cut off. The ends of cordeau detonant, if they are likely to be left in charges for a long period, should be protected. A good method is to cap them with a No. 8 detonator.

Right angled bends in cordeau detonant must be avoided as the continuity of the explosive is liable to be broken and failure result.

Junctions in cordeau detonant may be made either by means of a Spanish knot (*see* Plate **172**, Figs. 3 and 4) or by using a junction box filled with ammonal into which the ends of the cordeau are led ; each should be capped with a No. 8 detonator.

115. *Making up and fixing charges.*

1. **General principles.**—In making up and fixing charges the following points are important :—

>   i. All portions of the charge must be in close contact with each other.
>   ii. The charge must cover the whole surface to be destroyed, and be in close contact with it.
>   iii. The charge as a whole must be firmly fixed to the object to be destroyed.

2. **Connecting up primer, detonator and fuze.**—i. The safety fuze is cut to the length required to give time to get to a place of safety after it is lighted. The end to be inserted in the detonator is cut straight across ; the other end is cut on the slant to expose a larger surface for lighting.

ii. The straight cut end of the fuze is then carefully inserted in the open end of the detonator and pushed gently home so that the end of the fuze is in contact with the fulminate composition. The detonator should be held at the open end, and with the closed end pointing away from the body. The open end of the detonator is then gently pinched on to the fuze with a pair of pliers to make it grip and so prevent it being withdrawn. Care should be taken that no pressure is put on the closed end containing the fulminate.

iii. The primer should be tested to receive the detonator; if the hole in the primer is not large enough it must be enlarged by means of a "rectifier" (wooden tool made for the purpose); if the hole is too large, paper must be wrapped round the detonator to make it fit firmly. On no account is force to be used to get the detonator into position; screwing or twisting it is particularly dangerous.

3. Gun-cotton charges.—One dry primer with detonator will fire a gun-cotton charge of any reasonable length or size, provided contact between all the slabs is good. The primer should fit tightly into the tapered hole provided in the slabs for it.

In making up gun-cotton charges for demolishing walls, arches, &c., it is often convenient to lash the slabs to a board which can be fixed firmly to the object to be destroyed. A hole is drilled in the board to enable the detonator to be inserted from the opposite side.

Timber packing and mud or clay are useful to secure close contact in fixing gun-cotton charges for the destruction of girders.

The " bag, gun-cotton, waterproof," which is an article of store, is used for demolitions under water or in damp places. It is a rubber bag with a wide mouth, provided with a wooden clamp for sealing it. The wooden clamp has two grooves cut in it, to permit electric leads or safety fuze being passed into the bag. Waterproof bags of this description holding 25 gun-cotton slabs or 25 lbs. of ammonal, form part of the equipment of engineer field units.

A gun-cotton slab can only be detonated by cordeau detonant with the aid of a gun-cotton primer, with which it must be in close contact. A simple method of firing slabs with cordeau is that shown on Plate **172**, Fig. 2.

4. Ammonal charges may be fired by detonating—

    i. A slab of gun-cotton fixed in close contact with a portion of the charge.

    ii. Two or three turns of cordeau detonant wound round a tin of ammonal.

    iii. A detonator and primer buried in a portion of the charge. In the latter case, if the charge is in a damp place, a waterproof bag should be used as the container.

Ammonal re-acts chemically with copper and will gradually eat away the tube of a detonator rendering its withdrawal after any length of time a dangerous operation. Method (iii.) should be used only when it is intended to fire the charge at once.

In emergency a detonator without primer is sufficient for the detonation of ammonal, but the explosive effect obtained is not so great.

In large charges ammonal may be left in the tins in which they are issued. The ammonal will detonate through the thin walls of the tins, provided the latter are packed in close contact.

For small charges ammonal may be placed in sandbags or waterproof bags. For rails and girder demolitions bully beef or tobacco tins often make suitable containers. Placing the charge in a tube of strong canvas is a good method for certain demolitions; the flexibility of the charge is its chief merit. For bore-holes, stove-piping joined together and if necessary waterproofed makes an excellent container. A 6-inch pipe will take 10 to 11 lbs. of ammonal per foot run.

5. Dynamite charges.—A gun-cotton primer is not required for the detonation of dynamite; a detonator or commercial cap is sufficient. The paper covering containing the cartridge is unfolded at one end and a hole, with a piece of wood the same size and shape as the detonator, made in the end of the cartridge. The fuzed detonator is then inserted in the hole and the end of the paper tied firmly round the fuze to hold the detonator in position (*see* Plate **173**, Fig. 4).

If dynamite requires ramming, as in a bore-hole, each cartridge should be gently squeezed into place with a wooden rammer, the fuzed cartridge being placed in last. The use of an iron bar for ramming is dangerous.

6. Precautions in fixing detonators, &c.—In making up charges, precautions must be taken to prevent any strain to which the fuze may be subjected being transferred to the detonator in the charge, since this might cause displacement and consequent failure, or the premature firing of the detonator. In the case of charges not requiring a container, the fuze should be securely fixed to the charge 3 or 4 inches from the detonator. Where a bag container is used, a small piece of stick should be lashed at right angles to the fuzes just inside the bag. The fuze should be doubled back outside the bag, and lashed firmly to it (*see* Plate **173**, Figs. 2 and 3). Lengths of safety fuze leading to an exposed charge should be lightly fixed or weighted so that when lighted it may not curl up and set fire to the charge prematurely (*see* Plate **174**, Fig. 3).

The severe jerk set up in cordeau on detonation, tends to displace it. Cordeau should therefore be firmly secured throughout its length especially to the charges it is to detonate. Lengths of cordeau in contact or close to each other should be separated with a board, as one may cut the other without detonating it and cause failure.

7. **Lighting fuzes.**—The simplest method is to use matches ; the match head should be held against the powder in the end of the fuze which has been cut on a slant, and ignited by striking the box on it.

Patent friction lighters may be used, they easily get damp and are not therefore very reliable. They fit on to the fuze which in this case should be cut straight across. If several fuzes have to be lighted in quick succession a port-fire may be used. It is an article of store and consists of a stick of slow burning composition with a wooden handle. It can be lit with a match and to be put out should be knocked against the heel of the boot.

8. **Mobile charges.**—Conditions do not always permit of the charge being made up on the spot, as for instance in a raid. Where detailed information of the objective is available, the form of the mobile charge may be adapted accordingly. Normally, however, the charge should be made up in box form of size and weight (not exceeding 20 lbs.) suitable for one man to carry without difficulty. It should be provided with a stout handle and two or more separate means of ignition by fuze. Patent lighters, being quicker to operate than matches, are a suitable means of ignition for mobile charges but matches should be carried in addition in case of the failure of the former.

9. **Tamping.**—The tendency of all explosives is to act along the line of least resistance. Tamping is material placed round a charge in order to increase the explosive effect. Sandbags filled with earth are the material usually used for tamping. High explosives are so rapid in action that where they are in close contact with the objective, as in a rail or girder demolition, they accomplish their object without being enclosed on the exposed sides. Thus tamping in such cases, though it increases the effect, is not as a rule necessary. On the other hand in the case of mined charges tamping is essential as otherwise the main force of the explosion will pass down the gallery. The gallery in mined charges should therefore be tamped solid.

116. *Firing charges electrically.*

1. **General remarks.**—Charges on service are fired electrically by means of electric detonators, connected up by insulated cable, the usual source of energy being the service exploder.

In the following description of apparatus and methods employed, it is assumed that the theory of continuous current electricity is understood.

2. **The detonator, electric, No. 13, Mark III,** consists of a copper tube containing fulminate of mercury similar to the No. 8 detonator, but enlarged at its upper end to receive an ebonite plug. Two short copper leads, insulated with a rubber covering outside the detonator, are passed through this plug $\frac{1}{4}$ inch apart. The ends of the lead inside

the detonator are connected together with a piece of fine iridio-platinum wire just above the fulminate of mercury. On a current of sufficient strength (not less than 0·8 amperes) being passed through the copper leads, the fine wire is raised to a white heat and fuzes, thus igniting the fulminate of mercury and causing it to detonate. The head of the detonator is painted white, and the tube containing the fulminate of mercury red. The detonators are packed in tin cylinders (25 detonators in each) the upper halves of which are painted white and the lower red. Detonators fired by an exploder must always be connected up in series in a circuit. The electrical resistance of a No. 13 detonator at fusing point is 2·6 ohms.

3. **Insulated cables.**—Any insulated cable may be used provided its electrical resistance is not too great. The cable especially designed for demolitions in the field and normally available on service is the cable electric E_1, Mark II. It consists of 6 copper and 1 steel strand covered with vulcanized india-rubber and coated with compound. It is issued wound on wooden drums. The electrical resistance of the E. Mark II cable is 1·31 ohms. per 100 yards.

4. **The Exploder, Dynamo, Electric, Mark V** is contained in a wooden box 13 by 8 by 6 inches, painted white and fitted with a lid that can be locked. It consists of a dynamo operated by a handle, which converts mechanical energy into electrical energy. It is fully described in Military Engineering, Vol. IV.

To use the exploder, the handle is pulled up as far as it will go, and the leads of the circuit connected to the exploder terminals. The downstroke of the handle turns the dynamo and thus generates a current, which is at its maximum at the bottom of the stroke, when it flows through the leads. The handle of the exploder should be forced down as swiftly and smoothly as possible.

If in a good condition a Mark V exploder will fire No. 13 detonators in a circuit the total resistance of which (including that of the detonators) is 100 ohms. Where, however, an exploder has not been tested to ascertain the actual resistance through which it will fire, it is unsafe to rely upon it for a circuit of more than 40 ohms resistance. This figure is given in order to allow a good margin of safety for any defects in the working of the exploder.

Detailed tests for exploders are given in Military Engineering, Vol. IV.

The approximate resistance of a circuit can be estimated by adding the total cable resistance (1·31 ohms per 100 yards of E_1, Mark II cable) to the total resistance of the detonators at fusing point (2·6 ohms. per detonator). If the resistance of the circuit thus ascertained is beyond the power of the exploder, it must be reduced by using fewer detonators, duplicate cables, or cables of lower resistance.

To ascertain roughly if an exploder is in working order a No. 13 detonator should be connected up to the terminals and fired. The detonators should be placed in an iron box or under a sod so that it may do no harm when fired.

5. Jointing insulated cables.—The jointing of insulated cables is a most important operation since a badly made joint may be the cause of failure. It should be carried out as follows :—Strip off 2 inches of the insulation of each cable, open out the stranded wires and clean each thoroughly by scraping with the back of a knife ; take great care not to nick the wires in doing this. Cross the ends of the cables thus cleaned at right angles as shown on Plate **175**, Fig. 1, and bend them round each other as shown on Plate **175**, Figs. 2 and 3, making three or four complete and close turns with each end. Cut off the spare ends and pinch them close in with the pliers. Now cut off about 6 inches of indiarubber tape and warm it by rubbing it between the hands. Then bind the rubber tape round the joint (as on Plate **175**, Fig. 4). The tape should be stretched as it is applied. When the joint has been covered with one layer of tape, the rubber should be smeared with rubber solution and the tape wrapped on in successive layers until used up, each layer being smeared with rubber solution. No solution should be allowed to reach the bare wires of the cable.

Defects in the insulation of cables may be dealt with in a similar manner. Rubber tape and solution form part of the contents of the boxes jointing and testing carried by engineer field units.

6. Testing circuits.—The continuity test is normally sufficient except for very important charges. It consists in sending a small current through the circuit not large enough to fire the detonators but sufficient to deflect the needle of a detector or galvanometer placed in the circuit, thus proving that the electrical circuit is unbroken. A special test cell (cell, electric, dry, E) is used to furnish the current, so constructed that it can under no conditions give sufficient current to fire a detonator. The use of any other type of cell, furnishing a larger current than a test cell is highly dangerous. A test cell and a 3-coil galvanometer are provided in the " Box jointing and testing " ; they should be connected up in series to the ends of the circuit to be tested. The 2-ohm. coil of the galvanometer should be used in testing for continuity.

In all important work the detonators and cables should be tested separately for continuity before being connected up as well as the whole circuit when laid.

Detailed tests for electric firing circuits are given in Military Engineering, Vol. IV.

7. Connecting up and firing charges electrically.—The following, therefore, are the steps to be taken :—

    i. The detonator, previously tested for continuity, is placed in the charge. The precautions laid down in Sec. **115**, para. 2, as to fitting the detonator into a primer must be observed.

    ii. The cables, having been tested for continuity, are laid out from the charge to the selected firing point. The ends of the cables at the firing point should be placed in charge of an N.C.O. and the exploder box kept locked.

    iii. The cable ends at the charge are now connected up to the detonator leads.

    iv. The whole circuit is now tested for continuity as described in para. **6.**

    v. The exploder box is unlocked; the handle raised and the ends of the cables made fast to the exploder terminals.

    vi. To fire the charge the handle is pushed down swiftly and smoothly.

        Where several charges are to be fired simultaneously the procedure is similar. The detonators are connected up in series by lengths of cable as required.

Diagrams of circuits for testing and firing are shown on Plate **175**, Figs. 5 and 6.

8. Common causes of failure.

    i. Broken leads.—The leads used in demolitions carried out in the presence of the enemy should, if possible, be buried. Two feet of earth is adequate protection against bullets and small shell splinters, and 7 feet from shell fire. The cables should not be subjected to undue strain at any part of the circuit. Special care should be taken where they pass through the tamping or round corners.

    ii. Badly-made joint, causing a high resistance and thus preventing sufficient current flowing through the circuit to fire the detonators.

    iii. Bad insulation of cable or joints, causing leakage of current.

    iv. Faulty exploder.

    v. Defective detonator. The iridio-platinum wire may be broken. This, however, would be detected by the continuity test. Detonators may also be over or under-sensitive (*see* Military Engineering, Vol. IV). The No. 13 detonator is, however, carefully tested before issue and is most reliable.

117. *Alternative methods of firing.*

    1. The chief merit of safety fuze as a method of firing is its simplicity and the fact that, once the charge is laid, no apparatus other than a box of matches is necessary to fire it.

    The electrical method, however, enables the charge to be fired from a distance and at the precise instant desired, while at the same time it is admirably suited for the simultaneous firing of multiple charges. Moreover, the facility with which an electric circuit can be tested, at a distance from the charge it is to fire, is an added advantage, especially where the charge is not easily accessible for examination. On the other hand electric leads are very liable to be cut by shell fire ; they may be protected by burying them but this entails considerable labour—to be reasonably safe from damage, 7 feet of earth is necessary.

    Thus the selection of the safety fuze or electrical method will depend on the conditions. Electric firing is undoubtedly the surer method in deliberate demolitions provided the danger of the leads being cut is remote and that it is carried out under the direction of an individual who understands it. On the other hand in hasty demolitions firing by safety fuze owing to its simplicity is, as a rule, the more suitable method. For instance mobile charges will almost invariably be fired with safety fuze, while it is usually preferable to fire mined charges electrically. Whichever method is adopted all important charges should be provided with at least two means of ignition. In many cases a good plan is to use both methods, using the safety fuze as a stand-by in case of the breakdown of the electrical firing arrangements.

2. **Common causes of failure.**—i. The charge may have become too damp to detonate. Even gun-cotton slabs, if exposed to damp for a long period, may fail. Cases have occurred where owing to the deterioration of the paraffin wax covering primers, the dry gun-cotton has absorbed moisture from the wet gun-cotton slab surrounding it and failed to detonate.

ii. The charge may not entirely cover the object to be destroyed or has not been placed in close contact with it.

iii. There may not be close contact between the bulk explosive and primer, primer and detonator or detonator and fuze. The precautions laid down in Sec. **115**, paras. 1 and 2, should be carefully adhered to.

iv. The fuze may have deteriorated through dampness or age and become unreliable.

v. Failures from electrical firing are dealt with in Sec. **116**, para. 8.

vi. In firing charges in the presence of the enemy, men should be detailed to replace casualties, in order that failure may not result from this cause.

3. **Miss-fires.**—If a miss-fire occurs the longest possible time should be allowed to elapse (at least half an hour) before the charge is approached. In accessible places, the charge should be "killed" by detonating a fresh charge as close as possible to it. The charge should only be withdrawn when there is no alternative, as its removal will be a dangerous operation. In such cases, the tamping in proximity to detonators must be carefully removed, the whole being previously drenched with water, and the detonators withdrawn at the earliest opportunity.

118. *Demolitions—general principles.*

1. The main uses of demolitions in warfare are :—

    i. To delay the advance of an enemy by the destruction of communications over which he must pass or material which will fall into his hands (defensive).

    ii. Impairing an enemy's resistance by the destruction of captured communications which cannot be permanently held or materials that cannot be removed, as for instance in a raid (offensive).

2. The following points should be considered when selecting objectives for demolition :—i. The execution of a few complete demolitions at points in communications where there is no alternative route will delay the enemy more than a number of demolitions each of which can be quickly repaired or circumvented.

ii. Subject to the conditions stated in (i.) the following are the most suitable points of attack on communications (roads and railways) :—

(a) Bridges, culverts and tunnels.

(b) Cuttings and embankments.

(c) Road and railway junctions, level crossings and cross roads.

(d) Causeways passing over low-lying or marshy ground.

iii. The possibility of effecting destruction by means other than explosive should not be overlooked. This is especially important where the explosive available is limited, as for instance in a raid. Wooden bridges and stores may be burnt, certain materials rendered unserviceable with water, machinery disabled with crowbars or by the removal of indispensable parts, &c.

3. **Reconnaissance.**—The importance of thorough reconnaissance in all demolitions cannot be over-estimated. Haphazard and promiscuous methods without a clearly-defined plan cannot produce good results. Whenever demolitions are to be carried out on an extensive scale, a comprehensive and well-considered scheme should be drawn up, in which due weight is given to both tactical and technical features. Individual objectives should invariably, in so far as conditions permit, be carefully examined before the details of the method of destruction to be employed are decided on.

4. The extent to which demolitions are to be carried out in an operation will be laid down by the higher command. The responsibility for giving an order to fire the charge must be vested in an officer on the spot. Officers in charge of demolition parties must see that their orders include clear and definite instructions as to when the charge is to be fired (see Engineer Training, Sec. 56).

5. **Deliberate and hasty demolitions.**—To effect the complete demolition of a structure requires careful reconnaissance, ample time for the preparation of charges, and conditions that will permit of the fixing and firing the latter without serious enemy opposition. Where these conditions do not prevail, procedure on these deliberate lines cannot be carried out. It will then, as a rule, only be possible to aim at effecting partial destruction ; the most rapid and easily-executed method of attack having to be adopted in preference to that which will cause the most damage.

Demolitions may therefore be classified broadly under the headings of " deliberate " and " hasty." For example, the destruction of one or both abutments by mined charges is normally the most important operation in the deliberate demolition of a girder bridge, while time would only permit of the destruction of the main girders in the case of a hasty demolition. It must be borne in mind, however, that the quantity of explosives available may also prove a limiting factor as to the method of destruction adopted, the problem presented being in all cases that of effecting the maximum damage to the objective in the time available, and with the means at disposal.

119. *Calculation of charges.*

1. The weight of explosive (untamped) in pounds required for various demolitions can be calculated from the formulæ in the following table. They are equally applicable to gun-cotton, ammonal or dynamite. In these formulæ :—

B = length to be demolished in feet.

T = thickness to be demolished in feet.

t = thickness to be demolished in inches, in the case of steel or iron plate only.

Object attacked.	Lbs.	Remarks.
Masonry arch, haunch or crown ...	$\frac{3}{4}$ B.T^2	Continuous charge.
Masonry pier	$\frac{2}{3}$ B.T^2	
Masonry wall over 2 feet thick ...	$\frac{1}{2}$ B.T^2	Continuous charge. The length of breach B not to be less than the height of wall to be brought down. Walls under 2 feet thick require 2 lbs. per foot run.
Hard wood (rectangular section) ...	3 B.T^2	For soft woods these charges may be halved.
Hard wood (circular section) ...	3T^3	
Hard wood, auger hole	$\frac{3}{8}$ T^2	
Iron or steel plate	$\frac{3}{2}$B.t^2	*t* is in inches. (N.B.—A slab of gun-cotton will cut a steel plate 1 inch thick.)
Steel wire cable 4 inches in circumference and over	$\frac{C^2}{16}$	C is circumference in inches. Cables under 4 inches in circumference require 1 lb. to cut them.

Masonry includes concrete, stone or brickwork. Tamping will increase the effect of the above charges. Charges placed hurriedly under conditions, where they cannot be examined properly after being fixed, should be increased by a percentage (say 50 per cent.) to allow for bad contact, &c.

2. **Mined charges**.—The method of calculation of mined charges to produce craters of varying diameters in different soils is laid down in Military Engineering, Vol. IV. The following table, however, gives the weight of ammonal or similar high explosive required

in hard chalk. For ordinary demolitions it will be found sufficiently accurate for use in all soils; in softer ground larger craters will be made.

Ammonal charge in lbs.

Depth to centre of charge in feet.	Diameter of crater.					
	30 feet.	40 feet.	50 feet.	60 feet.	70 feet.	80 feet.
	lbs.	lbs.	lbs.	lbs.	lbs.	lbs.
5 feet	160	360	—	—.	—	—
10 feet	170	390	750	1,260	1,950	2,900
15 feet	190	420	800	1,350	2,050	3,000
20 feet	210	460	850	1,400	2,150	3,100
25 feet	—	490	900	1,470	2,250	3,250
30 feet	—	—	1,000	1,500	2,350	3,400

The formation of craters by means of mined charges makes a formidable obstacle in roads and railways, especially where deviation is difficult. For mined charges, however, explosives in considerable quantities must be available and their preparation entails the expenditure of much time and labour. They are thus essentially deliberate demolitions. The ruling principle as to the selection of points of attack is laid down in Sec. **118,** para. 2. The charge should be placed under the centre of the road or railway, and should be calculated to form a crater large enough to remove the whole width of the permanent way or metalled surface. The deeper the charge, provided adequate quantities of explosive are available, the greater will be the effect attainable. In many cases, however, the depth may be limited by the presence of water. To lay the mine a vertical shaft is sunk at the side of the road, and a horizontal or inclined gallery driven from the bottom of the shaft to the required depth under the centre of the road, where a chamber for the reception of the charge is constructed. It is best to place the chamber on one side of the gallery to increase the tamping effect. It should be of dimensions to correspond with the charge calculated. The charge is then laid and prepared for firing and the horizontal gallery tamped with sandbags. Where the charge is to be placed in an embankment the digging of a shaft is not as a rule necessary, a horizontal or inclined gallery being driven from the surface to the required position for the charge. In some cases, the cellar of a house at the side of a road may form a convenient point from which to drive a gallery. The method of constructing shafts and galleries and of laying mined charges are given in detail in Military Engineering, Vol. IV.

The priming charges containing the detonators should be placed in waterproof bags, even in apparently dry soils, whenever the mined charge is to be left in position for any length of time. In marshy or water-logged soils it is often not practicable to lay mined charges, owing to the difficulties presented in constructing the shaft and gallery and of keeping the charge waterproof.

In such cases waterproofed metal tubes filled with ammonal (as described in Sec. **115**, para. 4) may be fired in bore-holes made with earth augers or borers. This method is, however, not so effective as a mined charge.

120. Demolition of iron and steel work.

1. As gun-cotton is the most suitable explosive for such demolitions, its use only is considered, but the principles laid down will apply where other high explosives are used.

The formula $\frac{3}{2}$ B.t^2 should be used in calculating all charges for iron and steel work. The charge must extend along the whole breadth of the material to be cut. It thus follows that the minimum charge of gun-cotton is one slab per 6 inches of breadth to be cut. It is useful to remember that by the formula a slab of gun-cotton will cut metal 1 inch thick.

To cut a first-class steel rail one slab of gun-cotton is sufficient, a good method of fixing it is that shown on Plate **174**, Fig. 3.

In carrying out the destruction of metallic substances (guns, girders, rails, &c.) it should be remembered that fragments are liable to be blown 1,000 yards or more away from the spot where the demolition is being carried out.

2. Girders.—There are so many different forms of girders in use that it is impossible to lay down rules for their destruction which shall be applicable to all. The engineer must be prepared to use his own judgment.

In demolishing girders there is, as a rule, difficulty in obtaining proper contact between the charge and the metal owing to the presence of rivet heads. The best method of meeting this difficulty is to fill the spaces between the rivet heads with clay and include the depth of this layer in the thickness to be cut.

All girders are made up of a top and bottom " flange " connected by a " web " consisting of continuous plates in plate girders or of open cross bracing in braced girders.

With plate girders, the most economical method of destruction is to place continuous charges across the top and bottom flanges and the web. The weight of the charges, sufficient to cut through the metal in each case, is calculated from the formula $\frac{3}{2}$ B.t^2. This method, however, involves the simultaneous firing of three separate charges, moreover, difficulties in fixing them to the girder may often arise. Normally, therefore, the best method of destruction of ordinary plate girders is that shown on Plate **173**, Fig. 1. Separate charges are calculated for the top and bottom flanges and the web.

In calculating the charge for the web, only that portion between the angle irons is considered, as allowance is made in the flange charges for the destruction of that portion of the web which is thickened at the junction with the flanges. If the web does not exceed 1 inch in thickness, one slab per 6-inch length to be cut is necessary. Where the thickness exceeds one inch, the charge must be calculated from the formula, $\frac{3}{2}$ B.t^2. For the flanges the charge is calculated from the formula, t is taken as the maximum thickness of flange plus rivet head and B as breadth of flange in feet. To allow for the fact that the charge is not placed continuously along the flange and for the additional explosive required to cut the thick portion of the web at the junction with the flange, the charge arrived at from the formula must be doubled.

Example : Girder, Plate **173**, Fig. 1.

   i. *Web.*—Thickness of web = $\frac{3}{8}$ inch, *i.e.*, under 1 inch.
      Therefore one slab per 6 inches will suffice.
      Length of slab between angle irons = 24 inches.
      $\therefore$ No. of slabs required for web = 4.

   ii. *Top flange.*—Maximum thickness $t = \frac{1}{2}$ in. (flange) $+ \frac{1}{2}$ in.
      (angle iron) $+ \frac{1}{2}$ in. (rivet head) = $1\frac{1}{2}$ ins.

$$\text{Breadth B} = \frac{15}{12} \text{ feet.}$$

$$\therefore \frac{3}{2}\text{B.}t^2 = \frac{3}{2} \times \frac{15}{12} \times \left(\frac{3}{2}\right)^2 = 4 \cdot 2 \text{ lbs.}$$

      $\therefore$ Charge required = $8 \cdot 4$ lbs. or 9 slabs.

   iii. *Bottom flange.*—Maximum thickness $t = \frac{1}{2}$ in. (flange)
      $+ \frac{5}{8}$ in. (angle iron) $+ \frac{1}{2}$ in. (rivet head) = $1\frac{5}{8}$ ins.

$$\text{Breadth B} = \frac{15}{12} \text{ feet.}$$

$$\therefore \frac{3}{2}\text{B.}t^2 = \frac{3}{2} \times \frac{15}{12} \times \left(\frac{13}{8}\right)^2 = 5 \text{ lbs.}$$

      $\therefore$ Charge required will be 10 lbs. or 11 slabs.

In the case of braced girders a suitable point along the girder must be selected at which to cut through all the members. Charges must then be calculated separately from the formula for each member, *i.e.*, top and bottom flanges and the web bracing.

3. **Guns.**—To destroy a gun with high explosives a shell should be loaded in the ordinary way ; the charge necessary for the destruction of the gun should be packed behind it so as to be in close contact with the shell and with the sides of the chamber.

After the insertion of the firing arrangements the charge should be tamped with earth or other suitable material to keep it in position.

The breach should be closed as far as possible, just allowing room for the safety fuze or electric leads for firing the charge. A shell is not absolutely necessary for destroying a gun by this method, but increases the effect. The charge required is calculated by the following rule :—
" For a 3-inch gun use 2 lbs. and double the charge for every inch increase in calibre, *e.g.*, for a 4-inch gun use 4 lbs, and for a 5-inch gun, 8 lbs."

If explosive in bulk is not available, a gun may be destroyed by placing a high explosive shell in the breach and detonating it, after first blocking the bore.

121. *Demolition of buildings, stockades, &c.*

1. **Buildings.**—Buildings are best demolished by placing charges of explosive in the interior, preferably in the angle of the main supporting walls on the ground floor. All windows, doors and chimneys should be closed. For small brick houses and cottages charges of 10 lbs. per room will generally be sufficient and for more solidly built houses 20 lbs. per room. The charges should, if possible, be fired simultaneously. For large halls, theatres, &c., and buildings of exceptionally solid construction the demolition in detail of the main walls may often be the most effective method of destruction. Hollow towers with a solid base, such as those in the Indian North West frontier, can be demolished by firing a well tamped charge up to 100 lbs. in the centre of the base.

2. **Stockades.**—The charge of high explosive required to effect a breach in a stockade may be calculated roughly from the following data :—
> Stockade of earth between timber up to 3 feet 6 inches thick requires 4 lbs. per foot run of breach. A heavy rail stockade requires 7 lbs. per foot run of breach.

3. **Barbed-wire entanglement.**—For cutting a passage through a barbed-wire entanglement in hasty demolitions, a form of mobile charge, known as a Bangalore torpedo, may be used. It consists of an iron pipe, or one made of stout zinc or tin plate, filled with ammonal. The fuzed detonator and primer are inserted at one end. The pipe is closed at its extremities with wooden plugs, through one of which a hole is made for the safety fuze. The pipe should be laid on the strands of barbed wire and should be at least 2 inches in diameter if a clear passage is to be cut. With Bangalore torpedoes over 10 feet long, a piece of cordeau detonant should be run through the length of the charge to ensure detonation throughout.

4. **Timber posts, trees, &c.**—The most economical method of destroying posts, trees, &c., with explosive is by making an auger hole to just beyond the centre for the reception of the charge which is calculated from the formula $\frac{3}{8}$ T^2.

Piles which are to be cut off under water at their base are best dealt with in the following manner. The waterproofed charge is attached to the pile above the water by a piece of wire rope, wound round the pile sufficiently loose so that the whole will slide down the pile. A stick up which the fuze or leads are lead is attached to the rope ring, and the charge is pushed down in position below the water-level to where the pile enters the ground.

122. *Demolition of bridges.*

1. **Deliberate.**—Provided an adequate supply of explosives is available the deliberate demolition of a bridge will involve the destruction of the following :—

(*a*) One or both abutments.

(*b*) The intervening piers (if any).

(*c*) The main girders.

(*a*) *Abutments.*—If the abutments of a bridge are destroyed, the difficulty of repairing it is much increased. Thus their destruction is normally the most important operation in deliberate bridge demolitions, especially when no deviation is possible.

The usual method adopted is the laying of a mined charge as described in Sec. **119**, para. 2, sufficiently close to the abutment to blow it down, at the same time as the crater is formed. The most suitable position and the weight of the charge will be governed by the strength of the abutment walls and other conditions. The following rules, however, will be found to work well in most cases (*see* Plate **173**, Fig. 5).

    i. If " b " is the breadth of the abutment, the charge should be placed in the centre of the abutment at a distance " e " from the outside face of the abutment equal to $\frac{1}{2}$ " b."

    ii. The depth of the charge " h " should be between $1\frac{1}{4}$ and $1\frac{1}{2}$ times " e " and in the case of masonry bridges at least as deep as the springing of the arch at the abutment.

    iii. The weight of the charge may be calculated from the table given in Sec. **119**, para. 2, but in this case the depth of centre of charge is taken as " e," and the diameter of crater desired as " b." If ample supplies of explosive are available, the charge thus calculated may be increased with advantage up to 100 per cent. in order to produce a greater range of disruptive effect on the abutment foundations.

Example.—It is required to destroy the bridge shown on Plate **173**, Fig. 1, which is 28 feet wide :—

$$\therefore \text{ b } = 28 \text{ feet.}$$
$$\text{and e } = 14 \text{ feet.}$$

Depth of centre of charge—

$$= 1\frac{1}{4} \text{ to } 1\frac{1}{2} \text{ times 14 feet.}$$
$$= 17\frac{1}{2} \text{ to 21 feet (say 20 feet).}$$

From table of charges Sec. **119**, para. 2, a crater of 30 feet diameter (b = 28 feet) is produced at a depth of 15 feet (e = 14 feet) by a charge of 190 lbs. A suitable charge (increasing by 100 per cent.) would therefore be 380 lbs.

(*b*) *Piers.*—The formula for calculating charges for destruction of piers is given in Sec. **119**, para. **1**.

(*c*) *Girders.*—The main girders will be brought down by the destruction of the abutments and piers, but it is important to ensure that they are sufficiently damaged to render them useless for re-erection. They should therefore be cut with explosive charges as described in Sec. **120**, para. 2.

2. In carrying out all bridge demolitions it should be borne in mind that the destruction of the approaches to a bridge by means of mined charges may often be as important as the destruction of the bridge itself.

3. Light wooden trestle bridges with timbers up to 9 inches by 3 inches may be burnt by using petrol or tar. Heavy wooden bridges with timber of larger dimensions should be destroyed with explosives.

4. **Hasty.**—In hasty demolitions the destruction of the girders or arches of a bridge is, as a rule, all that can be attempted, while occasionally conditions may permit of more extensive damage being effected by the destruction of one or more piers.

The arches of a masonry bridge may be attacked at the haunches or the crown. The former method is much to be preferred as a larger gap will be made, but more explosive will be required than for a single charge placed at the crown. The charge may be calculated from the formula given in Sec. **119**, para. 1. If time and conditions permit a trench should be dug across the roadway down to the masonry of the arch, for the reception of the charge, which should be continuous and in close contact with the masonry (*see* Plate **173**, Fig. 5). The trench should be filled in after the charge has been laid as this tamping will increase its effect. When conditions do not permit of a trench being dug, the charge may be fixed to a board as described in Sec. **115**, para. 3, and the whole secured firmly to the under side of the arch so that the charge is in close contact with it throughout the whole width of the bridge.

Although time, when the demolition is a hasty one, will never permit of the destruction of bridge abutments by heavy mined charges, on the lines stated in para. 1, it may be possible under certain conditions to damage them considerably. The main difficulty is to obtain a hole in or behind the abutment in which to place the charge. This may sometimes be overcome by blowing a small initial charge, up to 50 lbs., to form a cavity for the main charge. Although the whole operation is thus performed in two stages, it need not take more than half an hour to carry out if the charges can be prepared beforehand. Another method, specially suitable where the approach is an embankment, is to make one or more bore-holes with an earth-auger behind the abutment masonry and to load the holes with ammonal in stove-pipes (*see* Sec. **115,** para. 4).

123. *Demolition of railways.*

1. The destruction of railway bridges and the blowing of craters at selected points in the permanent way have already been dealt with in Secs. **118, 119** and **122.** Tunnels may be destroyed by placing mined charges in the roof or walls. Ventilating shafts often form a suitable chamber for such charges. An alternative method of destruction requiring less explosive is to demolish a length of the arch-ring, the charge required being the same as that for destroying the arch of a bridge. Where the destruction of the tunnel is not desirable, an effective obstruction may be made by causing the derailment or collision of rolling stock in it; the removal of the wreckage within the cramped space of a tunnel is a difficult and lengthy proceeding.

Much damage to the permanent way, rolling stock and appliances of a railway can be effected without explosives. The method of attack must depend largely on the time at the disposal of the working party, its numerical strength and on the extent of the damage it is desired to carry out.

2. When a railway is to be interrupted, the first step in every case is to sever or block the main lines of rails. As soon as this has been done, points and crossings, as being the most important parts of the permanent way, should be destroyed or removed. The water supply should then be rendered useless. Pumps and tanks should be destroyed either with explosives or by knocking off rivets, &c., with a sledge hammer and so causing leakage. All signals, both electric and visual, should be destroyed.

Station buildings, as a rule, are not indispensable to traffic and, therefore, not worth destroying; but workshops and repairs shops should, if possible, be burnt out and their fittings and machinery and all other technical tools or apparatus removed or destroyed. Fuel should be removed or burnt. If rolling stock cannot be removed it can be rendered useless by burning, or trains may be derailed, preferably over an embankment or in a tunnel by turning a rail.

3. The simplest method of attacking the permanent way is to remove or destroy portions of the line or lines at intervals, especially curves. If sufficient explosive is available, destruction may be effected by firing a charge fixed as described in Sec. **120,** para. 1, at each rail joint. The rails will then be damaged beyond practical repair. Rails may be destroyed without explosives, by making fires with the sleepers, placing the rails upon them and twisting them when hot. If the rails are only bent, they can be straightened on the spot, but if twisted they must be sent to the mill to be re-rolled before they can be used again. To remove the rails the fish plate nuts should be unscrewed with a spanner; if one is not available

they can generally be broken off with a hammer. The chairs should be broken off with a sledge hammer. If time and conditions permit the permanent way may be taken up and removed bodily in trains, but this requires careful organization and large working parties and should be left to railway experts.

4. When it is desired to only disable rolling stock or instruments the guiding principle should be to destroy or remove the whole supply of one article essential to the working of the railway rather than to effect promiscuous but incomplete damage of several things. The adoption of this course of action prevents a few complete units being formed from the parts of damaged ones. Locomotives can be rendered useless but still repairable by taking off the injector, the connecting rods, piston or valve ; carriages can be similarly disabled by removing the springs so as to let the body of the carriage fall on the wheels and axle.

CHAPTER XVIII.

LAND MINES, TRAPS, &c.

124. *Types of mines and traps.*

1. **Land mines** are explosive charges laid in the ground with the object of delaying the advance of an enemy, by impairing his morale, destroying his personnel and transport, and interrupting his communications after the evacuated terrain has fallen into his hands.

The quantity of explosive used will depend upon the purpose for which the land mine is laid and may vary from a few pounds to several hundreds. High explosive shells and trench mortar bombs may often be suitably used for the charge in place of bulk high explosive.

Land mines may be divided into three classes according to the method by which they are set in operation.

    i. Contact.
    ii. Observation.
    iii. Delay action.

    i. **Contact mines.**—These normally consist of a small charge of explosive buried a few inches below the surface of the ground and contained in a specially designed box, or a shell, fitted with some form of contact firing arrangement. The latter is so constructed that pressure on the surface of the ground caused by troops or vehicles passing over it, sets it in operation and fires the charge. This firing arrangement in most forms functions by percussion or friction, the release of a striker firing a percussion cap (just as a cartridge is fired in a gun) or igniting friction composition. In some types, however, it may operate electrically, the pressure on the surface closing a circuit and thus firing the charge.

The designs of contact mines that may be met with are very numerous. A few representative types are described in Military Engineering, Vol. IV. Extensive fields or belts of such mines may be laid and there is much scope for the skilful selection of sites where traffic is likely to pass and yet where the detection of mines is difficult. The mines should be so spaced as to render it practically impossible for a wheeled vehicle or tank to pass through the belt without exploding one of them.

It is, as a rule, difficult to conceal contact mines in the metalled surface of roads. They may sometimes be placed with advantage on the edge of roads where traffic is still likely to pass and where the surface is, as a rule, more muddy and thus affords greater facilities for concealment. A ruse often adopted is to place an obstacle in the road and to lay a minefield on each side of it where a deviation would normally take place. A crater forms a specially satisfactory obstacle in such cases, as the debris scattered round it from its explosion serves to obliterate any traces on the surface of the existence of a minefield.

ii. **Observation mines** are land mines which can be fired by electricity from a distance when the enemy is seen to pass over them. They may be laid in front of a defended position in ground over which the enemy is likely to advance or mass for attack. Their use, however, as compared with contact mines is rather limited.

iii. **Delay action mines** are operated by a delay action fuze, by means of which the time of the explosion, after the charge has been laid, may be deferred for a period varying from a few hours to several weeks or even months. The simplest and most satisfactory type of delay action fuze hitherto invented depends on the dissolving of a fine steel wire by a corrosive liquid ; the length of the delay is regulated by varying the strength of the liquid. It is fully described in Military Engineering, Vol. IV.

Delay action mines will, as a rule, consist of charges of several hundred pounds laid at depths suitable to form large craters. They are specially suitable for laying in the permanent way of railway lines, bridge abutments, &c., with a view to causing intermittent interruption of road and rail communications, after the damage effected by ordinary demolitions prior to the retreat has been repaired by the enemy. They may also be laid with success in billets, dug-outs, &c., which the enemy is likely to occupy, and in abandoned shell dumps. In the latter case difficulty of detection may be increased by the container of the delay action fuze being constructed to resemble an ordinary shell fuze.

2. **Traps.**—Improvised contact mines and charges, placed with the object of making the occupation of dug-outs, buildings, &c., dangerous when abandoned to the enemy are commonly known as " traps." They are not, as a rule, very destructive to personnel but the atmosphere of uncertainty they produce has a considerable moral effect on advancing troops and may deter them from using much valuable shelter.

In principle their method of working is similar to land contact mines. The design of traps must be adapted to suit the local features of each particular case, and in general the more varied their form the more difficult will be their detection. There is ample field for ingenuity and cunning in constructing these devices.

The following are a few typical instances of traps that may be laid :— A loosened board so arranged that a charge is fired on the former being stepped on. An attractive souvenir or trinket so attached to a concealed charge that it fires the latter on being moved. A charge placed in a chimney so that an explosion occurs as soon as a fire is lighted. Charges may be so made up that they are fired on the following actions :—the opening of a door, window, cupboard or drawer ; switching on electric light, pulling the plug of a water-closet, cutting or tripping over a wire.

3. **General remarks.**—The making and laying of all land mines and traps is a dangerous operation, and should be carried out by experts. Wherever they are to be used on an extensive scale a considered scheme is essential ; careful records should be kept of the position and nature of all mines and traps laid.

125. *The detection of land mines and traps.*

1. **Mines and traps** laid by a skilful enemy are most difficult to detect, and their successful action can only be circumvented by a thorough and conscientious search. During an advance the country must be systematically examined, whenever the enemy is suspected to have employed these devices. Specially trained parties of engineers, acting in close co-operation with the infantry, should be used for this purpose. They should be equipped with probing bars, electric torches and wire-cutters.

In searching suspected localities, contact and delay action mines may be detected from the following :—

 i. Disturbed appearance of surface soil, breaks in the continuity of weeds, &c.
 ii. Small subsidences in the ground ; these are likely to be accentuated by rainy weather.
 iii. Presence of spoil, explosive wrappings, boxes, &c.
 iv. Foot prints in soil foreign to the surface of the ground, *e.g.*, chalk marks where no chalk exists on the surface.
 v. Pegs or other marks placed in the ground without any obvious reason.

Delay action mines, since they require no contact making device near the surface, are particularly difficult to discover.

Where the enemy is using shells for the explosive charge, the deflection of a magnetic needle in the presence of iron may sometimes be a valuable aid to detection, especially in searching walls of dug-outs and buildings. In such cases search parties should be provided with compasses and dip needles.

Search parties should be carefully instructed in the various types of mines and traps that are likely to be encountered. When any new form is discovered, it should be immediately reported and a description rendered of its salient features. By this means all troops can be quickly warned and search parties placed on the look out for devices of the same type.

2. **The removal or rendering harmless** of mines and traps is a dangerous operation and should only be carried out by experts. In some cases the extraction of the firing device and detonator may be effected without danger and the bulk explosive left *in situ ;* in others, it may be advisable to remove the whole charge carefully and explode it in a safe place. Where the charge is fired electrically the wires of the circuit should be cut and the electric detonator removed. With delay action mines the fuze should be removed immediately it is discovered. Particular care must be exercised in doing so as the slightest jar may send it off. Explosives should not be thrown in ponds or down wells as they may poison the water.

APPENDIX I.

SPECIMEN OF A WORKING PARTY TABLE.

WORKING PARTY TABLE: NIGHT 6TH/7TH OCTOBER, 1917, "A" FIELD COMPANY, R.E., NORTH BRIGADE.

Serial No.	Date.	Number of working men.	Unit* providing party.	Rendezvous.	Time.	Guides.	Tools.	Task.†	Remarks.
1	6th	50		(Map Ref.) Bank R.E. Dump	18.00	Field Coy., Sapper Shovel	Tools from Bank Dump	Deepening Engineer Avenue : time work 4 hours.	I. Serial Nos. 1, 3, 4, 5, 6: All officers and N.C.Os. not digging to bring 6-foot sticks marked in feet. II. Instructions for brigade guide (serial No. 2) attached. III. Up route, Serial Nos. 3, 4, 5, 6, Railway Avenue ; Down Route, Engineer Avenue. IV.—Serial Nos. 3 and 4, pick up tools at rendezvous. V.—Serial Nos. 3 and 4 dump their tools on the work for Serial Nos. 5 and 6, who will bring their tools back to rendezvous.
2	6th	25		Do.	18.30	Brigade	Nil	Carrying trench boards for Engineer Avenue two journeys.	
3	6th	50		(Map Ref.) junction Railway Avenue and main road	18.00	Field Coy., Sapper Wire	Every man 1 shovel, every second man 1 pick in addition	Digging Png trench. Task work estimated 3½ hours	
4	6th	50		Do.	18.15	Do., Sapper Pick	Do.	Do.	
5	7th	50		Do.	00.15	Do., Sapper Post	Tools from Serial No. 3 party	Do., 2nd relief	
6	7th	50		Do.	00.30	Do., Sapper Screw.	Tools from Serial No. 4 party	Do.	

* This column is for use of general staff.
† State whether task or time work and probable duration. If duty is carrying stores, state number of journeys between dump and site of work.

Copies—
2 ... ... North Brigade.
1 ... ... C.R.E.
1 ... ... O.C. Field Company.

O.C. "A" Field Company.

APPENDIX IA.

SPECIMEN OF INSTRUCTIONS TO A UNIT PROVIDING A WORKING PARTY.

A

O.C. 28th Black Watch.

Please provide the following party for work as detailed below :—

Party Off.	Party O.R.	Tools to be brought.	Rendezvous Date.	Rendezvous Time.	Rendezvous Place.	To report to.	Nature and place of work.	Probable time to complete.	Remarks.
2	100	65 shovels 35 picks	11.2.18	19·00	Cross roads X. 23 a. 03	2nd Lieut. Smith, R.E.	Excavation of Highland Avenue	4 hrs.	Task work.

N.B.—Number of men required means actual working men; a proper proportion of N.C.Os. and stretcher bearers must be sent in addition.

This form is to be given to the guide who meets the party at the rendezvous in exchange for a similar form marked B, which will be handed in at the battalion orderly room on return from work.

D. McLEOD,
Brigade-Major,
226th Infantry Bde.

Date—10.2.18.

B

O.C. 28th Black Watch.

Please provide the following party for work as detailed below :—

Party Off.	Party O.R.	Tools to be brought.	Rendezvous Date.	Rendezvous Time.	Rendezvous Place.	To report to.	Nature and place of work.	Probable time to complete.	Remarks.
2	100	65 shovels 35 picks	11.2.18	19·00	Cross roads X. 23 a. 03	2nd Lieut. Smith, R.E.	Excavation of Highland Avenue	4 hrs.	Task work.

N.B.—This form is to be given to the officer in command of the party in exchange for a similar form marked A.

Copy to O.C. 690th Field Coy., R.E.

D. McLEOD,
Brigade-Major,
226th Infantry Bde.

Date—10.2.18.

APPENDIX II.

TABLE OF TIME, MEN AND TOOLS REQUIRED FOR THE EXECUTION OF CERTAIN FIELD WORKS.

It is assumed that :—

    i. All tracing and marking out has been done beforehand.

    ii. Materials are on the site of the work, except when provision for carrying is made.

    iii. The labour is ordinarily trained infantry working parties.

    iv. Rain is not falling.

    v. The march to work does not exceed 1½ hours.

Nature of work.	Unit or party.	Time.	Amount of work.	Tools per party.	Remarks.
I.—ENTRENCHING.					(a) Soil average easy; increase by 50 per cent. for very easy soil; decrease by 30 per cent. for very difficult soil.
1. Excavation ...	1	1 hour ...	20 cubic feet ...	1 Pick 1 Shovel	(b) Decrease by 30 per cent. for very dark nights.
	1	4 hours ...	60 cubic feet ...	Do.	(c) Maximum throw 12 feet and lift 4 feet, or maximum lift only 9 feet. When these maxima are exceeded, one shoveller, with one shovel, is required for every two diggers.
					(d) When depth of trench exceeds 4 feet, one shoveller, with one shovel, is required for every two diggers, to clear berms and level parapet; parados must be left heaped and uneven.

	No.	Time	Quantity	Tools	Remarks
					(e) In heavy clay provide sticks as scrapers; handles of entrenching tools are not to be used.
					(f) Officer in charge of work is responsible for the provision of any special tools, such as crowbars, spare pick handles, spades, axes, billhooks, &c., required by the nature of the soil.
					(g) One pick between two diggers or one pick between three diggers will suffice in certain soils.
2. Moving earth 25 yards, depositing and return	1	2 mins. ...	1 cubic foot ...	2 wheelbarrows ...	(a) Spare wheelbarrows or stretchers being filled while the others are being emptied.
	2	2 mins. ...	1 cubic foot ...	2 stretchers ...	
	1	2 mins. ...	½ cubic foot ...	1 sandbag ...	(a) Sandbags ready filled—dumped, but not emptied.
3. Filling sandbags ...	3	1 min. ...	1 bag ...	3 shovels ...	(a) Sandbags to be three-quarters filled.
II.—REVETMENTS, &c.					
1. Sandbag revetment—					
i. Filling sandbag ...			See I. 3, above.		
ii. Carrying sandbags to site, &c.			See I. 2, above.		
iii. Building sandbags	2	2 mins. ...	1 square foot of revetment	2 filled bags ... 1 flat beater	(a) Size of sandbag 20 by 10 by 5 ins. (b) Alternate courses of headers and stretchers. (c) Flat beater may be a billhook or a spade.

Nature of work.	Unit or party.	Time.	Amount of work.	Tools per party.	Remarks.
II.—REVETMENTS, &c.—*continued.*					
2. Sod revetment:—					
i. Cutting sods	3	3 mins. ...	5 sods	3 sharp spades ...	(a) Size of sod 18 by 9 by 4 ins.
ii. Carrying sods to site, &c.			*See* I. 2, above.		(b) 1 sod to be taken as ¾ cubic foot.
iii. Building sods	2	3 mins. ...	1 square foot of revetment	1 shovel or spade ...	(c) Allow 5 sods, each 18 by 9 by 4 ins. per square foot of surface revetted 18 ins. thick.
3. Sheeting and anchored pickets	10	30 mins....	10 feet run of revetment	2 mauls ... 1 saw 1 billhook 1 pair pliers 2 shovels 1 pick 1 crowbar	(a) Trench cut to section. (b) Sheeting consisting of C.G.I. sheets, planking, hurdles or X.P.M. (c) Wire, sharpened pickets, and sheeting distributed at frequent intervals on site. (d) 1 anchorage picket to 2 main pickets. (e) 4 strands of wire from each picket to anchorage. (f) Anchorage pickets 8 feet from main pickets. (g) With angle iron pickets use sledge hammers instead of mauls. (h) 2 men on anchorage pickets. 2 men on main pickets. 2 men sheeting. 2 men wiring. 2 men filling and trimming.

Task		No.	Time	Unit	Tools	Remarks
4. Sheeting and small "A" frames	...	7	15 mins....	10 feet run (of trench)	2 picks 2 shovels 2 mauls 2 spades 1 saw 1 hammer nails	...	(a) Trench cut to section. (b) Materials distributed at frequent intervals. (c) Sheeting consisting of C.G.I. sheets, planking, hurdles or X.P.M. (d) Allow 15 mins. per corner. With rounded corners time includes bending C.G.I. sheets as required. (e) 2 men supplying materials. 2 men trimming and packing. 3 men placing frames.
5. Gabions, placing and filling	...	1	5 mins. ...	1 square foot revetted	1 shovel 1 pick	...	(a) Gabions 2 ft. wide by 2 ft. 9 ins. high; area revetted 5¼ sq. ft. Contents 5½ cub. ft. Earth to be excavated.
6. Picket trestles and laying trenchboards on same	...	5	15 mins....	6 feet run (of trench)	1 maul 2 saws 2 hammers nails	...	(a) Pickets distributed on site. (b) Trestles at one foot intervals.
7. Trenchboards laying on "A" frames	...	3	10 mins....	10 feet run (of trench)	1 saw 1 hammer nails	...	(a) 1 man supplying material; 2 men laying and fitting. (b) Allow 10 mins. per corner.
8. Hurdles, rough, making	...	3	20 mins....	1 hurdle	2 billhooks 2 knives 1 mallet 1 pair pliers	...	(a) Materials: 75 lbs. brushwood and 60 ft. of wire or yarn per hurdle, 6 ft. by 2 ft. 9 ins. (b) Weight of each complete, about 56 lbs. (c) Brushwood ready cut on site.
9. Fascines, making	...	4	1 hour ...	1 fascine	3 billhooks 1 handsaw 1 pair pliers 2 knives 1 maul 1 choker	...	(a) Materials: 200 lbs. of brushwood and 60 ft. of wire or hoop-iron (40 ft.) per fascine, 18 ft. long by 9 ins. diam. (b) Weight complete about 140 lbs. (c) Cradle for making requires ten pickets, 6 ft. 6 ins. by 3 ins. diam. (d) All materials on site.

Nature of work.	Unit or party.	Time.	Amount of work.	Tools per party.	Remarks.
III.—CUTTING AND FELLING.					
1. Trees, felling ...	1	1 min. ...	1 in. in diam. ...	1 felling-axe or saw	(a) Up to 12 ins. diam. If over 12 ins. diam., allow time in mins. = $\frac{d^3}{144}$ where "d" = mean diam. in inches. (b) If only hand-axes are available allow twice the time as calculated in both these rules.
2. Woods, clearing of brushwood and small trees	10	2½ mins....	20 sq. yards (up to 12 ins. diam.)	10 billhooks ... 4 felling-axes 2 saws 2 saws 2 saws 1 grindstone 2 whetstones	(a) All hands felling at first, then a proportion detailed for collecting and removing according to purpose in view. (b) Produce: about 5 lbs. brushwood per 1 sq. yard.
3. Hedges, cutting stems	2	5 mins. ...	1 yard run (up to 2 ins. diam.)	1 billhook or hand-axe 1 saw 3 fathoms of rope	(a) Average stiff thorn hedge. (b) If necessary use rope to expose lower stems to the cutting tool.
4. Brick walls, cutting loopholes in	1	30 mins....	1 loophole	1 pick or 1 crowbar...	(a) Up to 18 ins. thick. (b) If possible, obtain a mason's chisel and hammer.
5. Brick walls, notches in	1	10 mins....	1 notch ...	1 pick or 1 crowbar	(a) Up to 18 ins. thick. (b) If possible, obtain mason's chisel and hammer.

IV.—WIRE ENTANGLEMENTS.

1. Standard French wire	10	Day, 10 mins. ... Night, 20-30mins.	50 yards	1 pair pliers ... / 98 windlassing sticks / Gloves, if desired / 26 long pickets / 4 anchorage pickets / 6 coils French wire / 24 staples / 3 coils barbed wire / 3 spirals barbed wire	*(a) Two coils barbed wire for thickening entanglement, if spirals are not available.
2. Standard double belt of concertinas	8	Day, 20 mins. ... Night, 30-35 mins.	50 yards	1 pair pliers ... / 7 windlassing sticks / Gloves, if desired / 34 long pickets / 4 anchorage pickets / 16 concertinas / 2 coils barbed wire	
3. Standard low (or knee-high) wire entanglement	8	Day, 30 mins. ... Night, 1-1¼ hours	50 yards	1 pair pliers ... / 7 windlassing sticks / Gloves, if desired / 54 medium pickets / 2 coils barbed wire * / 3 coils barbed wire (50 yards) / 4 spirals	*(a) Four coils barbed wire for thickening entanglement, if spirals are not available.
4. Standard double apron fence	10	Day, 30 mins. ... Night, 45-60 mins.	50 yards	1 pair pliers ... / 9 windlassing sticks / Gloves, if desired / 16 long pickets / 32 anchorage pickets / 2 coils barbed wire * / 10 coils barbed wire (50 yards)	
5. Entangling hedges, &c. rough abatis	8	20 mins....	50 yards	2 billhooks ... / 2 pairs pliers / 5 coils barbed wire (50 yards)	

* Coils of barbed wire are 100 yards coils, unless otherwise stated.

APPENDIX III.

TABLE GIVING LOADS FOR MAN, G.S. WAGON AND 3-TON LORRY, FOR ENGINEERING STORES IN GENERAL USE IN THE FIELD.

[NOTE.—The loads are based on fair conditions only—*i.e.*, a man at 30 lbs., a G.S. wagon at 1,900 lbs., and a 3-ton lorry at 5,600 lbs. : under good conditions, the load of a man may be increased to 40 lbs. and the loads of a G.S. wagon and a 3-ton lorry by one-sixth. Man-loads to be " bundled " beforehand, whenever possible.]

Item.	Article.	Description.	One man load.	G.S. wagon load.	3-ton lorry load.
1	Sandbags	Bales of 250 ; weight, 96 lbs.	50 to 75	20 bales	60 bales
2	Coil barbed wire ...	100 yards ; weight, 28 lbs.	1	70	200
3	Screw pickets, long	5 ft. 7 ins. long, 4 eyes ; weight, 6 lbs.	3	200	600
4	Screw pickets, medium	3 ft. 9 ins. long, 2 eyes; weight, 4½ lbs.	5	300	900
5	Screw pickets, short	2 ft. 1 in. long, with loop: weight, 2½ lbs.	8	550	1,600
6	Angle iron pickets, long	5 ft. 10 ins. long; weight, 14 lbs.	2	150	400
7	Angle iron pickets, medium	3 ft. 6 ins. long ; weight, 8 lbs.	4	250	700
8	Posts, wooden, 5 ft.	3½ to 4 ins. diameter ; weight, 9 lbs.	3	200	600
9	Posts, wooden, 2 ft. 6 ins.	2½ to 3 ins. diameter; weight, 3 lbs.	10	600	1,800
10	French wire, coil. (English manufacture)	Issued in bundles of 5 coils ; weight of coil, 14 lbs.	2 coils	150 coils or 30 bundles	400 coils or 80 bundles
11	Staples, French wire	Boxes of 300 ; weight, 160 lbs.	—	12 boxes	35 boxes
12	Plain wire, coil ...	Weight, 56 lbs.; 100 lbs. per mile	½ coil	40 coils	100 coils
13	Rabbit netting, roll	3 ft. wide, 50 yards in roll ; weight, 80lbs.	20 yards	24	70
14	X.P.M., sheets ...	6 ft. 6 ins. long by 3 ft. wide ; weight, 27 lbs. In cases of 20 sheets	1	70	200
15	Corrugated iron, 6-ft. sheets	Width, 2 ft. 9 ins. ; weight, 16 lbs.	2	120	350
16	Corrugated iron, 7-ft. sheets	Width, 2 ft. 9 ins. ; weight, 18½ lbs.	1½	100	300
17	Corrugated iron, 9-ft. sheets	Width, 2 ft. 9 ins. ; weight, 28 lbs.	1	70	200

Item.	Article.	Description.	One man load.	G.S. wagon load.	3-ton lorry load.
18	Felt, roll	3 ft. wide, 25 yards in roll; weight, 85 lbs.	10 yards	22	66
19	Canvas, Hessian, roll	3 ft. wide, 110 yards in roll; weight, 70 lbs.	50 yards	27	80
20	Canvas, rot-proof, roll	5 ft. wide, 120 yards in roll; weight, 130 lbs.	30 yards	14	40
21	"A" frames (small)	Weight, 30 lbs. ...	1	40	120
22	Trench board ...	6 ft. long; weight, 35 lbs.	1	35–40	120–150
23	Timber, 4-in. by 2-in.	F.R.; weight, 2½ lbs.	—	750 F.R.	2,250 F.R.
24	Timber, 9-in. by 3-in.	F.R.; weight, 9 lbs.	—	200 F.R.	600 F.R.
25	Planking. 1-in.	Supplied by the foot run	Dependent on width of planking. Weight of soft timber may be taken as 40 lbs. per cubic foot.		
26	Planking, 1½-in.				
27	Planking, 2-in.				
28	Pit prop, 9 ft. long ...	6 ins. diameter; weight, 90 lbs.	—	25	75
29	Pit prop, 9 ft. long ...	9 ins. diameter; weight, 180 lbs.	—	12	36
30	Cement, cask ...	400 lbs.	—	5	14
31	Sand	—	½ sandbag	¼ cub. yd.	2½ cub. yd.
	Gravel	—	,,	,,	,,
	Chalk...	—	,,	,,	,,
	Earth	—	,,	,,	,,
32	Corrugated steel shelter, large	*See* Plate **143** for quantities	...	½	1
33	Corrugated steel shelter, small	*See* Plate **141** for quantities	...	1 in 1 G.S. wagon; 3 in 2 G.S. wagons	5
34	R.S.J. 9 ft. by 5 ins. by 3 ins.	100 lbs.	¼	19	56
35	Nails—				
	1-in.	800 to 1 lb.			
	2-in.	122 to 1 lb.			
	3-in.	52 to 1 lb.			
	4-in.	30 to 1 lb.			
	5-in.	20 to 1 lb.			
	6-in.	14 to 1 lb.			
36	Staples, No. 8 S.W.G.	50 to 1 lb.			
37	Shovels	Weight, 5 lbs. ...	6	400	1200
38	Picks	Weight, 8 lbs. ...	4	240	700
39	Tapes, tracing ...	In 50-yard rolls ...	12	—	—

APPENDIX IV.

PRINCIPAL TOOLS, MATERIALS, AND STORES SUITABLE FOR USE IN
FIELD ENGINEERING.

*The tools and stores provided for the peace instruction of troops in field
engineering are as laid down in the Regulations for the Equipment of the
Army, Part I, 1909, para. 323 and Appendix VI. The tools and stores
forming war equipment of units are similarly detailed in the various
sections of Part II, Equipment Regulations, and in Mobilization Store
Tables (A.F.G. 1098).*

*The following Tables of tools, materials, and stores are intended as a
guide for the selection and preparation of articles, suitable for use in war,
for such operations of field engineering as are described or indicated in
this manual.*

*The method of obtaining supply of such articles will follow the instructions
laid down in F.S. Regulations, and Ordnance Manual (War),
paras. 36 and 37.*

*The special equipment required for the following engineer services is
not included in these tables, except in so far as certain articles comprised
therein may be suitable for general field engineering purposes :—*

Electrical instruments and electric light stores.
Railway tools, plant and armoured trains.
Survey instruments and stores.

Demand and issue, except where otherwise stated, are " per article."

The tables in this Appendix are :—

1.—Tools, entrenching.
2.—Tools, cutting.
3.—Tools and stores, miscellaneous.
4.—Sandbags, canvas, &c.
5.—Corrugated shelters and iron sheets.
6.—Rolled steel joists and rails.
7.—Posts and pickets.
8.—Materials supplied for camouflage.
9.—Timber.
10.—Cordage.
11.—Bridging and boat stores.
12.—Wire and wire rope.
13.—Bolts, dogs, nails and spikes.
14.—Water supply stores.
15.—Demolition stores.

TABLE 1.—*Tools, entrenching.*

Designation.	Detail.	How issued.
Axes, pick, heads	4½ lbs. and 8 lbs.	
,, ,, helves	36 ins. ferruled...	
Barrows, hand, double	6 ft. 7 ins. long	
,, wheel, entrenching ...	Steel tubular frame	
Crowbars, chisel and claw ends ...	6 ft. 37 lbs., 5 ft. 6 ins. 31 lbs., 4 ft. 6 ins. 20 lbs., 3 ft. 6 ins., 12 lbs. 2 ft. 3 ins., 7 lbs.	
Picks, miners	22½ ins. 6 lbs.(special short for cramped work)	
,, push	30 ins. 3 lbs. 6 ozs. (heart shaped, straight stabbing)	
Shovels, G.S.	32 ins. helve, 3½ lbs.	
,, R.E.	32 ins. helve, 5 lbs.	
,, miners	30 ins. 6 lbs., and 5 ft. long	
Spades, Mark III	32¾ ins. helve, 5¾ lbs.	

TABLE 2.—*Tools, cutting.*

Designation.	Detail.	How issued.
Adzes, carpenters, handled ...	4½ lbs.	
Axes, felling	32 ins. helve, 6 lbs. 7 ozs.	
Axes, hand	16 ins. helve, 2 lbs. 3 ozs.	
Chisels, brick	18 ins., 1¼ ins. end.	
Chisels, hand, cold ...	Metal cutting, 1 in., ⅞ in. and ¾ in. wide.	
Chisels, firmer	Wood-cutting, blades 3 ins. to 1/16 in. wide.	
Grindstones, F.S.	18 ins., 76 lbs. 10 ins., 25 lbs.	
Hooks, bill	1 lb. 13 ozs.	
Hooks, reaping	1 lb.	
Pliers, side-cutting ...	8 ins. and 5 ins. long	prs.
Saws, cross-cut	5 ft. blade, 6¼ lbs.	
Saws, folding, in leather case ...	3 ft. 9 ins. blade, 2 handles, 1 lb. 12 ozs.	
Saws, hand	26 ins. and 20 ins.	
Sets, cold, large	15 ins. handle; for cutting steel wire rope, &c.	
Stones, rag	For reaping hooks, &c.	doz.
Wire-cutters	9½ ins. long, 1¼ lbs.	

TABLE 3.—*Tools and stores, miscellaneous.*

(Not included in Tables 1 or 2)

Designation.	Detail.	How issued.
Anvils	1 cwt.	
Bars, boring (steel chisels) ...	For rock, 3 ins., 1½ ins., and 1¼ ins. wide, up to 4 ft. long.	
Bars, jumping (chisel each end) ...	3 ins., 1½ ins., and 1¼ ins. wide, 5 ft. 6 ins. and 7 ft. long.	
Bars, pinching (spike and lever)	2½, 3, 3½ and 4 ft. long.	
Blocks, tackle, G.S., cast iron, galv.	Single, double, treble, and snatch (and size of cordage).	
Blowers, rotary, Mk. IV	With hose and wrenches; for ventilating mines, &c.	
Buckets, miners	14 lbs. 7 ozs. for raising earth from shafts.	
Candlesticks, miners	With bottom and side spikes.	
Chokers, fascine	2 four ft. levers and 4 ft. chain.	
Crabs, hoisting, iron	Hand power winches, to lift 1, 25, and 50 tons.	
Forges, field, G.S.	276 lbs. and poker, slice, tongs and vice.	
Grapnels, iron	2, 3, 16, 40, and 50 lbs.	
Hammers, claw	20 ozs.	
Hammers, masons	10 lbs. chisel point.	
Hammers, miners, boring ...	5 and 7 lbs.	
Hammers, miners, sledge ...	14 lbs.	
Jumpers, steel	For post holes, 2 ft. 9 ins. × 2 ins., 27¾ lbs.	
Ladders, field telegraph	16 ft. 6 ins. in 2 lengths.	
Ladders, rope, miners	20, 30, and 50 ft.	
Lamps, acetylene	Land and portable and calcium carbide.	N.I.V.
Lamps, electric	Land and portable and span accumulators.	,,
Lamps, hurricane	Oil or candle.	
Levels, F.S.	4 ft., 3 lbs. 7 ozs.	
Mauls, G.S.	14 lb.	
Rods, measuring, common ...	Wood, 10 ft. and 6 ft. marked 3 ins., 5 ft. marked ⅛ in.	
Scoops and scrapers	For clearing bore-holes; 3 ins., 1½ ins. and 1¼ ins. × 6 ft. 6 ins.	
Scrapers, earth	7 ft. long; 3 lbs.; also with 3 ft. handle.	
Spanners, adjustable	15 ins.	
Spanners, McMahon	9 ins.	
Tapes, measuring	In leather case, 100 ft.	
Tapes, tracing	50 yds.; 1½ ins. white web.	
Trucks, miners, elm	69 lbs.; for removing earth in saps and mines.	
Vices, standing, 36-lb.	Jaws 4 ins. wide.	

TABLE 4.—*Sandbags, canvas, &c.*

Designation.	Detail.	How issued.
Bags, sand, common	Bales of 250	
Bags, guncotton, waterproof ...	Canvas, to hold 2, 5, and 25 lbs. ...	
Canvas, rot-proof	5 ft. wide ; 120 yds. in roll	Yards.
Canvas, Hessian	3 ft. wide ; 110 yds. in roll	,,
Cloth, union, anti-gas	54 ins. wide; 74 yards in roll... ...	,,
Covers, sailcloth, waterproofed ...	Sizes in feet : 30 × 30, 30 × 20, 24 × 18, 20 × 16, 18 × 15, 15 × 10, 12 × 10.	
Felt, roofing	3 ft. wide ; 25 yards in roll	Roll.
Sheeting, corrugated iron ...	2 ft. 9 ins. wide ; in 6-ft., 7-ft. and 9-ft. lengths.	
Tarpaulins	Sizes in ft. : 30 × 30, 30 × 20, 30 × 16, 24 × 18, 20 × 16, 20 × 10, 18 × 15, 15 × 10, 12 × 10, 10 × 6.	

TABLE 5.—*Corrugated shelters and iron sheets.*

Designation.	Detail.	How issued.
Straight sheets, black or painted...	6 ft. to 9 ft. by 2 ft., 22 or 24 gauge, 10 sheets to bundle.	Bundles.
Straight sheets, black or painted, galvanized.	7 ft. by 2 ft., 22 gauge, 10 sheets to bundle.	,,
Curved sheets for bivouac shelter	9 ft. by 2 ft. 2 ins., 18 gauge, bent to 7 ft. radius. Carries one layer of sandbags or equivalent weight. 5 sheets to bundle.	,,
Curved sheets for bivouac shelter	9 ft. by 2 ft. 2 ins., 22 gauge, bent to 9 ft. radius. Will carry no weight. 10 sheets to bundle.	,,
Corrugated steel shelter, large ...	17 ft. 9 ins. by 9 ft. 6 ins. by 6 ft. 2½ ins. inside dimensions. Made up of 21 curved plates 7 ft. by 2 ft. 6 ins., each with six 5-in. corrugations.	
Corrugated steel shelter, small ...	12 ft. 9 ins. by 5 ft. 3 ins. by 3 ft. 8 ins. inside dimensions. Made up of 10 curved plates 5 ft. 6½ ins. by 2 ft. 6ins. each with five 6-in. corrugations.	
Troughing	6 ft. or 9 ft. long, 3 ft. 3 ins. wide ...	

TABLE 6.—*Rolled steel joists and rails.*

Designation.	Detail.	How issued.
Joist steel, rolled—		
12 ins. by 5 ins.	Length 22 ft., 32 lbs. per foot run ...	
10 ins. by 5 ins.	Length 20 ft., 30 lbs. per foot run ...	
9 ins. by 4 ins.	Length 18 ft., 21 lbs. per foot run ...	
8 ins. by 4 ins.	Length 16 ft., 18 lbs. per foot run ...	
5 ins. by 3 ins.	Length 9 ft., 11 lbs. per foot run ...	
Rail, steel, bullheaded—		
40 lbs.	Height $3\frac{1}{2}$ ins., $13\frac{1}{2}$ lbs. per foot run...	
60 lbs.	Height $4\frac{1}{4}$ ins., 20 lbs. per foot run ...	
80 lbs.	Height 5 ins., $26\frac{1}{4}$ lbs. per foot run ...	

TABLE 7.—*Posts and pickets.*

Designation.	Detail.	How issued.
Pickets, angle iron, long	5 ft. $10\frac{1}{2}$ ins. long	
,, ,, medium ...	3 ft. 6 ins. long...	
,, screw, long ...	5 ft. long, 4 eyes	
,, ,, medium ...	3 ft. 6 ins. long, 3 eyes	
,, ,, short ...	1 ft. 6 ins. long, 2 eyes	
Posts, wire entanglement, wood ...	Various lengths and diameters ...	

TABLE 8.—*Materials supplied for camouflage.*

FOR CONCEALMENT OF FIELD WORKS, BATTERIES, &c.

1. Fish netting in 30 feet by 30 feet squares, or wire netting in rolls 30 feet by 6 feet, garnished with canvas knots, with or without irregular islands of scrim (an open mesh form of canvas), mainly for use in open country.

2. Fish netting (30 feet by 30 feet) or wire netting (30 feet by 12 feet), furnished only with large islands of scrim, for use in broken country. Fish nets (10 feet by 10 feet) with raffia.

3. Irregular patches of scrim, with or without bare rolls of wire netting 30 feet by 6 feet, to be used for supplementing, or actually making up material mentioned in para. 2 *in situ.*

4. Scrim sheets, 30 feet long by 6 feet or 12 feet wide, for covering spoil, sandbags, &c., or any other light-toned objects under material 1, 2 or 3.

5. Posts 2 inches by 2 inches, of varying lengths, pickets and wire, for supporting camouflage.

SNIPER'S REQUISITES.

1. "Symien" pattern, consisting of loose-fitting jacket, with hood attached, separate legs, rifle cover and gloves.

2. Dummy heads for locating enemy snipers.

TABLE 9.—*Timber.*

(NOTE.—Timber may be either felled and trimmed on the spot, collected from timber stores in adjacent towns or villages, obtained by dismantling structures containing timber, or demanded from the engineer parks and dumps. The following table gives the ordinary sizes in which timber may be expected to be available).

Designation.	Size.	
Planking	$\frac{3}{4}$ in. ... ... 1 in. ... ... $1\frac{1}{2}$ ins. ... ... 2 ins.	In various widths and lengths. Demand by the F.R.
Scantlings	3 ins. by 3 ins. ... 4 ins. by 2 ins. ... 4 ins. by 3 ins. ... 4 ins. by 4 ins. ... 6 ins. by 3 ins. ... 6 ins. by 4 ins. ... 6 ins. by 6 ins. ... 9 ins. by 3 ins. ... 9 ins. by 4 ins. ... 9 ins. by 6 ins. ... 9 ins. by 9 ins. ...	In various lengths. Demand by the F.R., giving a minimum "piece length" for 6 ins. by 6 ins., 9 ins. by 6 ins., and 9 ins. by 9 ins.
Baulks	10 ins. by 10 ins. 12 ins. by 6 ins. 12 ins. by 12 ins. 14 ins. by 12 ins. 16 ins. by 8 ins.	In various lengths. Demand by the piece.
Spars	—	In various sizes and lengths. Demand by the piece, specifying length and diameter.
Pit props	—	In various sizes and lengths. Demand by number, specifying diameter and length.

TABLE 10.—*Cordage.*

Designation.	Detail.	How issued.
i. Cordage. hemp, hawser, 3-strand.	Service cordage in general use ; either *tarred* or *white*; in the following sizes, circumference in inches:— 9, 7, 6, 5, 4, $3\frac{1}{2}$, 3, $2\frac{1}{2}$, 2, $1\frac{1}{2}$, 1. *Tarred* cordage is weaker, but will stand exposure to weather better than *white*.	Fathom [coils of 113 fms.]
ii. Cordage. manilla, hawser, 3-strand.	A stronger cordage, in the following sizes, circumference in inches:—5, 4, $3\frac{1}{2}$, 3, $2\frac{1}{2}$, 2, $1\frac{1}{2}$, 1, $\frac{3}{4}$, $\frac{1}{2}$.	,,
iii. Cordage, coir, hawser, 3-strand.	A coarse, light, elastic, cordage, which will float upon water, but has only one-sixth the strength of hemp cordage of same size. Sizes :—9, 7, 6, 5, 4, and $2\frac{1}{2}$ ins.	,,

Designation.	Detail.	How issued.
iv. Lashings, falls, guys, &c. ...	Cordage, as in i. above, of sizes as under :— 3 ins. Footropes, 9 fms.; cables, 30 fms.; falls, 50 fms.; guys, 30 to 36 fms. 2½ ins. Slings for cask piers, 6 fms. 2 ins. Falls, 50 fms.; lashings, 6 and 9 fms. 1½ ins. Braces, 3 fms.; breast lines, 10 fms.; lashings, 6 fms. 1 in. Buoy lines, 10 fms.; lashings, 3 and 6 fms.	Fathom [coils of 113 fms.]
v. Small cordage, yarn, twine, &c. Cordage, spun yarn, hemp ...	3-thread, tarred, rough	Cwt.
Lines, Hambro	150 ft., strong and light	

TABLE 11.—*Bridging and boat stores.*

(*See also* Military Engineering, Vol. III.)

Designation.	Detail.	How issued.
Anchors, boat	1 cwt. and ½ cwt.	
Bailers, pontoon	Tin, with handle.	
Baulks, Mk. III, tapered	15 ft. 9¾ ins. by 3¼ ins. to 1½ ins. by 6 ins.; 56 lbs.	
Baulks, shore end, inside	3 ft. 7 ins. long; 3 to set; 15½ lbs. each.	
Baulks, shore end, outside ...	3 ft. 6¼ ins. long; 2 to set; 21½ lbs. each.	
Beams, saddle, Mk. II. ...	In two pieces; 58 lbs. pair.	
Boats, collapsible	Bow and stern sections; 6 ft. 1½ ins. long; 9-ft. oars.	
Buoys, pontoon, iron	For anchors; 5 lbs.	
Chalk, prepared	White or coloured; 144 pieces ...	Box.
Chesses, Mk. II.	10 ft. by 12 ins. by 1½ ins.; 45 lbs.	
Drivers, pile, Swiss...	With iron guide rod; about 130 lbs.; hand power.	
Hooks, boat...	18 ft., 11 ft. 7½ ins., and 6 ft. long.	
Life-belts, cork.		
Life-buoys, Mk. IV	Reindeer hair, covered canvas.	
Oars, ash	20 ft. to 8 ft. long, in sizes increasing by 1 foot. 12 ft. for pontoons.	
Pontoons, bipartite, Mk. II. ...	Bow and stern pieces; 1,008 lbs. per pair.	
Ribands, Mk. II.	15 ft. 9 ins. by 3¼ ins. by 6 ins.; 79 lbs.; can be used as baulks.	
Sticks, rack	With 6 ft. of 2-in. lashing; 1¾ lbs.	
Transoms, shore end, Mk. III ...	11 ft. 6 ins. long; 73 lbs.	
Trestles, bridging, Mk. III ...	With 2 tackles, differential, 10 cwt.; weight, 816 lbs.	

For lashings and wire rope, *see* Tables 10 and 12.

TABLE 12.—*Wire and wire rope.*

Designation.	Detail.	How issued.
Rope, galvanized, steel, wire ...	In coils of 100 fms.	fms.

Sizes, circumference, inches	4	3	2½	2	1½	1¼	1⅛	1	Sizes 12 to 5 ins. areN.I.V.
Approx. weight, lbs. per fm.	12	7	4½	2¾	1¾	1	⅞	¾	
Safe load (9c²) cwt.	144	81	56	36	20	14	11	9	

Rope, steel, ·65 in....	For use with collapsible boats, 2 to 2½ tons breaking strain ; ·42 lb. fm.	100 yds.
Wire, galvanized iron, No. 14 S.W.G.	In 28 lb. and 56-lb. coils	Coil.
Wire, barbed	In 28-lb. (100 yds.) or 15-lb. (50 yds.) coils.	,,
Wire, French	One coil=16 yds. In bundles of 5 coils. Staples in boxes containing 300.	,,
Rabbit wire netting	3 ft. wide, in rolls containing 50 yds....	Roll.
X.P.M. (expanded metal) ...	Sheet 6 ft. 6 ins. long by 3 ft. wide. In cases of 20 sheets.	Sheet.

The table below gives the properties, weight, &c., of new iron wire. New steel wire may be taken as twice the strength given, otherwise similar in size, &c ; galvanized wire is heavier.

Size, S.W.G.	1	2	3	4	5	6	7	8	9	10	11	12	13
Diam., inches ..	·300	·276	·252	·232	·212	·192	·176	·160	·144	·128	·116	·104	·092
Yards, per cwt...	155	183	220	260	311	380	452	546	675	854	1040	1293	1653
Lbs. per mile ..	1268	1073	895	758	633	518	436	369	292	231	190	152	119
Approx. breaking strain, lbs. ..	3804	3219	2685	2274	1899	1554	1308	1080	876	693	570	456	357

Size, S.W.G.	14	15	16	17	18	19	20	21	22	23	24	25	26
Diam., inches ..	·080	·072	·064	·056	·048	·040	·036	·032	·028	·024	·022	·020	·018
Yards, per cwt...	2186	2699	3416	4462	6073	8745	10796	13663	17846	24290	28908	34978	43184
Lbs. per mile ..	90	73	58	44	32·5	22·5	18·2	14·4	11	8·1	6·8	5·6	4·6
Approx. breaking strain, lbs. ..	270	219	174	132	98	68	55	43	33	24	20	17	14

TABLE 13.—*Bolts, dogs, nails and spikes.*

Designation.	Detail.	How issued
Bolts, with nuts, hexagon head.	Principal store sizes ; length and diam. in inches 14 × 1 or ¾, 12 × ¾ or ⅝, 8 × ¾ or ⅝, 6 × ⅝ or ½, 5 × ⅜. Other sizes prepared as required.	
Bolts, drift ...	¾ in. × 24 ins. and ⅝ in. × 20 ins.	
Dogs, railway and sawyers, Mk. II ...	Straight. 15 and 12 ins. long, with 6-in. teeth... ...	

Nails, iron, spike (*quote store No.*)							
Length, inches :—		10	9	8	7	6	5
Nails in 1 cwt. (app.):—		114	155	193	294	430	590
Army Store No. :—		*187*	*186*	*185*	*184*	*183*	*182*

Nails, wire, iron, grooved	Length, ins.:	6	5	4	3	2½	2	1¾	1½	1¼	1	lb.
	Nails in 1 lb. (approx.) :-	14	20	50	70	100	150	200	300	400	600	

Staples, No. 8 S.W.G	Approximately 51 per lb.	,,

TABLE 14.—*Water supply stores.*

Designation.	Detail.	How issued.
Hose, canvas	3 and 4 ins. diam.	Yards.
Hose, delivery, canvas, 2¾-in. ...	In lengths 100, 50, 30, ft., with screw unions.	Lengths.
Hose, delivery, 2¾-in.	In 30-ft. lengths ; prepared for pump L. & F. Mk. IV.	,,
Hose, suction, 2-in.	In 12-ft. lengths ; prepared for pump L. & F., Mk. IV.	,,
Hose, suction, syphon, with cap ...	3 ins. and 2½ ins. in 10 ft. 3 in. lengths ; 2 ins. in 10-ft. lengths.	,,
Pails, iron, galvanized	3 and 4 gallons	
Pumps, deep well	50 ft. 3-in. bore ; and 100 ft., 3 and 4-in. bore.	Sets.
Pumps, lift and force, Mk. IV ...	With four 12-ft. lengths of suction, and 30 ft. of delivery hose, to lift 60 ft. ; weight 84 lbs. ; with hose 216 lbs.	
Pumps, steam, portable, Merry-weather	Small "Valiant" on wheels, 8½ cwts. ; to raise 1,500 gallons per hour 250 ft.	
Tanks, iron, galvanized, rectangular	In sizes from 20 to 1,000 gallons ...	
Tanks, steel, corrugated, galvanized, circular	25, 50, 100 to 1,000 gallons ; with taps and covers.	N.I.V.
Tanks, waterproof, 2,300 gallons, open	16 ft. 9 ins. × 16 ft. 9 ins. with stores as M.E., Vol, V.	
Tanks, waterproof, 1,500 galls., closed	Octagonal ; no extra stores required...	
Troughs, waterproof, 600 gallons...	With standards, 10 to a set, to water 16 animals at one time.	Sets.
Windlasses, well, light	With large drum	

TABLE 14—*continued.*

Description.	Detail.	How issued.
Pipe, wrought iron, **galvanized** ...	In sizes 6, 4, 3, 2, 1½, 1 and ¾ inches internal diameter.	Ft. run.
Bends, elbows, ties, connectors, reducing sockets, crosspieces, plugs, caps, nipples, back nuts, screwed flanges.	For pipes of sizes as above	
Valves air. sluice or reflex ...	For 6, 4, and 3-in. pipes	
Valves, ball, stop or reflex ...	For 2-in. pipes	
Valves, ball	For 1½ and 1-in. pipes	
Cocks, stop	For 1¼ and 1-in. pipes	
Cocks, bib	For 1 and ¾-in. pipes	
Stocks and dies, sets	For 6, 4, 3, 2, 1½, 1 and ¾-in. pipes ...	Set.
Pipe-cutters, pipe vices, pipe wrenches.	For sizes of pipe as above	

TABLE 15.—*Demolition stores.*

Designation.	Detail.	How issued.
	BULK EXPLOSIVES.	
Gun-cotton, wet, slabs, field, 15-oz.	Slabs 6 by 3 by 1¾ ins., 1 perforation for 1-oz. primer ; packed in waterproof boxes containing 16 slabs	Lbs.
Ammonal	Packed in 25 and 50-lb. waterproof tins	Lbs.
Dynamite ... ... ⎫ Blasting gelatine ... ... ⎭	Not an ordnance store, but, as a rule, obtainable from this source ; manufactured in various grades and strengths, usually in 2-oz. cartridges packed in boxes of 5 lbs. and 50 lbs.	Lbs.
	AUXILIARY EXPLOSIVES, &c.	
Cable, electric, E₁, Mark II ...	For electric firing ; weight 5·7 lbs. per 100 yds.	Yards.
Caps, copper, blasting. (Commercial caps)	The commercial equivalent of the No. 8 detonator ; manufactured in eight standard strengths, sizes 3 to 10	
Detonators No. 8, Mark VII ...	For use with safety and with detonating fuze ; packed in tins of 25	
Detonators, electric, No. 13, Mark III	For use in electric firing ; packed in tins containing 25 detonators and a rectifier	
Exploders, dynamo, electric, quantity, Mark V	For firing charges electrically ; size 13½ by 8½ by 6¼ ins. ; weight 27 lbs.	Each
Fuze, safety, No. 11	In tin cylinders of 8, 24 or 50 fathoms	

Designation.	Detail.	How issued.
Fuze, detonating, Bickford's ...	Wound on drums	Yards
Fuzes, electric, No. 14, Mark III	For use with low explosives fired electrically; packed in tins of 25	
Gun-cotton, dry, primers, field, 1-oz.	Packed in sealed tin cylinders containing 10 primers	
Matches, vesuvian	Fuzees; 20 in a box	
	TESTING APPARATUS, ELECTRIC FIRING.	
Boxes, testing and jointing, filled	Tin box in leather cover; size 14 by 8 by 5½ ins.; weight 12 lbs. 3 ozs.; the contents are as follows:—	
	For testing:—	
	1 box of resistance coils (100 ohms)	
	1 "Q" and "I" detector ...	
	1 cell, electric, dry, "E"... ...	
	2 reels of iridio-platinum wire ...	
	1 chamois leather...	
	1 box of plate powder	
	For jointing:—	
	1 pair 5-in. side-cutting pliers ...	
	2 tubes of indiarubber solution ...	
	4 cylinders of indiarubber tape ...	
	¼ lb. of cotton waste	

APPENDIX V.

TABLES GIVING STRESSES IN, AND SIZES OF ROPES, SPARS, SCANTLING, GIRDERS AND RAILS FOR USE IN SIMPLE FIELD STRUCTURES.

1.—Stresses in derricks, sheers and gyns.

2.—Sizes of spars and baulks for derricks, &c.

3.—Tackles.

4.—Power required on falls of tackles in tons.

5.—Sizes of road-bearers and beams for light bridges, dug-outs, &c.

6.—Safe nett distributed loads in lbs. which can be supported by girders or rails over different spans.

Table 1.—*Stresses in derricks, sheers and gyns.*
(Stresses are stated in terms of the weight to be lifted. Allowance has been made for the weight of tackle, &c.)

Standing Derrick.		Sheers.	
Spar	1·5 W	Leg with leading block	·9 W
Running guys	·5 W	Other legs	·7 W
Other guys	·3 W	Back guy	1·3 W
Gyn.		**Swinging Derrick.**	
Spar with leading block ...	·6 W	Upright spar	2·0 W
Other spars	·4 W	Swinging arm	1·3 W
		Struts	1·0 W
		Connecting tackle	1·7 W
		Guys	1·5 W

In the case of the swinging derrick, the length of the upright spar and the swinging arm are assumed to be about equal. Any alteration in these proportions will affect them. The size of spars and guys can be found from Tables 2 and 3.

This table gives the maximum stresses which are likely to occur in practice, for the ordinary conditions under which these machines are used. In special cases the stresses should be worked out graphically.

Table 2.—*Size of spars and baulk for derricks, &c.*

Mean diameter of spar in inches.	Length of spar in feet from point of attachment of main tackle to the ground.												Size of square baulk in inches.
	5	10	15	20	25	30	35	40	45	50	55	60	
6	4	1¾	⅝	—	—	—	—	—	—	—	—	—	5½
7	6	2¼	1¼	¾	—	—	—	—	—	—	—	—	6
8	10	4	2	1¼	¾	½	—	—	—	—	—	—	7
9	15	6	3	1¾	1¼	¾	½	—	—	—	—	—	8
10	20	8	4½	2½	1¾	1¼	⅞	⅝	½	—	—	—	9
11	26	12	6½	4	2½	1¾	1¼	1	¾	½	—	—	10
12	33	16	9	5½	3½	2½	1¾	1½	1	⅞	—	—	11
13	41	20	12	7½	5	3½	2½	2	1½	1¼	1	—	11½
14	50	26	15	10	6½	4¼	3½	2¾	2¼	1¾	1	—	12½
15	—	33	20	12½	8½	4½	3½	2¾	2¼	2	1½	1½	13½
16	—	40	25	16	11	8	5¾	4½	3½	3	2½	2¼	14½
17	—	50	30	20	14	10	7¼	5¾	4½	3¾	3	2¾	15
18	—	—	36	25	17	12	9	7	5¾	4¾	4	3¼	16
19	—	—	43	30	20	15	11	8½	7	6	5	4¼	17
20	—	—	50	35	24	18	14	11	8	7	6	5	18
Inches.	Safe load in tons.												Inches.

This table is derived from Gordon's formula $P = \dfrac{rA}{1+a\left(\dfrac{l}{d}\right)}$ *taking* $r = 1{,}000$ *lbs. per square inch,* $a = \dfrac{1}{48}$ *for round and* $\dfrac{1}{62}$ *for square timber.*

TABLE 3.—*Tackles.*

Minimum size in inches of unselected cordage and steel wire rope to be used in main lifting tackles with leading block. The figures are illustrated on Plate **108**.

Weight to be lifted in tons.	Type of tackle and theoretical gain of power.					
	Fig. 4. W=P	Fig. 3. W=2P	Fig. 5. W=3P	Fig. 6. W=4P	Fig. 8. W=5P	Fig. 9. W=6P
	CORDAGE.					
¼	2½	2	2	—	—	—
½	—	2½	2½	2½	2	2
¾	—	—	3	2½	2½	2½
1	—	—	—	3	3	2½
1½	—	—	—	—	—	3
	STEEL WIRE ROPE.					
1	2	1½	1½	1	1	1
1¼	2	1½	1½	1½	1	1
1½	2	1½	1½	1½	1½	1
2	2½	2	1½	1½	1½	1½
2½	3	2	2	1½	1½	1½
3	—	2½	2	2	2	1½
4	—	2½	2½	2½	2	2
5	—	3	2½	2½	2	2
8	—	—	3	3	2½	2½
10	—	—	—	3	3	3
11	—	—	—	—	3	3
12	—	—	—	—	—	3
Tons.	Inches.					

This table is derived from the formula $P = \dfrac{W}{G}(1 + \mu n)$ *where* — P = *Power required,* W = *Weight to be lifted,* G = *Theoretical, gains* μ = *Coefficient friction.*

If $\mu = \dfrac{1}{8}$ *(an average value) then* $P = \dfrac{W}{G}\left(1 + \dfrac{n}{8}\right)$. *The safe stress in cordage has been taken as* $\dfrac{C^2}{20}$ *tons and of steel rope as* $\dfrac{9C^2}{20}$ *tons.*

TABLE 4.—*Power required on falls of tackles in tons.*

Size of rope.	Cordage.	Steel wire rope.
1½-inch.	—	1 ton.
2 ,,	1/5 ton.	2 ,,
2½ ,,	1/3 ,,	3 ,,
3 ,,	½ ,,	4 .,

Man-power equals a pull of 56 lbs. (1/40 ton) per man. Field capstan gives a gain of about 10 to 1.

TABLE 5.—*Sizes of road-bearers and beams for light bridges, dug-outs, &c.*

1. Table showing the number of road-bearers of different scantlings required for various spans.

Width of roadway has been taken as 9 feet, and the decking must be 3 inches thick.

	Span in feet.					
	5	7	9	11	13	15
Scantling on edge (in inches).						
6 × 2	9	13	—	—	—	—
9 × 2	5	7	8	10	11	13
7 × 2½	6	8	10	12	—	—
6 × 3	7	9	11	—	—	—
8 × 3	5	6	7	8	10	11
9 × 3	5	5	6	7	8	9
4 × 4	10	—	—	—	—	—
5 × 4	7	10	12	—	—	—
7 × 4	5	6	7	8	10	11
9 × 4	5	5	5	6	6	7
5 × 5	6	8	10	12	—	—
7 × 5	5	5	6	7	8	9
9 × 5	5	5	5	5	5	6
6 × 6	5	5	6	8	9	10
8 × 6	5	5	5	5	6	6
8 × 8	5	5	5	5	5	5
9 × 9	5	5	5	5	5	5
Round poles aver. diam.						
5 inches	9	12	—	—	—	—
6 ,, ...	6	8	10	12	—	—
7 ,, ...	5	5	7	8	9	11
8 ,, ...	5	5	5	6	7	8
R.S.J. (in inches).						
5 × 3 or larger ...	5	5	5	5	5	5
4 × 3	5	5	5	5	6	7
4 × 1¾	5	6	8	10	11	—
3 × 3	5	5	6	7	9	10
3 × 1½	7	10	—	—	—	—
20-lb. Decauville rail ...	6	8	10	12	—	—
80-lb. rail	5	5	5	5	5	5

This table is calculated for a working stress in timber of $\frac{1}{2}$ ton per square inch, and in steel of $7\frac{1}{2}$ tons per square inch, and allows for a maximum moving load of 2 tons on one axle.

2. To use the above table for beams for dug-out roofs, divide 16,000 lbs. by the number given in the table for the scantling or pole at the required span. The result gives the safe distributed load in lbs. which can be carried by one beam. Then proceed as in Table 6. Example :—

9-inch by 3-inch joists will safely carry $\dfrac{16,000}{5} = 3,200$ lbs. per beam

in a roof of 7-foot span, 5 being the figure given in the table for 9-inch by 3-inch timber over a span of 7 feet.

TABLE 6.—*Safe nett distributed loads in lbs. which can be supported by girders or rails over different spans.*

(If the load is concentrated in the middle of the girder or rail only half these loads are safe.)

Span in feet.	8-in. × 4-in. "I" girder, weight 18 lbs. per ft.	5-in. × 3-in. "I" girder, weight 11 lbs. per ft.	40-lb. steel rail, weight 13¼ lbs. per ft.	60-lb. steel rail, weight 20 lbs. per ft.	80-lb. steel rail, weight 26¾ lbs. per ft.
4 ...	37,500	12.700	9,720	17,770	26,800
5 ...	30,000	10.150	7,775	14,200	21,450
6 ...	25,000	8.440	6,480	11,500	17,875
7 ...	21,400	7,200	5,550	10,150	15,320
8 ...	19,900	6,300	4,860	8.880	13,400
9 ...	16,600	5,600	4,320	7,890	11,900
10 ...	15,000	5,000	3,890 ·	7,100	10,725
11 ...	13,500	4,600	3,530	6,460	9,750
12 ...	12,5C0	4 2C0	3,240	5,900	8,940

The weight per foot run is given for purposes of comparison. Thus, the 8-inch by 4-inch girder weighing 18 lbs. per foot is the most suitable for roofs of dug-outs, and is twice as strong as a 60-lb. rail weighing 20 lbs. per foot.

Steel rails are described by their weight per yard, and may be recognized by their measurements :—

40-lb. rail is 3½ inches high.
60-lb. rail is 4¼ inches high.
80-lb. rail is 5 inches high.
Other weights in proportion.

To find the weight of earth in lbs. which is supported by one girder, multiply the span in feet by the distance apart of the girders in feet by the depth of earth in feet by 100 (a cubic foot of earth weighs roughly 100 lbs.).

To find the suitable spacing for the girders of a roof. Take the weight of 1 foot width of roof, and compare with the table above.

Thus for the roof illustrated on Plate **143**, the weight of one foot width of roof will be 9 (span in feet) × 8½ (depth of roof covering in feet) × 100 = 7,650 lbs., and if 8-inch by 4-inch girders are used a suitable spacing will be $\frac{16,600}{7,650}$ feet, or approximately 2 feet centre to centre. If 5-inch by 3-inch girders are used $\frac{5,600}{7,650}$ will be the safe spacing, or approximately 9 inches.

APPENDIX VI.

TABLES OF WEIGHTS AND MEASURES.

Table—
1.—Linear measure.
2.—Square measure.
3.—Cubic measure.
4.—Liquid measure.
5.—Measures of weights.
6.—Area and contents of certain figures.

TABLE 1.—*Linear measure.*

	Ins.	Ft.	Yds.	Pls.	Chs.	Furs
Foot	12	1				
Yard...	36	3	1	1		
Rod, pole or perch ...	198	16½	5½	1		
Chain	792	66	22	4	1	
Furlong	7,920	660	220	40	10	1
Mile	63,360	5,280	1,760	320	80	8
Fathom	72	6	2			
Nautical or geographical mile.	72,960	6,080	2,026⅔			
Cable's length	7,296	608	202⅔			
Millimetre	·039	·003	·001			
Centimetre	·394	·033	·011			
Decimetre	3·937	·328	·109			
Metre	39·37	3·28	1·094			
Kilometre	39,370·79	3,280·90	1,093·633			

1 kilometre is approximately ⅝ mile ; to convert miles to kilometres multiply by 1·609.

To convert yards to metres multiply by ·914.

(B 14783)Q   H

TABLE 2.—*Square measure.*

	Ins.	Ft.	Yds.	Pls.	Chs.	R.	A.
Square foot ...	144	1					
Square yard ...	1,296	9	1				
Rod, pole or perch.	39,204	272¼	30¼	1			
Square chain	627,264	4,356	484	16	1		
Rood ...	1,568,160	10,890	1,210	40	2½	1	
Acre	6,272,640	43,560	4,840	160	10	4	1
Square mile ...	—	—	3,097,600	102,400	6,400	2,560	640
Centiare ...	1,550·059	10·764	1·196				
Area (100 sq. metres).	—	1,076·430	119·603	—	—	—	·025
Hectare ...	—	—	11,960·333	—	—	—	2·471

To convert acres to hectares, multiply by ·405.

TABLE 3.—*Cubic measure.*

1,728 cubic inches = 1 cubic foot.
27 cubic feet   = 1 cubic yard.
1 cubic metre   = 35·3156 cubic feet.
To reduce cubic feet to cubic metres, multiply by ·028.

1 cubic foot of fresh water weighs 1,000 ozs., or 62½ lbs.
1 cubic foot of fir weighs 40 lbs. approximately.
1 cubic foot of oak weighs 59 lbs. approximately.
1 cubic foot of beech weighs 43 lbs. approximately.
1 cubic foot of earth weighs 80 to 100 lbs. approximately.
1 cubic foot of brickwork or concrete weighs 120 lbs. approximately.

TABLE 4.—*Liquid measure.*

	Pints.	Quarts.	Gallons.
4 gills	1	—	—
Quart	2	1	—
Gallon	8	4	1
Firkin	72	36	9
Kilderkin	144	72	18
Barrel	288	144	36
Hogshead	432	216	54
Puncheon	576	288	72
Butt	864	432	108
Litre	1·759	·880	·22
Hectolitre	175·976	87·988	21·997

To convert gallons to litres, multiply by 4·56.
A gallon of fresh water weighs 10 lbs.

TABLE 5.—*Measures of weights.*

(Avoirdupois weight.)

	Ozs.	Lbs.	Qrs.	Cwts.	Grains.
16 drachms	1	—	—	—	= 437·5
Pound	16	1	—	—	= 7,000
Stone	—	14	—	—	—
Quarter	—	28	1	—	—
Hundredweight	—	112	4	1	—
Ton	—	2,240	80	20	—
Gram	·032	·022	—	—	—
Kilogram	35·26	2·204	—	—	—

To convert lbs. to kilograms, multiply by ·454.

TABLE 6.—*Areas and contents of certain figures.*

Circle   radius $(r) = \frac{1}{2}$ of diameter (d).

circumference $= 2\pi r = \pi d$. $\pi = 3\cdot14159$ or $\dfrac{22}{7}$ nearly.

area $= \pi r^2$.

Triangle   area $= \frac{1}{2}$ base $\times$ perpendicular from apex to base.

Cylinder   of height h — content $= \pi r^2 h$.

Cone   of height h — content $= \frac{1}{3}\pi r^2 h$. $r =$ radius of base.

Pyramid   of height h — content $= \frac{1}{3}h \times$ area of base.

INDEX.

A.

229

C.

E.

F.

H.

I.

K.

L.

N.

O.

252

257

U.

V.

W.

Z.

Printed under the Authority of His Majesty's Stationery Office by
HARRISON & SONS, LTD., 44–47, St. Martin's Lane, W.C. 2.

Plate 1.

Fig. 3.

Fig. 2.

Fig. I.

USE OF PICK.

Plate 2.

Fig. 2.

Fig. 1.

USE OF SHOVEL.

8921. 1:08+ PP 3115/987

Malby & Sons, Lith.

Plate 3.

USE OF SHOVEL.

8921. 11084. PP3115/387.

Malby & Sons. Lith.

Plate 4.

USE OF SHOVEL.

8921. 11084. PP 9115/387.

Plate 5.

Fig. 2.

Fig. 1

USE OF SPADE.

8921. 11084. P P 3115/387.

Malby & Sons. Lith.

Plate 6.

FASCINES.

WITHES.

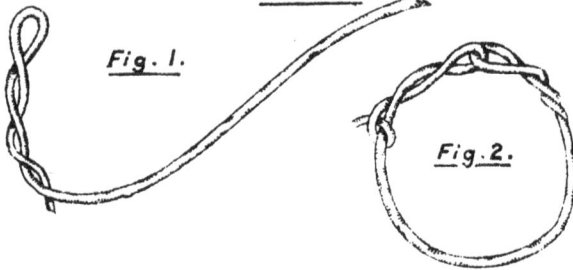

Fig. 1.

Fig. 2.

Fig. 3.

FASCINE TRESTLE.

Fig. 4.

Stakes
about 6½ feet long

FASCINE CHOKER
For compressing brushwood

18-FT FASCINE   Trestles 4 apart

Fig. 5.

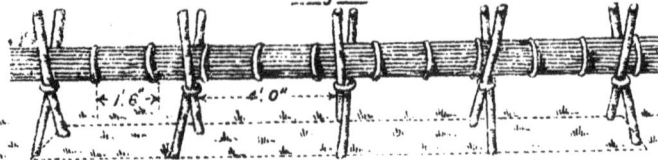

Suitable tracing rectangle measures 16′. 0″ × 4′. 0″

8921. 11084 P.F 3115/387.

Malby & Sons. Lith.

Plate 7.

HURDLES.

Rough Hurdle.

Binding Wire. Binding Wire. Binding Wire.

6' 0"

2' 9"

Fig. 3.

ELEVATION.

PLAN.
(Shewing pickets and binding wire).

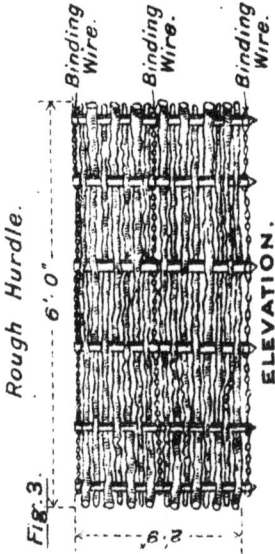

3" Expanded Metal turned over at each end in order to obviate tendency to pull out.

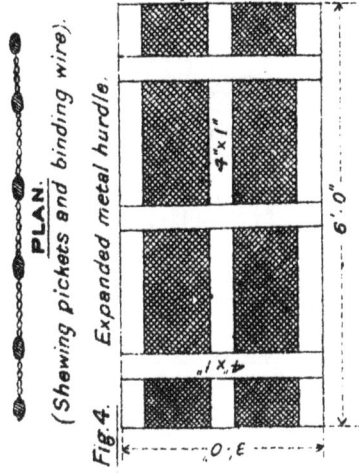

Expanded metal hurdle.

4" x 1"

4" x 1"

6' 0"

3' 0"

Fig. 4.

3 Men, 2½ hours. Tools; 2 Bill Hooks, 2 Gabion Knives, 1 Mallet, 1 six f⁺ Rod, 1 pair Pliers (if sewn in wire), 1 Gauge (made on ground).

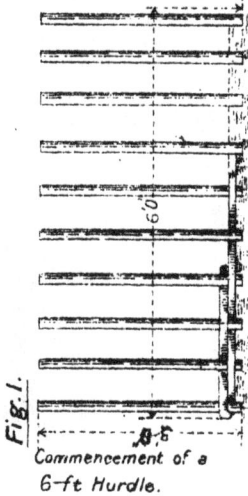

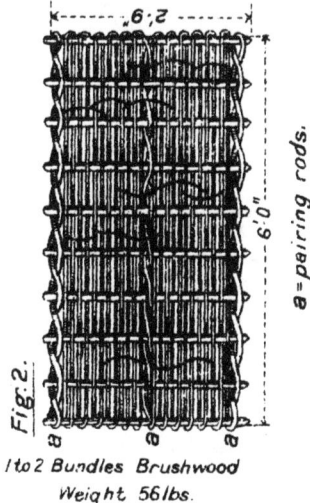

6' 0"

2' 9"

6' 0"

a = pairing rods.

Fig. 1.

Fig. 2.

Commencement of a 6-ft Hurdle.

1 to 2 Bundles Brushwood Weight 56 lbs.

a   a   a

Plate 8.

COLLAPSIBLE WIRE NETTING GABION.

Wire Netting Omitted.

FIG. 1.

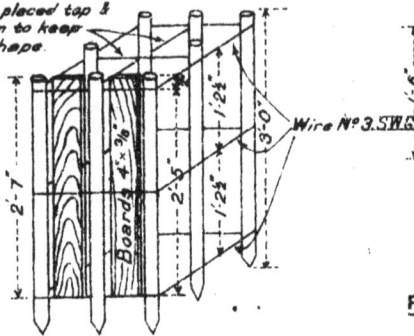

Wires placed top & bottom to keep the shape.

Boards 4"×⅜"

Wire Nº 3.S.W.G.

2'-7"

2'-5"

1'-2½"

1'-2½"

3'-0"

ISOMETRIC VIEW.

FIG. 2.

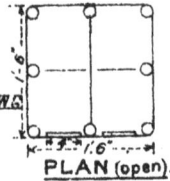

1'-6"

1'-6"

1'-6"

PLAN (open).

PLAN (closed).

LIGHT COLLAPSIBLE GABION OF CANVAS.

FIG. 3.

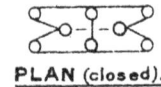

Waterproof Canvas.

3"×1 Timber

Spreader ½" Bolt

3"×3 Timber

4'-0"

2'-0"

PLAN.

FIG. 4.

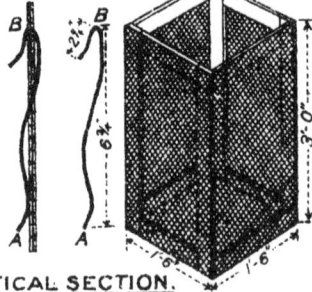

Spreader

2'-6"

ELEVATION.

1. The fastener prevents the raw edges of the X.P.M. bulging outwards.

2. The end A is inserted just far enough to pass through the overlapped sheets and the fastener is then pressed vertically downwards by tapping it with a hammer. This causes the hook end B to engage also through the overlapped sheets. The hook is driven down as far as it will go. It is impossible to withdraw without tools.

N.B. Wooden framework is not essential.

EXPANDED METAL GABION.

FIG. 6.

B

B

2¼"

6¼"

A

A

VERTICAL SECTION.

FIG. 5.

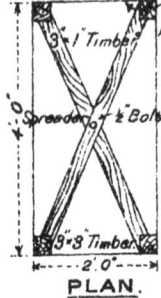

3'-0"

1'-6"

1'-6"

Malby & Sons. Lith.

Plate 9.

Fig.1. SHORT REVETTING FRAME.

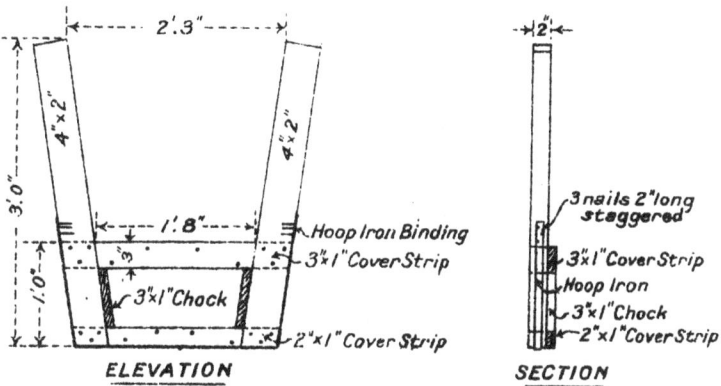

2'.3"

4"×2"

4"×2"

3'.0"

1'.8"

Hoop Iron Binding

3"×1" Cover Strip

3"

3"×1" Chock

1'.0"

2"×1" Cover Strip

ELEVATION

2"

3 nails 2" long
staggered

3"×1" Cover Strip

Hoop Iron

3"×1" Chock

2"×1" Cover Strip

SECTION

Fig.2. TRENCH BOARD.

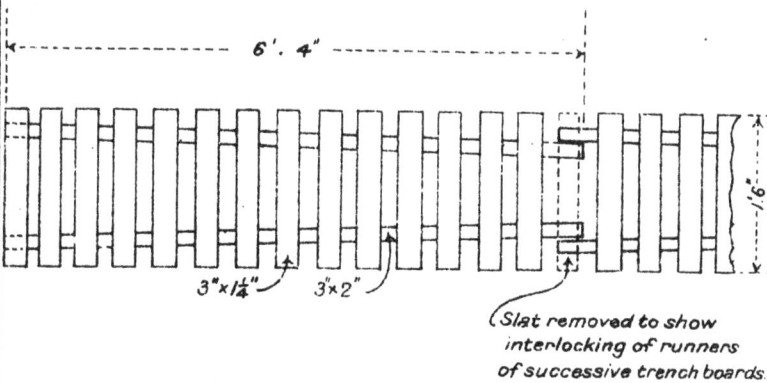

6'. 4"

1'.6"

3"×1¼"

3"×2"

Slat removed to show
interlocking of runners
of successive trench boards.

Plate 10.

FIELD GEOMETRY.

Fig. 2.

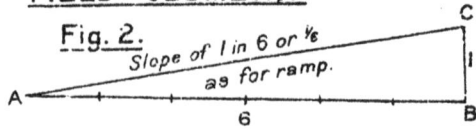

Slope of 1 in 6 or ⅙
as for ramp.

Fig. 1.

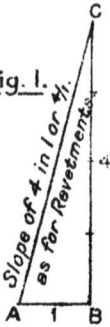

Slope of 4 in 1 or 4/1
as for Revetment.

Fig. 3.

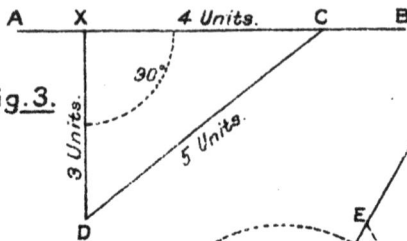

A   X   4 Units.   C   B

30°

3 Units.

5 Units.

Fig. 5.

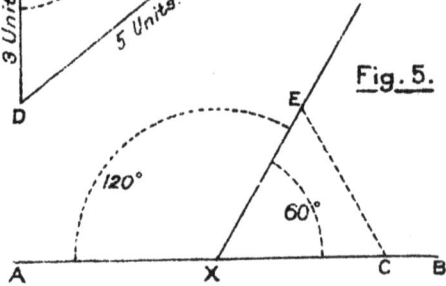

120°

60°

Fig. 6.

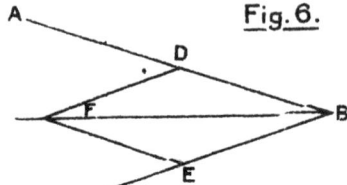

Fig. 4.

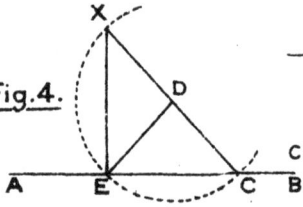

Fig. 7.

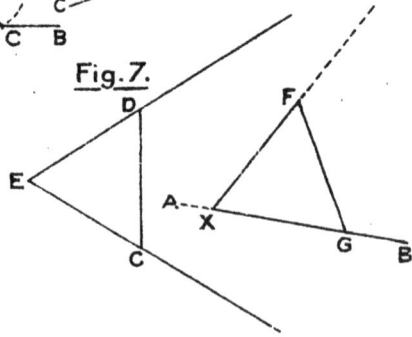

Fig. 8.

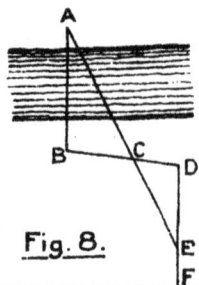

8921. 5084 PP 3115/387   Malby & Sons. Lith.

Plate 11.

FIELD LEVEL.

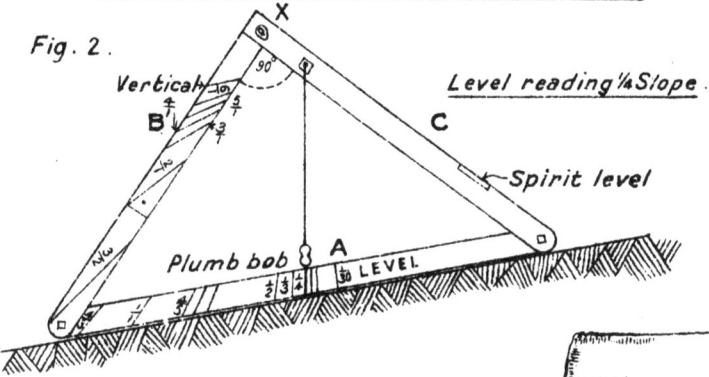

Fig. 2.

Vertical

B

C

Level reading ¼ Slope.

Spirit level

Plumb bob   A

LEVEL

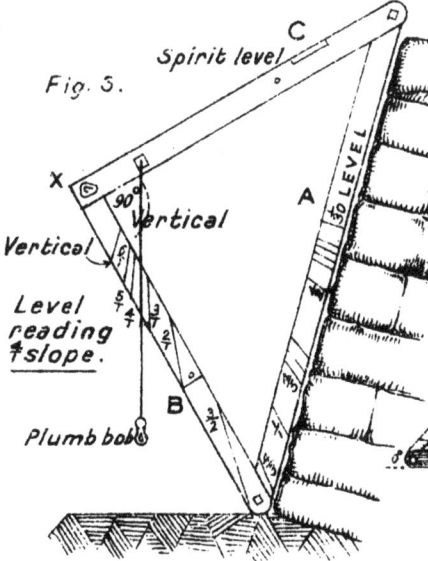

Fig. 5.

Spirit level   C

X   Vertical

Vertical

Level reading ⅛ slope.

Plumb bob

B

A   LEVEL

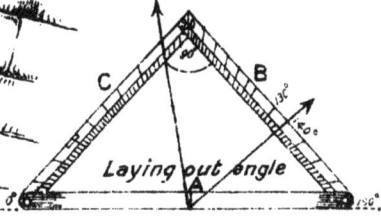

Fig. 1.

C   B

Laying out angle

Fig. 4.

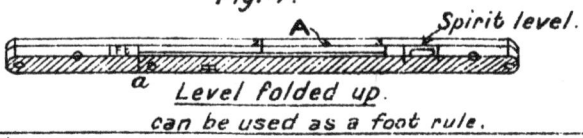

A   Spirit level.

a   Level folded up.
can be used as a foot rule.

IMPROVISED LEVEL

Fig. I.

Glass Tube 8"

Approx. 5 yds. centre to centre when stretched.

Water level

Glass Tube 8"

⅜" Rubber tubing

Fig. 2.

Parapet of Trench.

← 5 yds. → ← 5 yds. → ← 5 yds. → ← 5 yds. → ← 5 yds. → ← 5 yds. → ← 5 yds. → ← 5 yds. →

I   H   G   F   E   D   C   B   A

2'.0"

1'.0"

Bottom of Trench

SIDE ELEVATION.

6921 11084 P.P 3115/387.

Malby & Sons. Lith.

Plate 13.

THE SITING OF TRENCHES.

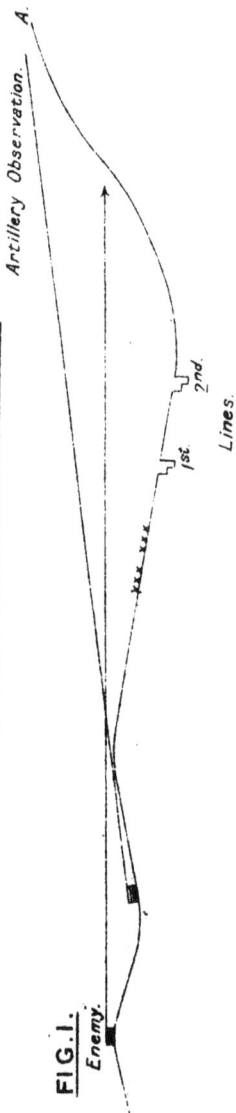

FIG. I.

Enemy.

Artillery Observation. A.

1st. 2nd.

Lines.

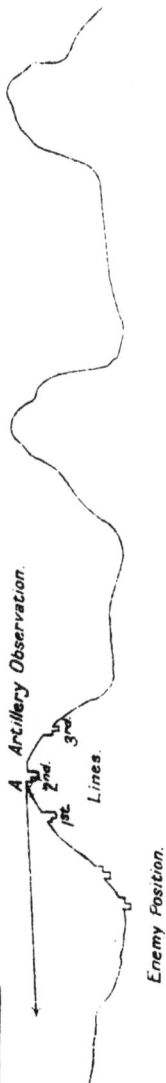

FIG. 2.

A. Artillery Observation.

1st. 2nd. 3rd.

Lines.

Enemy Position.

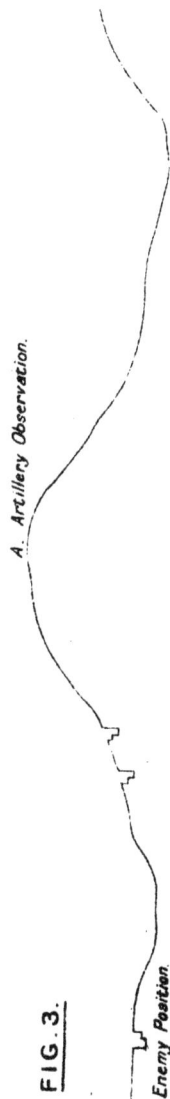

FIG. 3.

A. Artillery Observation.

Enemy Position.

Plate 14.

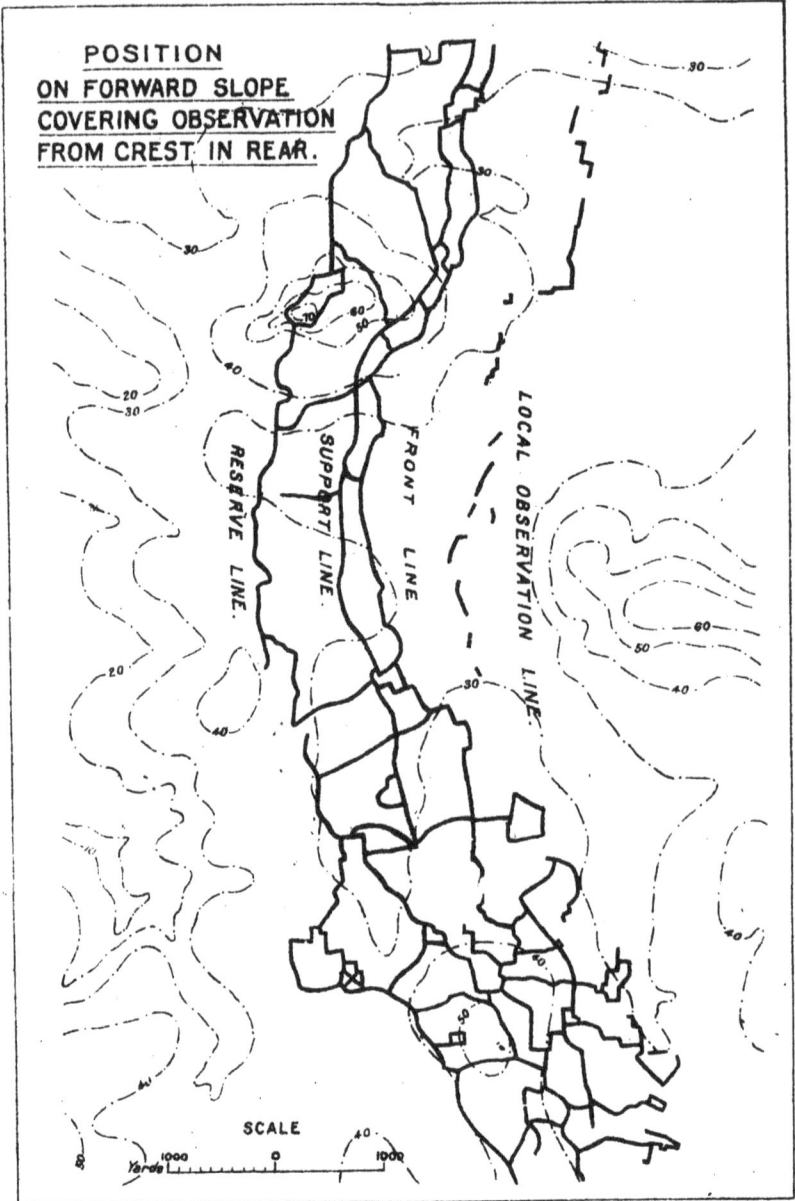

POSITION
ON FORWARD SLOPE
COVERING OBSERVATION
FROM CREST IN REAR.

RESERVE LINE.

SUPPORT LINE.

FRONT LINE.

LOCAL OBSERVATION LINE.

SCALE

Yards 1000   0   1000

Plate 15.

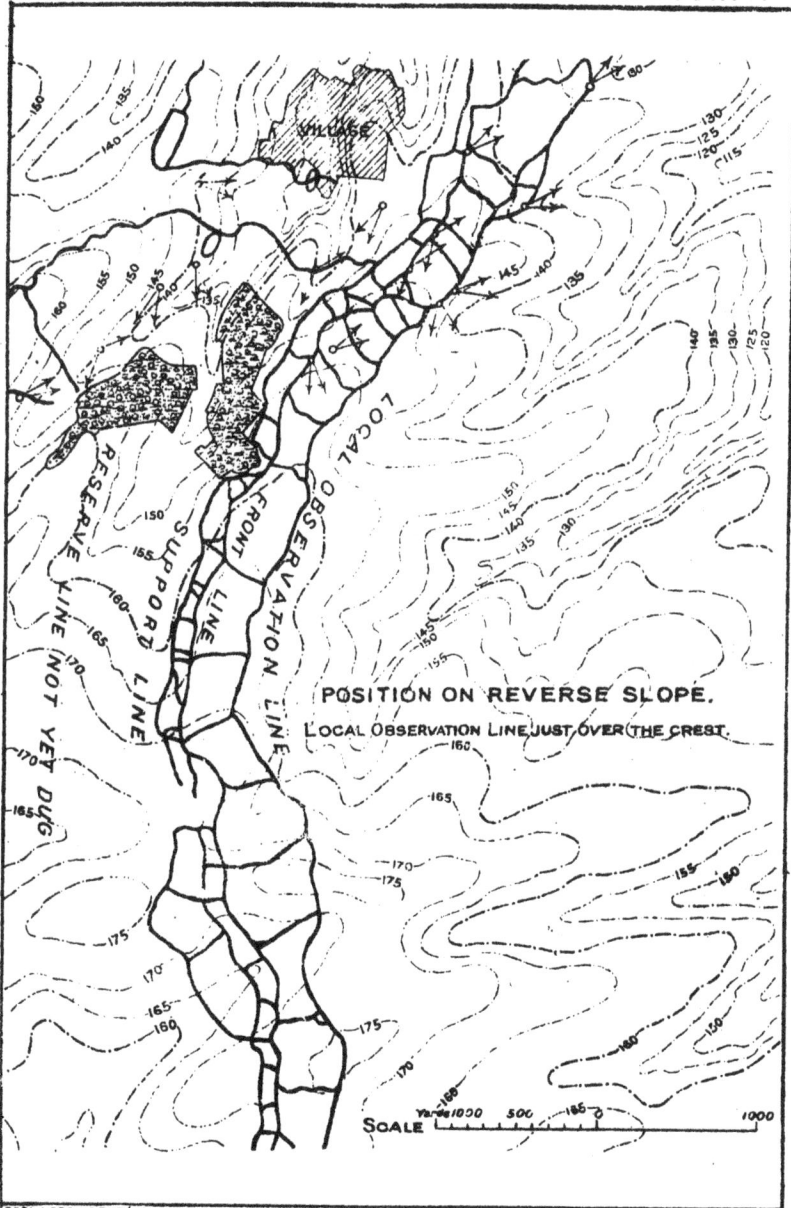

POSITION ON REVERSE SLOPE.

LOCAL OBSERVATION LINE JUST OVER THE CREST.

VILLAGE

LOCAL OBSERVATION LINE

FRONT LINE

SUPPORT LINE

RESERVE LINE (NOT YET DUG)

SCALE Yards 1000 500 0 1000

8921. 11084. PP 3113/387.

Malby & Sons. Lith.

Plate 16.

OBSTACLES IN A WOOD.

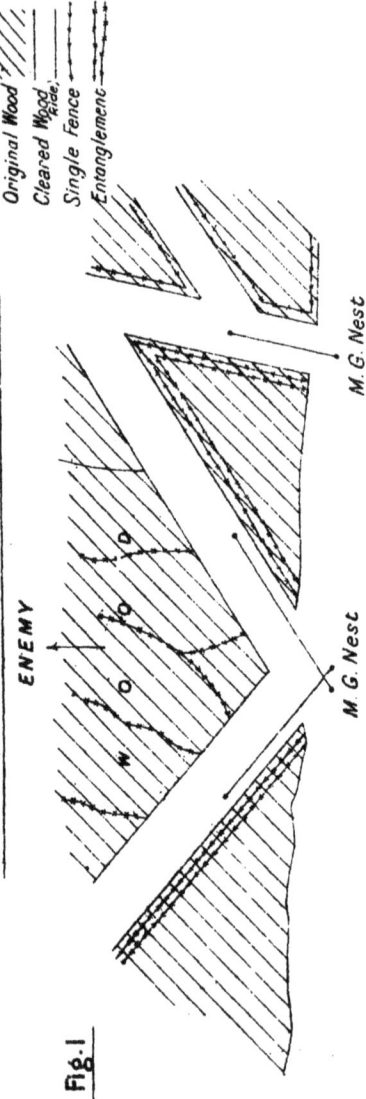

Fig.I

Original Wood
Cleared Wood
Ride.
Single Fence
Entanglement

ENEMY

W O O D

M.G. Nest

M.G. Nest

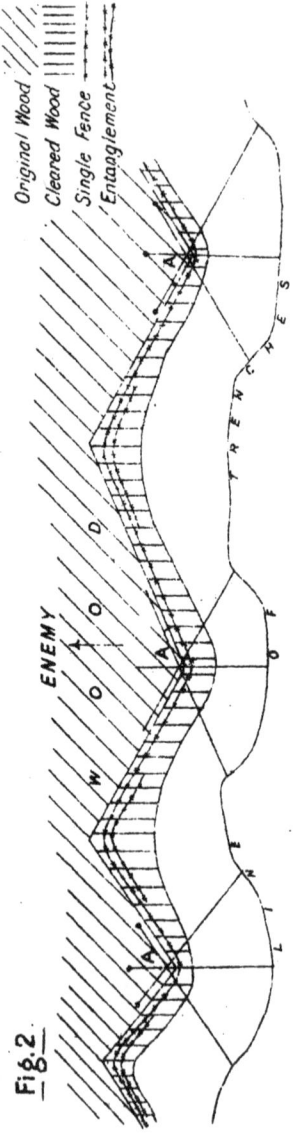

Fig.2

Original Wood
Cleared Wood
Single Fence
Entanglement

ENEMY

W O O D

A

A

A

A

LINE OF TRENCHES

Malby & Sons Lith.

Plate 17.

MINIMUM DIMENSIONS OF
M.G. EMPLACEMENTS.

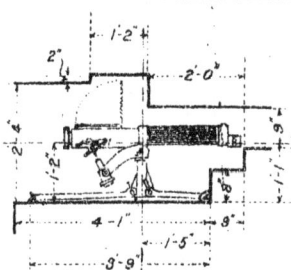

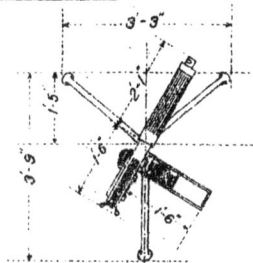

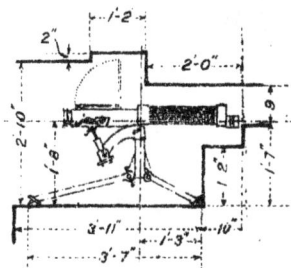

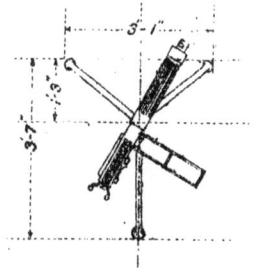

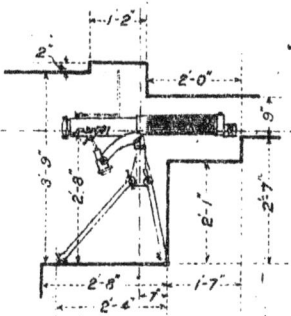

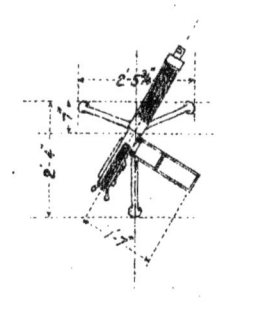

Plate 18.

SHELL-HOLE M.G. EMPLACEMENT.

Fig. 1

Fig. 2

Fig. 3

Fig. 4

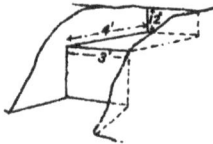

8921. 11084 PP3115/587

Malby & Sons. Lith

Plate 19.

CONCEALMENT OF M.G. EMPLACEMENT.

FIG. 1.

FIG. 2.

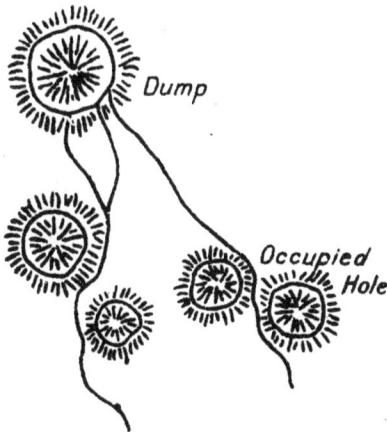

Dump

Occupied Hole

FIG. 3.

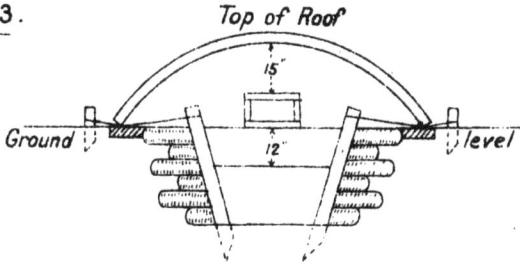

Top of Roof

15"

Ground level

12"

Emplacement with Splinter-proof Roof

(Material on top of Roof not shown.)

Malby & So Lith

Plate 20.

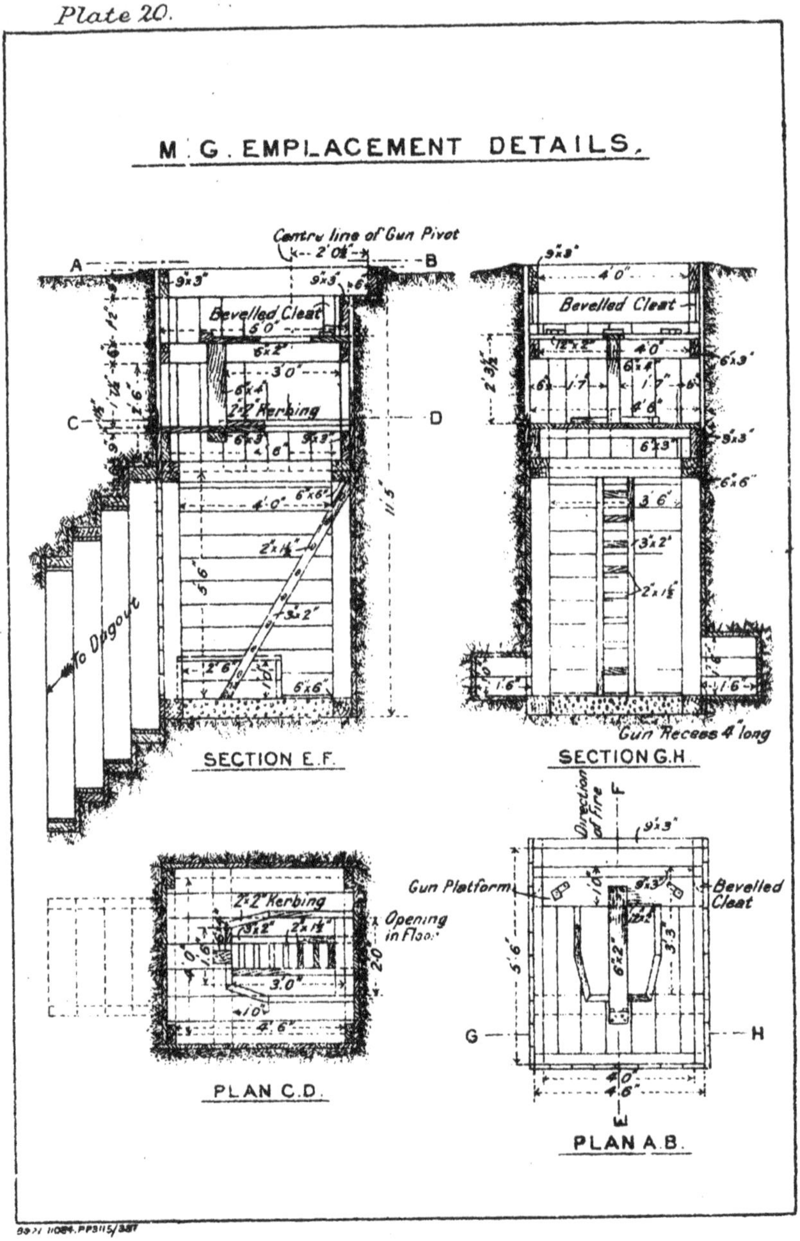

M.G. EMPLACEMENT DETAILS.

Centre line of Gun Pivot

A — B

Bevelled Cleat

C — D

SECTION E.F.

SECTION G.H

Gun Recess 4' long

PLAN C.D.

Gun Platform

Bevelled Cleat

Direction of Fire

Opening in Floor

PLAN A.B.

Plate 21

DETAILS OF "T-BASE" FOR AN OPEN MACHINE GUN PLATFORM.

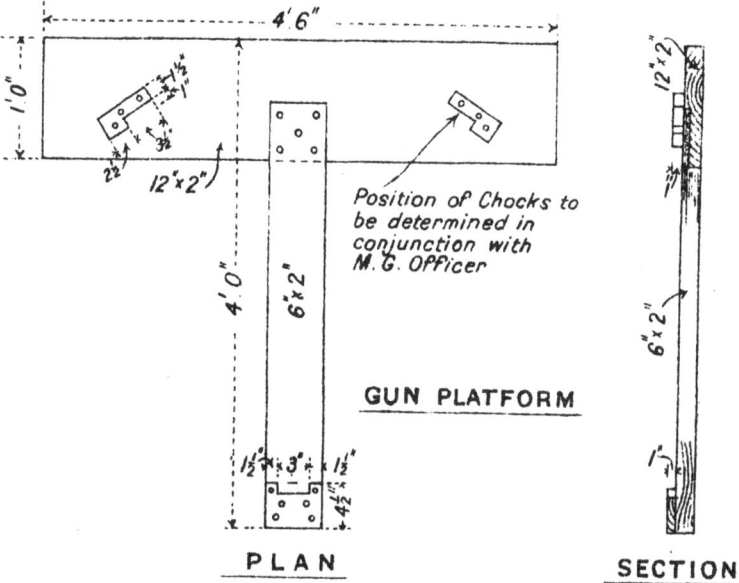

4' 6"

1' 0"

2½ × 1½ × 1"

3½

2¼

12" × 2"

Position of Chocks to be determined in conjunction with M.G. Officer

4' 0"

6" × 2"

GUN PLATFORM

1½" × 3" × 1½"

4½" × 2"

PLAN

12" × 2"

6" × 2"

1"

SECTION

SKETCH
Showing Gun being fired from reverse position.

8921 - 11084 PP.9115/987

Malb & Sons Lith

Plate 22.

THE "MOIR" PILL-BOX, SHOWING MAIN DETAILS OF CONSTRUCTION, DIMENSIONS AND WEIGHTS.

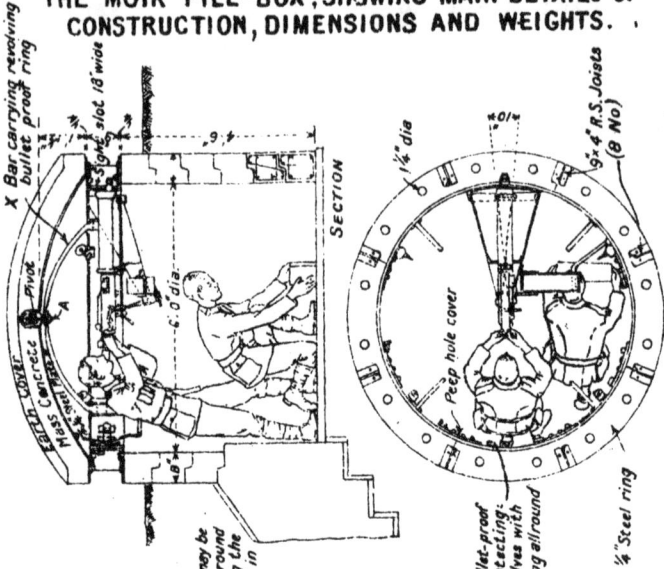

X Bar carrying revolving bullet proof ring

Sight slot 18"wide

SECTION

EP9in Cover

Mass Concrete

pivot

6'0 dia

6'0 dia

8"

Ground Line

NOTE:—
Bullet-proof ring may be lowered to give allround view by unscrewing the nut "A" (lettered in Section)

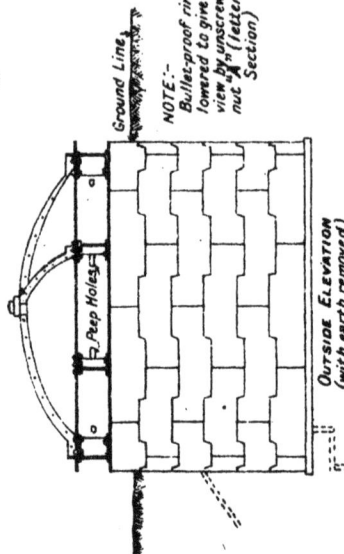

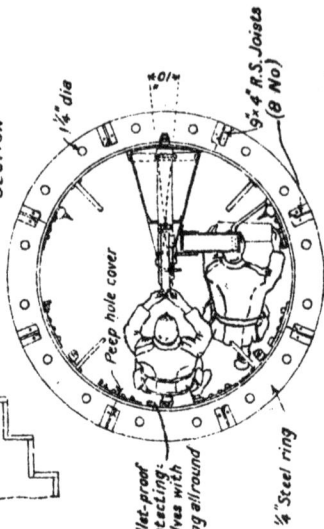

¼" dia

1'0½"

9"×4" R.S. Joists
(8 No)

Peep hole cover

PLAN (with cover removed)

12 7⁄16 bullet-proof steel protecting ring revolves with gun giving allround fire

¼" Steel ring

OUTSIDE ELEVATION
(with earth removed)

Peep Holes

Description	Weight	No	Total Weight
			lbs
Concrete block	200 lbs	48	9600
¾" Steel cover	132 "	4	528
¼" Steel ring	35½ "	4	142
R.S. Joist	11½ "	8	92
¼" Bullet-proof ring	76 "	4	304
X Bar	55 "	1	55
Sundries	77 "	1	77
Suspension gear	70 "	1	70
Vickers gun	28¼ "	1	28¼

Plate 23.

ACTUAL EXAMPLE SHOWING PROTECTIVE AND TACTICAL WIRE.

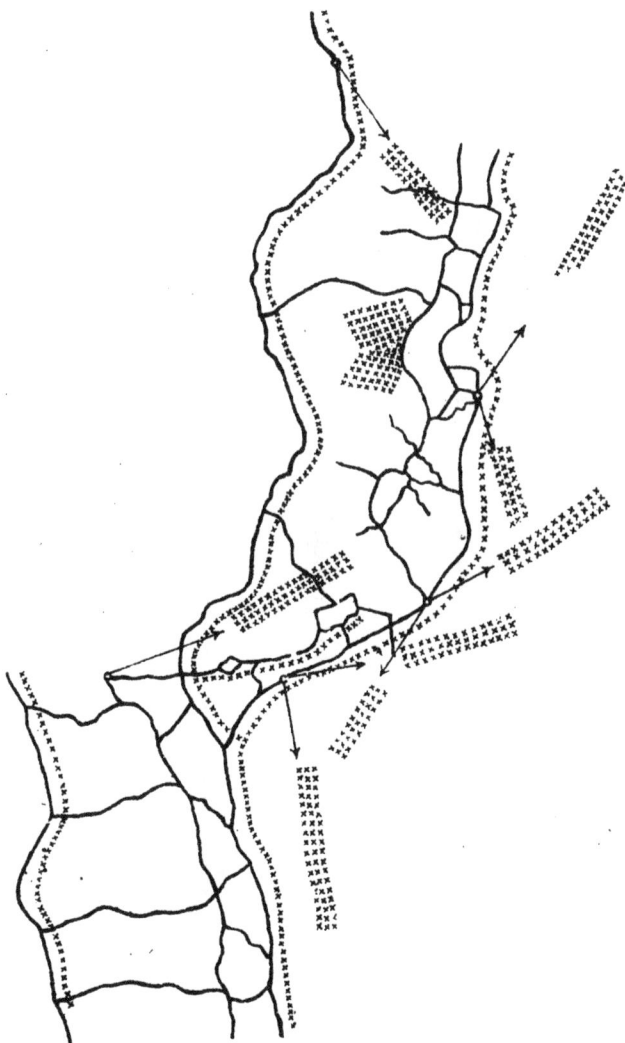

8921.11084.FP9115/387

Malby & Sons Lith

Plate 24.

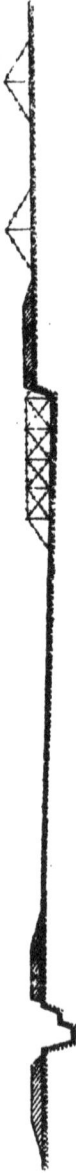

DUG IN WIRE.

Fig. 1

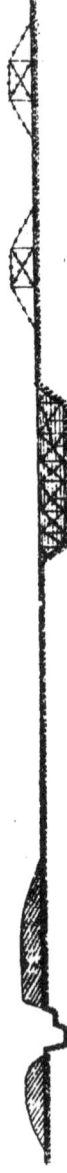

BORROW-PIT FILLED WITH WIRE AND FLOODED.

Fig. 2

892(.11084. PP3H5/387

Malby & Sons. Lith.

Plate 25.

DIAGRAM SHOWING HOW TO FINISH OFF,
THE ENDS OF THE ENTANGLEMENT AT A GAP.

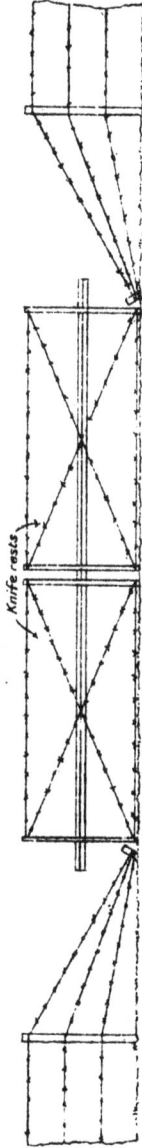

Knife rests

WRONG.

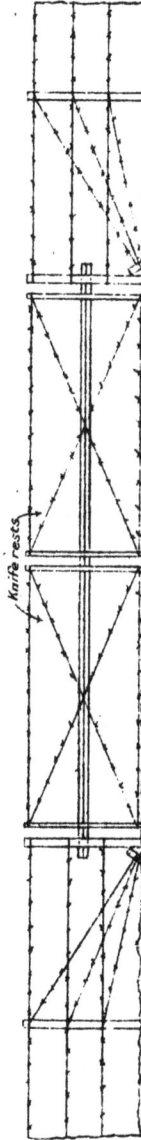

Knife rests

RIGHT

NOTE.   Knife Rests must be securely anchored to the entanglement and to each other.

Plate 26.

BLOCKING OF ROADS.

TWO BELTS OF WIRE CROSSING A ROAD DIACONALLY.

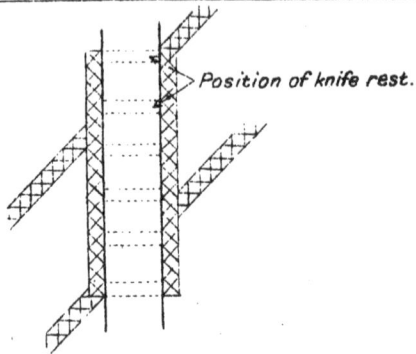

Position of knife rest.

ROAD OPEN PREPARED FOR BLOCKING.

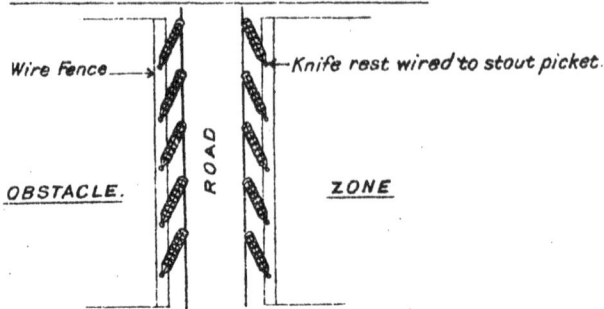

Wire Fence⟶

⟵Knife rest wired to stout picket.

ROAD

OBSTACLE.   ZONE

ROAD CLOSED.

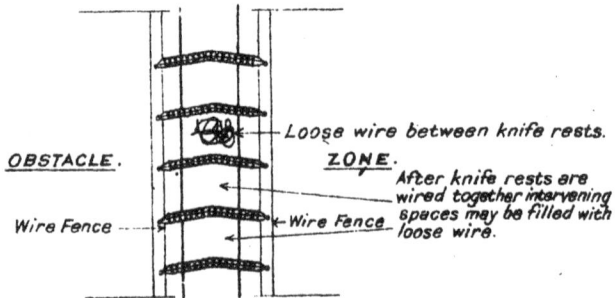

OBSTACLE.   ZONE.

⟵Loose wire between knife rests.

Wire Fence⟶   ⟵Wire Fence

After knife rests are wired together intervening spaces may be filled with loose wire.

Plate 27.

STANDARD FRENCH WIRE ENTANGLEMENT.

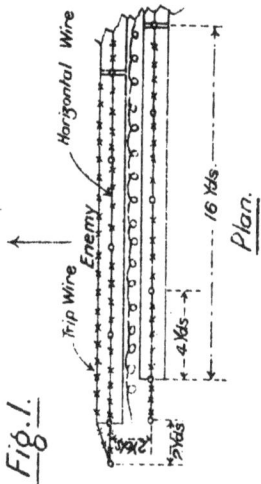

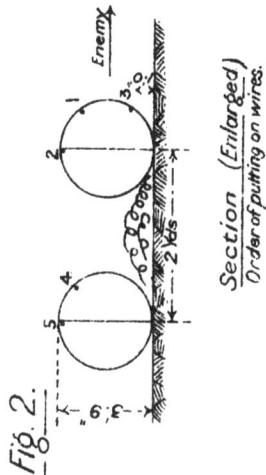

Fig. I.

Horizontal Wire

Enemy

Trip Wire

16 Yds.

4 Yds.

2 Yds.

Plan.

Fig. 2.

Enemy

Section (Enlarged)
Order of putting on wires.

2 Yds.

3' 9"

DOUBLE BELT OF CONCERTINAS.

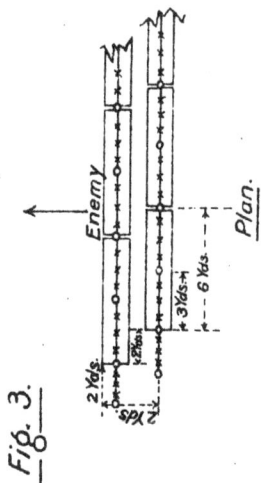

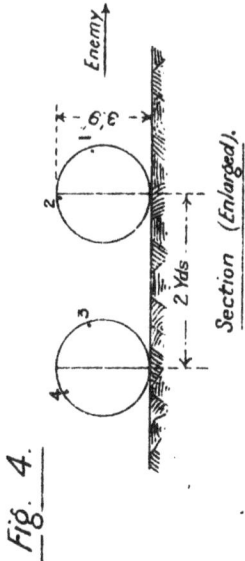

Fig. 3.

Enemy

6 Yds.

3 Yds.

2 Yds.

2 Yds.

Plan.

Fig. 4.

Enemy

Section (Enlarged).

2 Yds.

3' 9"

Plate 28.

STANDARD LOW WIRE ENTANGLEMENT. STANDARD DOUBLE APRON ENTANGLEMENT.

FIG. 1.

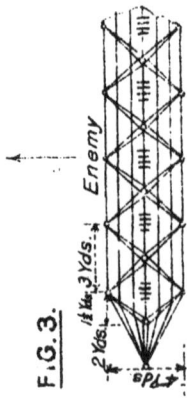

PLAN.

FIG. 2.

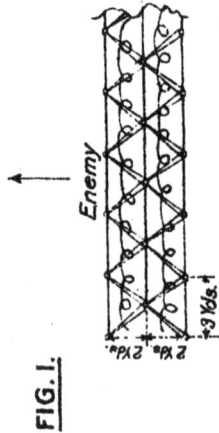

SECTION (Enlarged).
Order of putting on the wire.

F.G. 3.

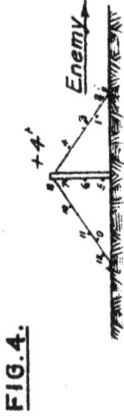

PLAN.

FIG. 4.

SECTION.
Order of putting on wires.

Plate 29.

Plate 29.

DOUBLE APRON FENCE.

AS FRAMEWORK FOR WIDE OBSTACLES.

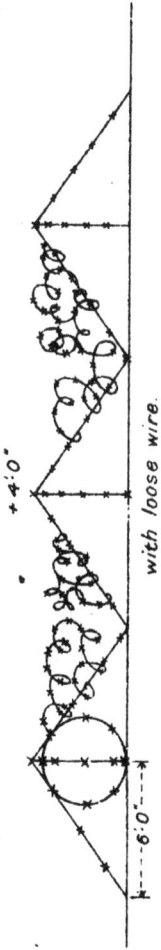

with loose wire.

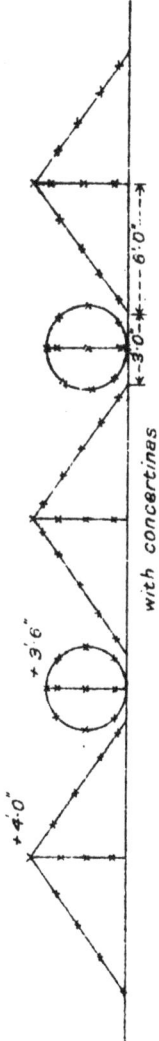

with concertinas

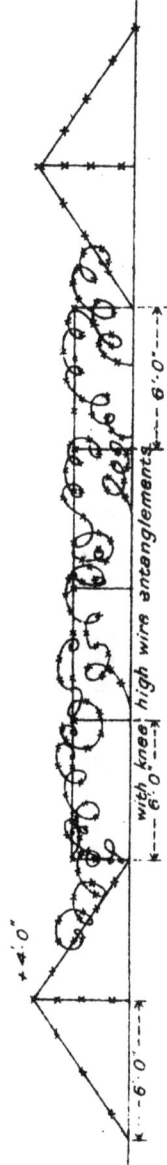

with knee high wire entanglements.

Malby & Sons. Lith.

892/. 1084 PP 3115/387

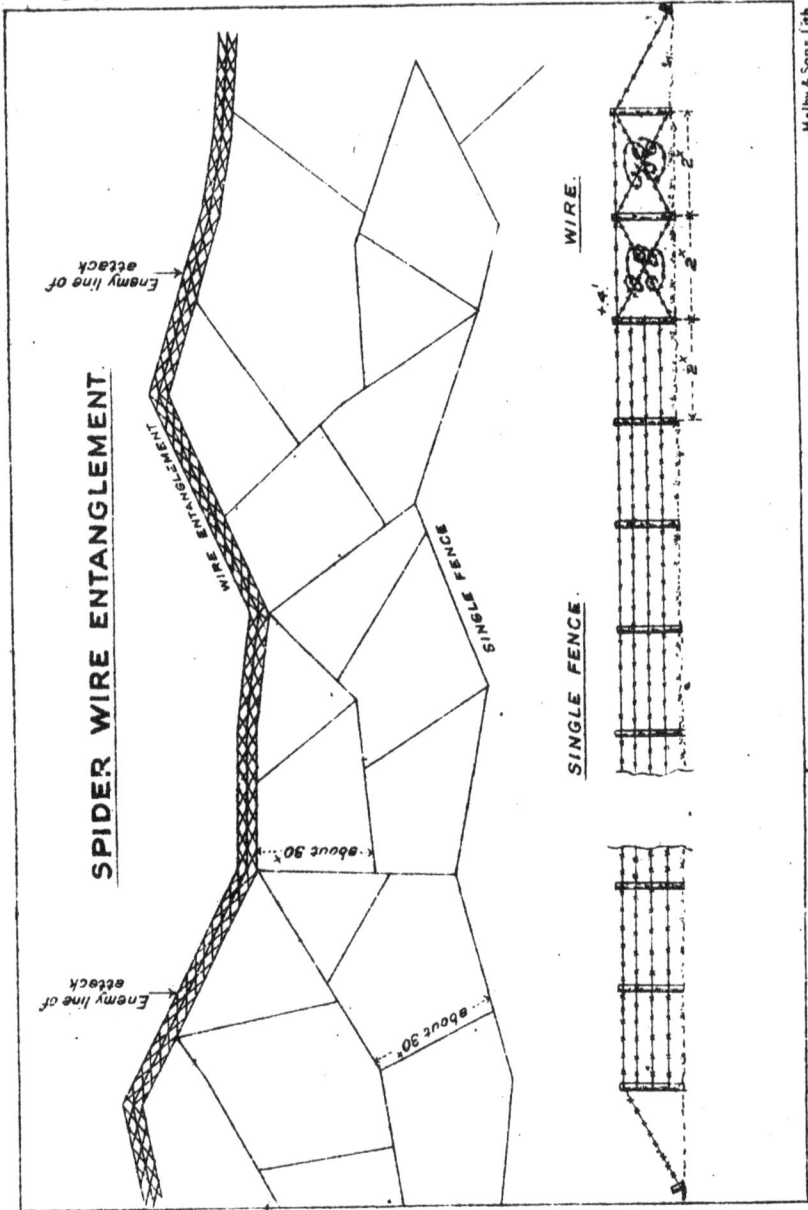

Plate 30.

SPIDER WIRE ENTANGLEMENT

SINGLE FENCE.

WIRE.

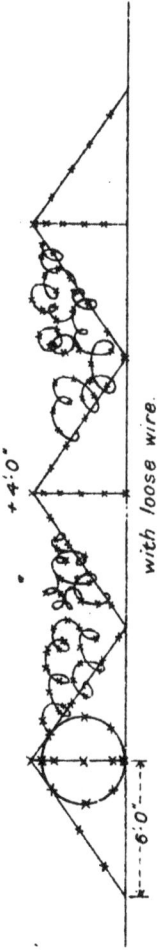

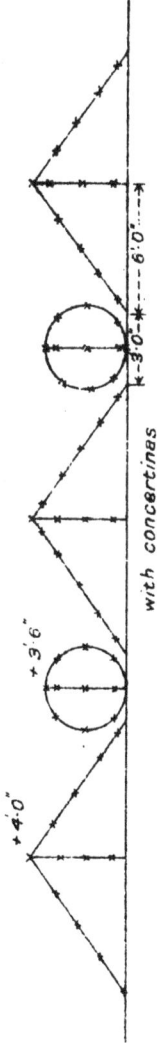

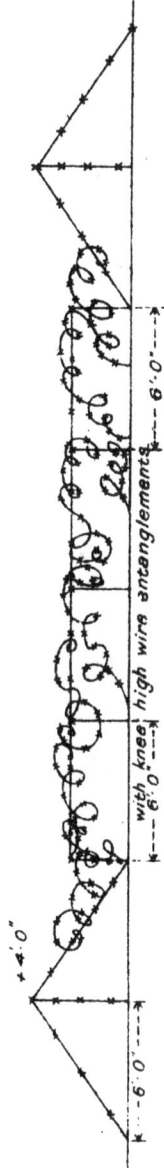

Plate 29.

DOUBLE APRON FENCE.

AS FRAMEWORK FOR WIDE OBSTACLES.

with loose wire.

with concertinas

with knee high wire entanglements

Malby & Sons. Lith.

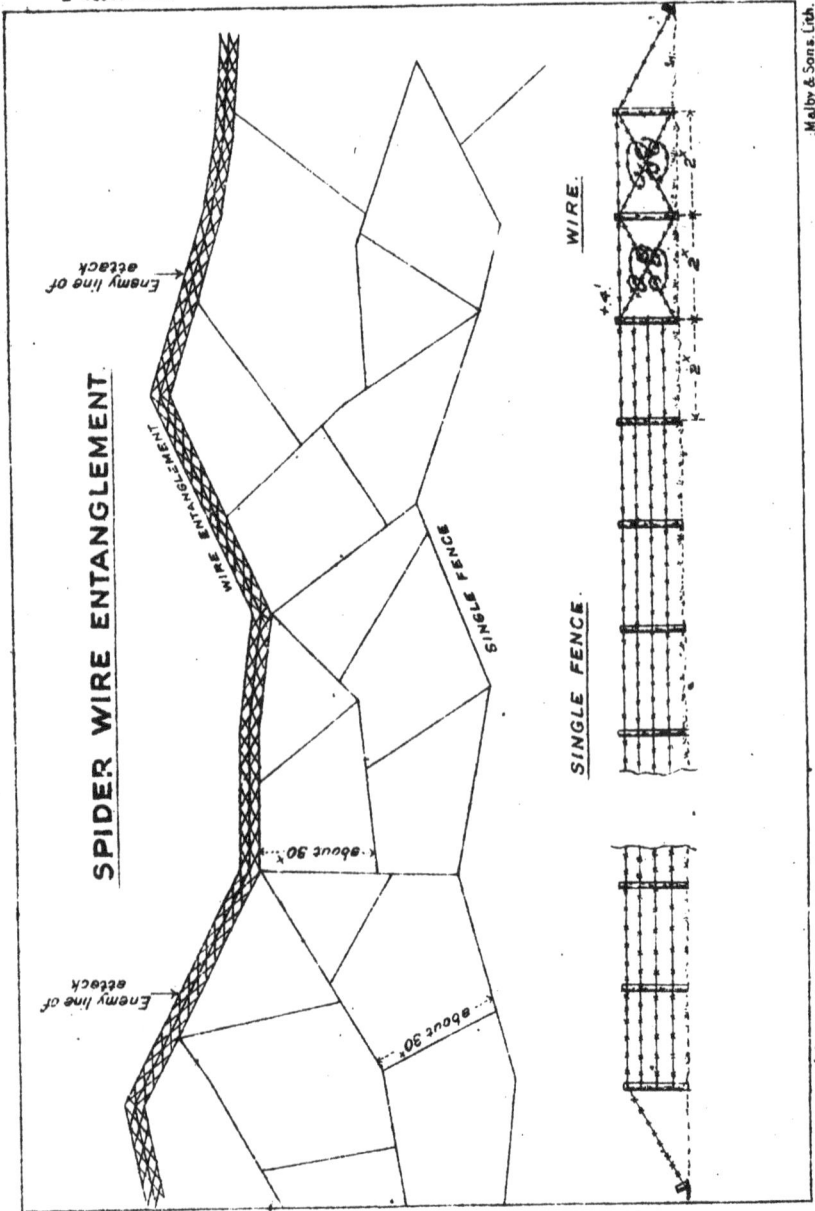

Plate 30.

SPIDER WIRE ENTANGLEMENT

Enemy line of attack

WIRE ENTANGLEMENT

SINGLE FENCE

about 30'

Enemy line of attack

about 30'

WIRE.

SINGLE FENCE.

Plate 31

SCREW POSTS FOR WIRE ENTANGLEMENTS.

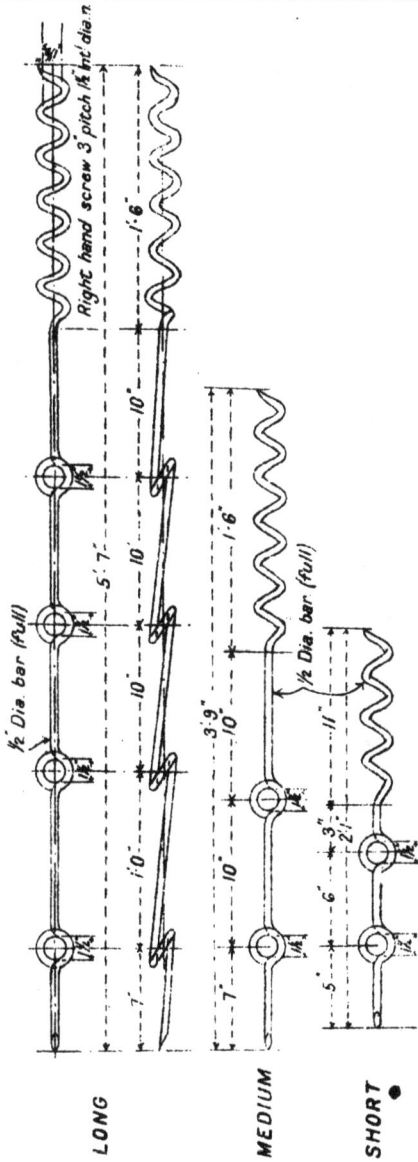

LONG

MEDIUM

SHORT

½ Dia. bar (full)

Right hand screw 3" pitch 1½ int' dia n.

½ Dia. bar (full)

1'·6"

1'·6"

10"

10"

10"

5'·7"

1'·0"

7"

3'·9"

10"

10"

7"

11"

6"

5"

3"

2½"

Plate 32.

METHOD OF FIXING WIRE TO PICKETS.

FIG. 1.

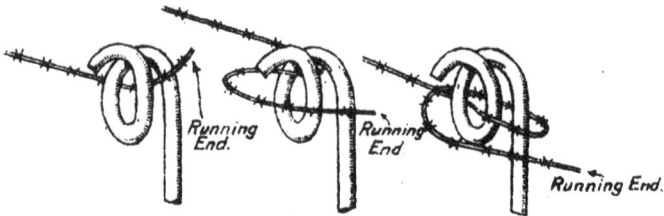

Running End.   *Running End*   *Running End.*

Sketch of Top eye show-   2nd Operation showing   Showing wire in the eye
ing wire forced up into it.   running end passing down   and a turn taken below
       over the point of the eye.   the eye.

FIG. 2.

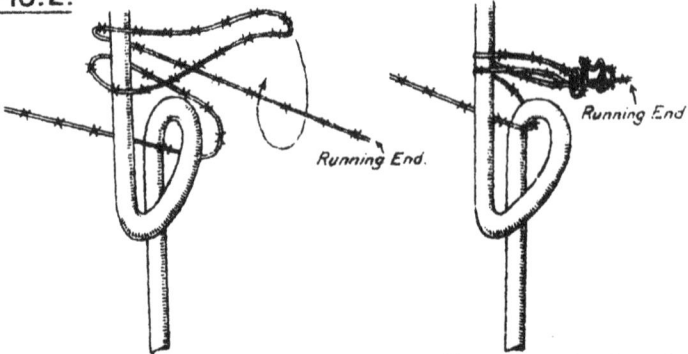

Running End.   *Running End*

Sketch showing a wire forced up into   Showing bight finished off
a lower eye of picket and the bight   on the running end.
taken round the picket above the eye.

FIG. 3.

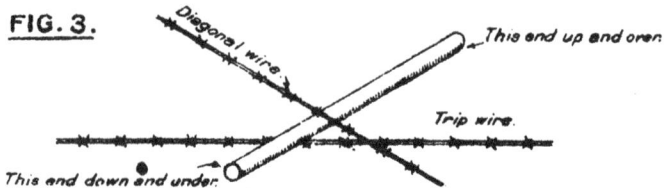

Diagonal wire   This end up and oven.

Trip wire.

This end down and under.

Sketch showing commencement of Windless.

Malby & Sons Lith

Plate 33.

BARBED WIRE CONCERTINAS.

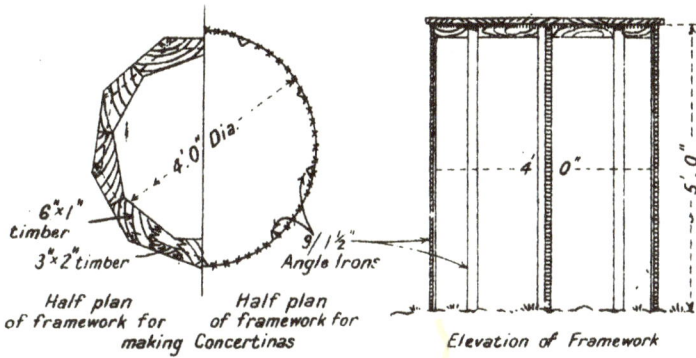

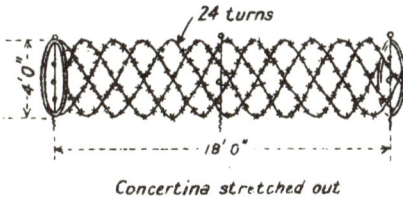

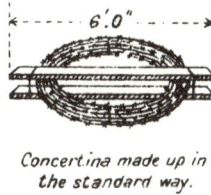

4'0" Dia.

6"x1" timber

3"x2" timber

9/1½" Angle Irons

Half plan of framework for making Concertinas

Half plan of framework for making Concertinas

4' 0"

5' 0"

Elevation of Framework

24 turns

4'0"

18' 0"

Concertina stretched out

6'0"

Concertina made up in the standard way.

Sketch showing one man carrying Concertina on shoulder

Sketch showing two men carrying Concertina.

Plate 34

KNIFE RESTS.

LARGE KNIFE REST MADE OF WOOD

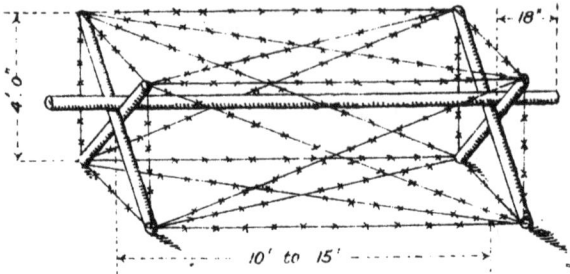

18"

4' 0"

10' to 15'

KNIFE REST MADE OF IRON PICKETS.

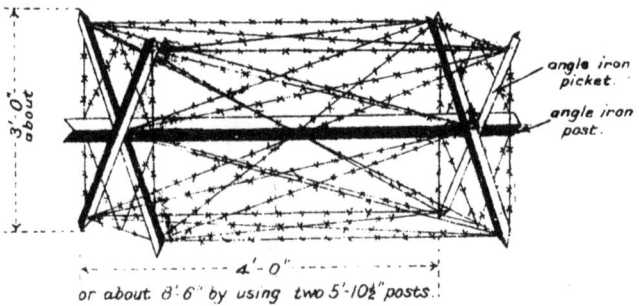

angle iron
picket.

angle iron
post.

3' 0"
about

4' 0"

or about 8' 6" by using two 5'-10½" posts.

BARRICADES.

FIG. 1.

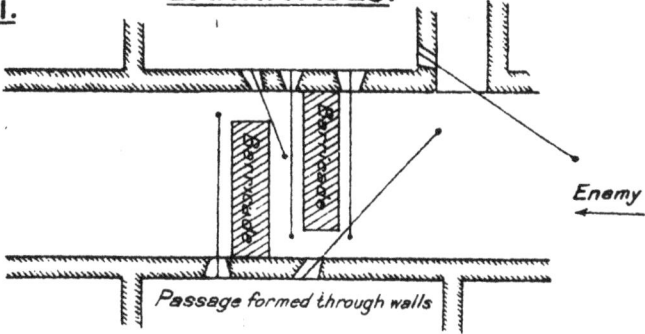

Enemy

Passage formed through walls

FIG. 2.

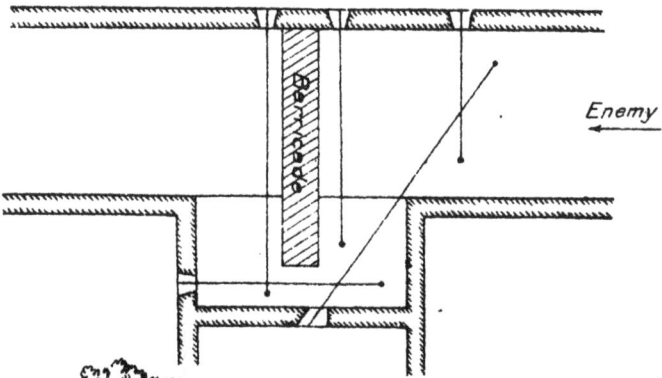

Enemy

FIG. 3.

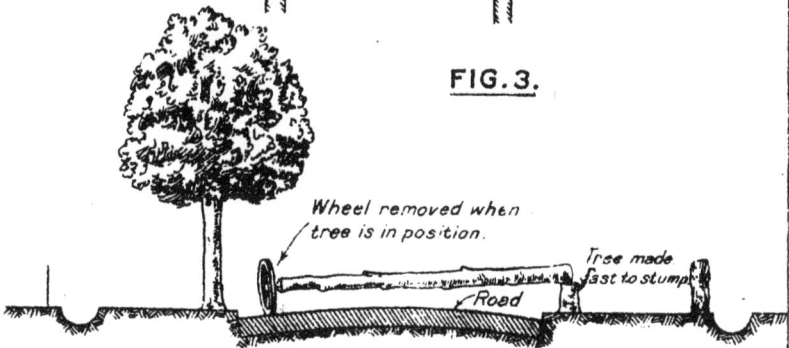

Wheel removed when
tree is in position.

Tree made
fast to stump

Road

Malby & Sons' Lith.

Plate 36.

TYPE OF OBSTACLE ON ESPLANADE RETAINING WALL.

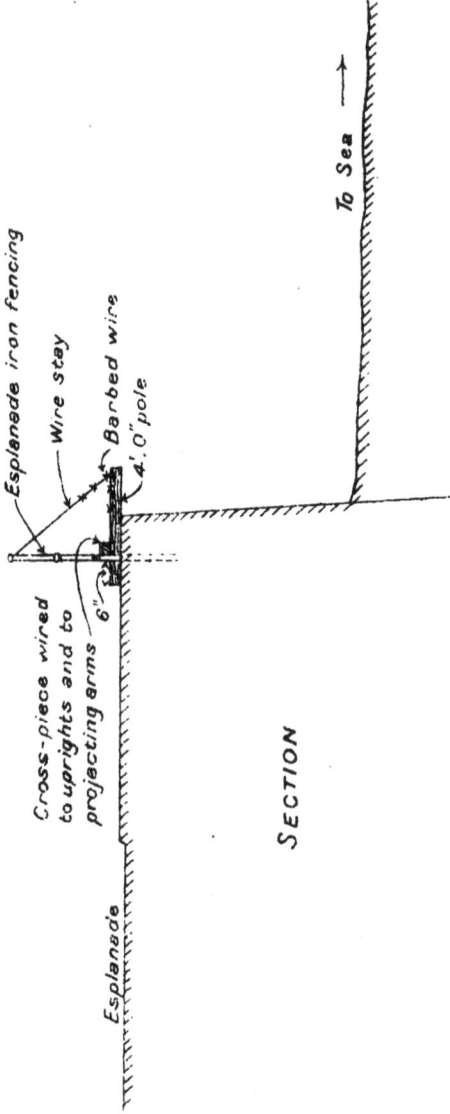

Esplanade iron fencing

Wire stay

Barbed wire

4' 0" pole

Cross-piece wired
to uprights and to
projecting arms

6"

Esplanade

SECTION

To Sea →

Maiby & Sons, Lith.

Plate 37.

TRACE OF TRENCHES.

FIG.1.

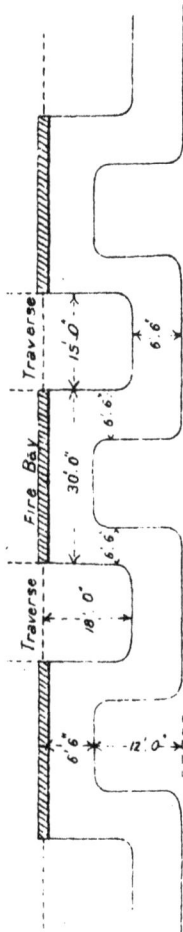

FIG.2.

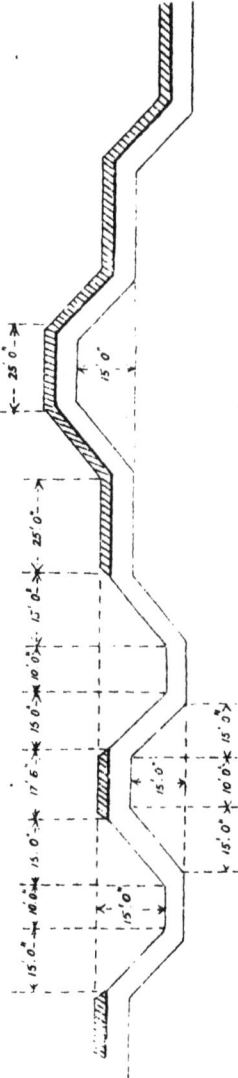

Fire steps to be arranged to suit number of garrison

Plate 38.

TRACE OF TRENCHES.

FIG. 1.

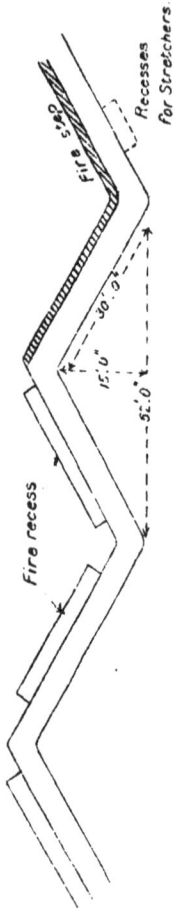

Fire recess

Fire step

Recesses
for Stretchers.

30'.0"

15'.0"

51'.0"

FIG. 2.

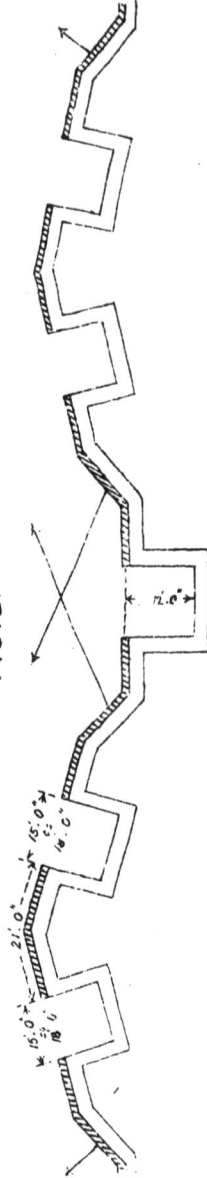

15'.0"

18'.0"

21'.0"

15'.0"

18'.0"

11'.0"

Showing application of normal trace to the ground.

Malby & Sons, Lith.

892. 11084 p 9 9115/987.

Plate 39.

ALTERNATIVE TRACE OF TRENCHES.

FIG.1.

OCCASIONAL FORWARD TRAVERSE

Provides a good position for a snipers post or position for a machine gun for flanking fire

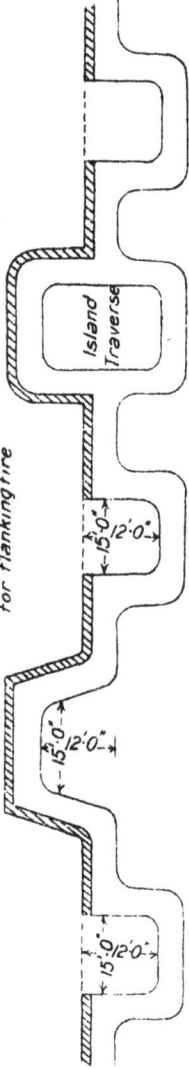

Island Traverse

15'0" 12'0"

15'0" 12'0"

15'0" 12'0"

FIG.2

Fire Recess
18'0 to 21'0'

Fire Recess
16'0 to 21'0'

CURVED WITHOUT TRAVERSES

A traversed trench is the better but takes longer to dig

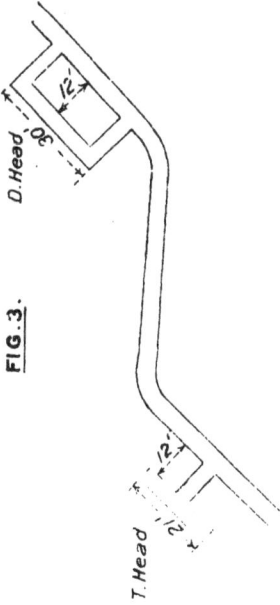

FIG.3.

D. Head

12'

30'0"

T. Head

12'

2'6"

Malby & Sons Lith.

Plate 40.

ALTERNATIVE TRACE OF TRENCHES.

FIG. I.
DOG LEGS.

18' to 30' Fire Step

Fire Step

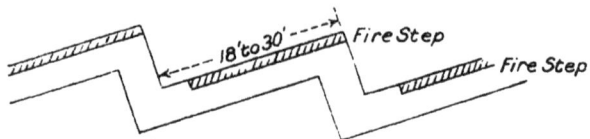

FIG. 2.
EXAMPLE OF USE OF DOG LEGS.

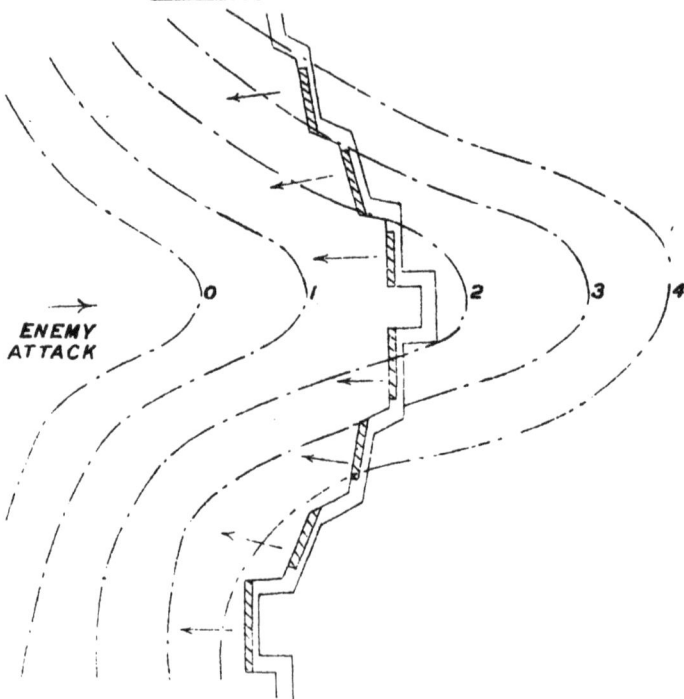

0   1   2   3   4

ENEMY
ATTACK

Malby & Sons. Lith

Plate 41.

FIG.I.   **NORMAL SECTION OF FIRE TRENCH.**

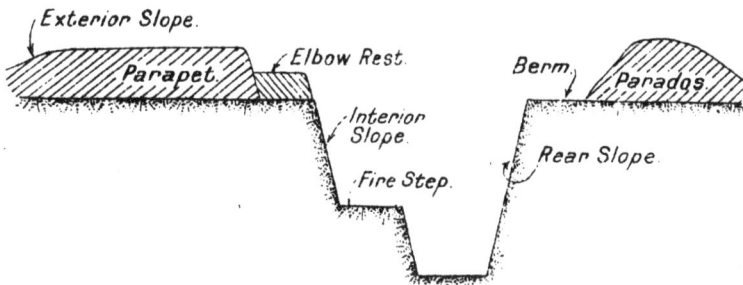

Exterior Slope.
Parapet.
Elbow Rest.
Berm.
Parados.
Interior Slope.
Fire Step.
Rear Slope.

FIG.2.   **ORDER OF WORK.**

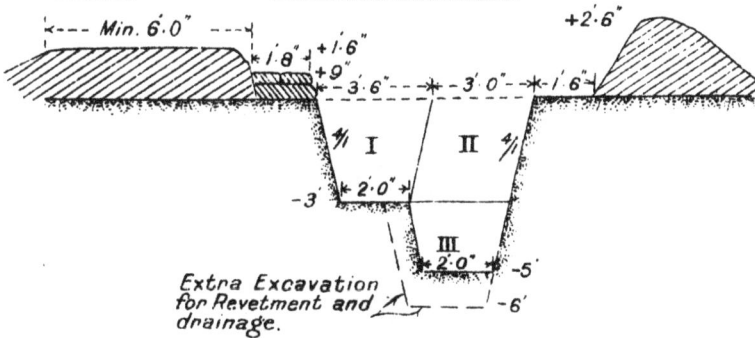

--- Min. 6ʹ0ʺ ---
1ʹ8ʺ
+1ʹ6ʺ
+9ʺ
+2ʹ6ʺ
3ʹ6ʺ
- 3ʹ0ʺ -
1ʹ6ʺ
I
II
-3ʹ
2ʹ0ʺ
III
2ʹ0ʺ
-5ʹ

Extra Excavation for Revetment and drainage.
- 6ʹ

FIG.3.   **FINAL SECTION OF TRENCH.**

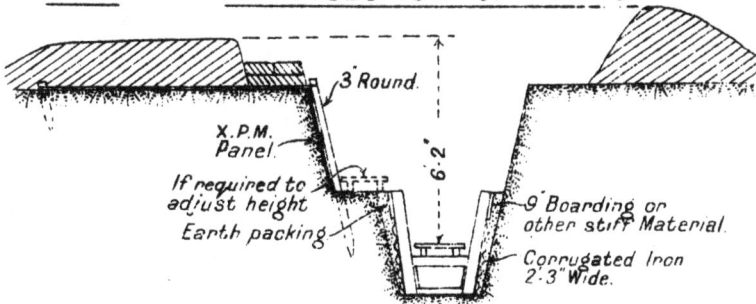

3ʺ Round.
X.P.M. Panel.
6ʹ2ʺ
If required to adjust height
Earth packing
9ʺ Boarding or other stiff Material.
Corrugated Iron 2ʹ3ʺ Wide.

Plate 42.

TRACE & SECTION OF COMMUNICATION TRENCH.

FIG.1.
TRACE

Corners rounded

Fixing Line

5 to 7 Yards

10 to 15 Yards

10 to 15 Yards

5 Feet

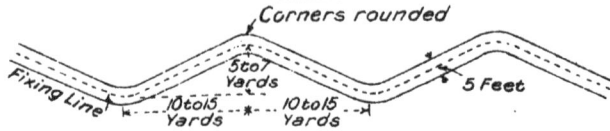

The bends in the trench must conform to the ground so as to get the best advantage in cover, but roughly the distance between bends should not be more than 15 yards up to support trenches, and not more than 10 yards between support and fire trenches.

FIG.2.
REVETTED SECTION

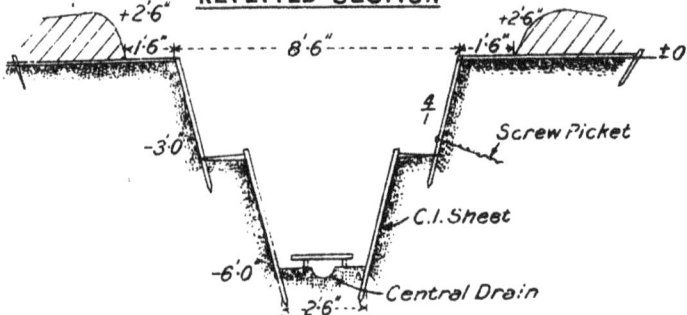

+2'6"

+1'6"

8'6"

+2'6"

+1'6"

±0

$\frac{4}{1}$

-3'0"

Screw Picket

C.I. Sheet

-6'0"

Central Drain

-2'6"

Note   In shifting & wet soils revetting frames as in Plate 41. Fig. 3. should be used.

FIG.3.
UNREVETTED SECTION

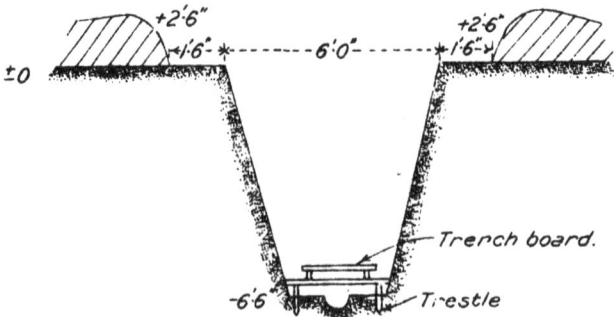

+2'6"

+1'6"

6'0"

+2'6"

+1'6"

±0

Trench board.

-6'6"

Trestle

8921. 11084. PP3115/3B7.   Malby & Sons. Lith

Plate 43.

METHODS OF DEFENCE OF COMMUNICATION TRENCHES.

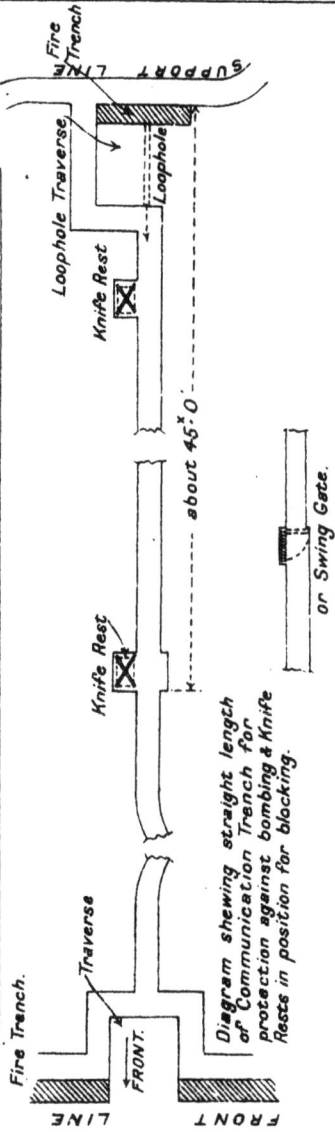

Fig. 1.

Fire Trench.
Traverse
FRONT.
FRONT LINE
Knife Rest
Knife Rest
Loophole Traverse
Loophole
Knife Rest
SUPPORT LINE
Fire Trench
FIRE LINE
about 45'·0"
or Swing Gate.

Diagram shewing straight length of Communication Trench for protection against bombing & Knife Rests in position for blocking.

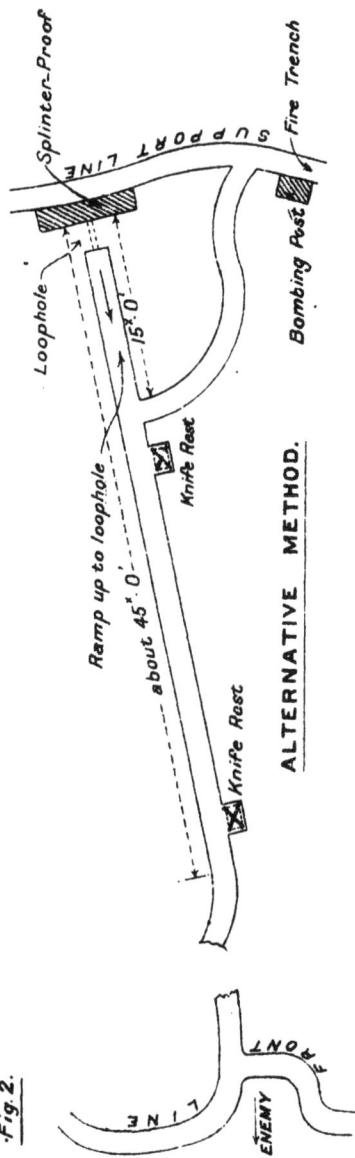

Fig. 2.

ENEMY
FRONT LINE
Knife Rest
about 45'·0"
Ramp up to loophole
Loophole
Knife Rest
15'·0"
Splinter-Proof
SUPPORT LINE
Bombing Post
Fire Trench

ALTERNATIVE METHOD.

Malby & Sons, Lith.

837l. 11084. p⁵ 3115/587.

Plate 44.

DEFENCE OF COMMUNICATION TRENCH BY "D" HEADS, "T" HEADS AND RECESSES.

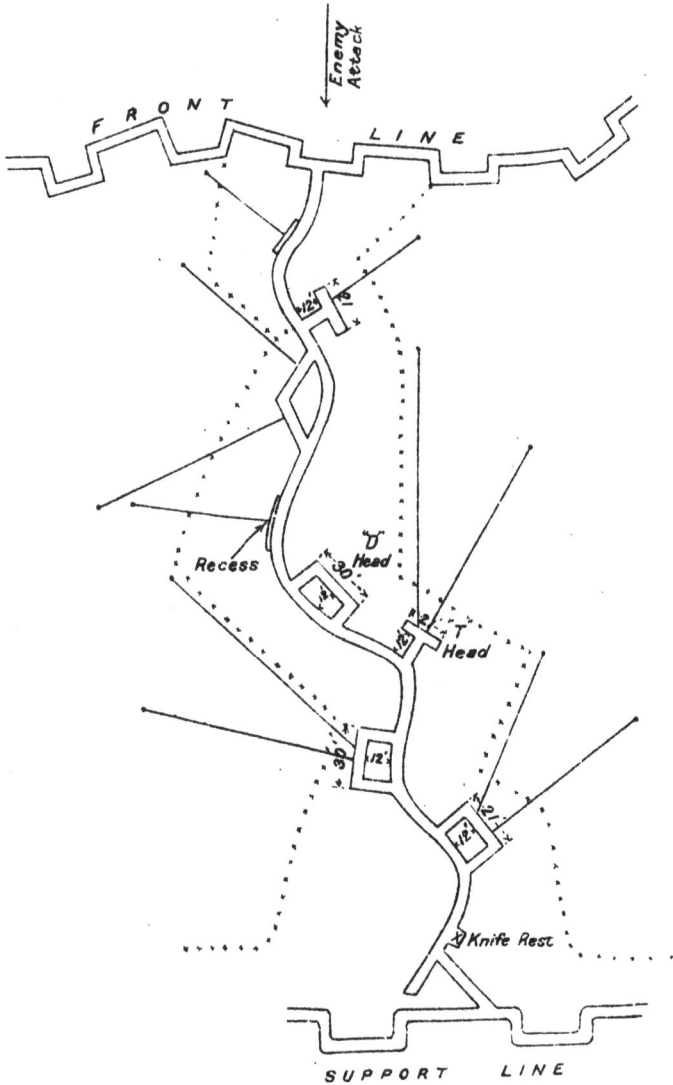

Plate 45.

FIG. I. SECTION OF BREASTWORK IN WET SOIL.

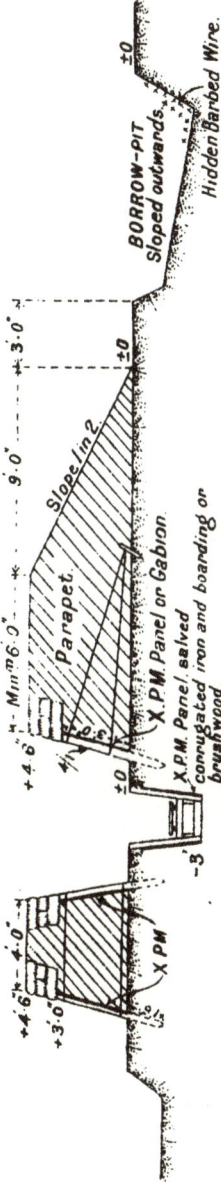

Parapet.

Slope 1 in 2.

BORROW-PIT
Sloped outwards.

Hidden Barbed Wire.

X.P.M. Panel or Gabion.

X.P.M. Panel solved
corrugated iron and boarding or
brushwood.

X.P.M.

Note:- Depth of bottom to vary according to ground. Set frames as low as possible
to save breastwork. Drainage at grade of 1/100 to lower ground essential.

+4.6" — 9.0"

Min. 6.0"

+4.6" — 4.0"

+3.0"

±0

+3.0"

±0

-3'

FIG. 2. COMMUNICATION TRENCH
CROSSING A FIRE TRENCH.

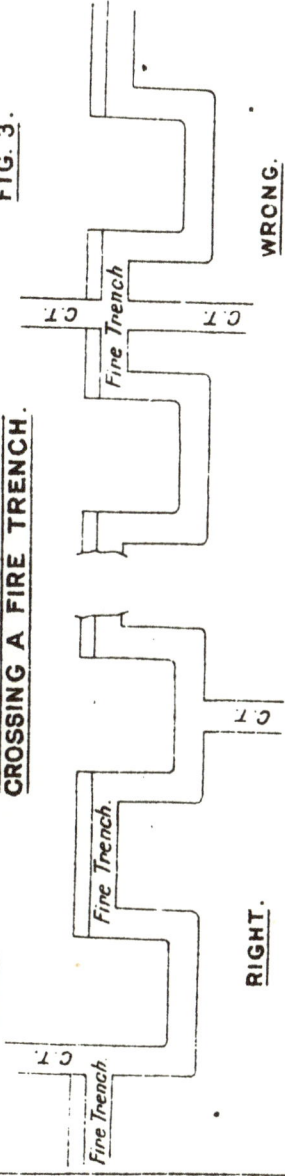

FIG. 3.

Fire Trench.

Fire Trench.

Fire Trench.

Fire Trench.

C.T.

C.T.

C.T.

C.T.

RIGHT.

WRONG.

Plate 46

SHELL SLITS.

TO GIVE QUICK IMPROVISED SHELTER.

UNSTRUTTED.

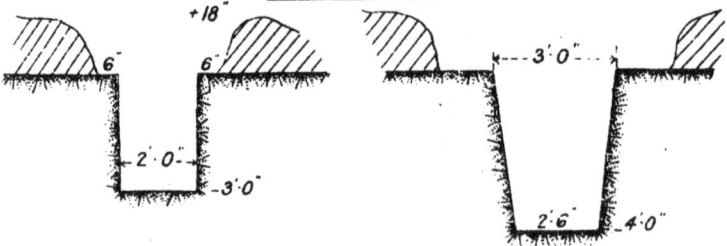

STRUTTED.

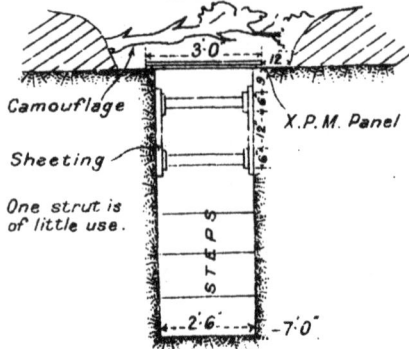

Camouflage

X.P.M. Panel

Sheeting

One strut is
of little use.

STEPS

PLAN OF STRUTTED TRENCH.

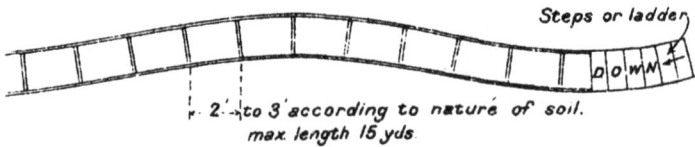

Steps or ladder

DOWN

2' to 3' according to nature of soil.
max. length 15 yds.

Malby & Sons. Lith.

Plate 47.

SAPPING.

FIG. I.

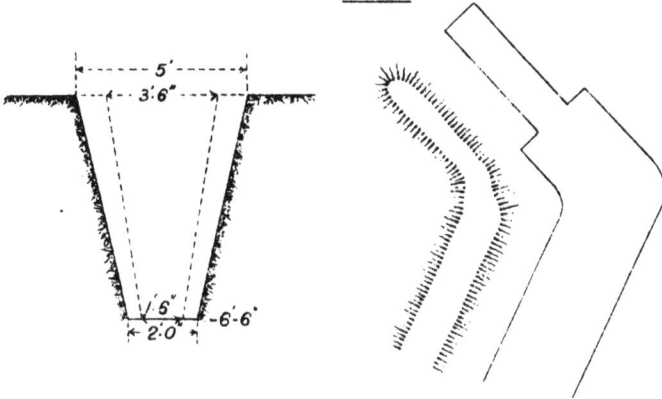

5'
3'6"
6"
2'0"
6'6"

FIG. 2.

RUSSIAN SAP.

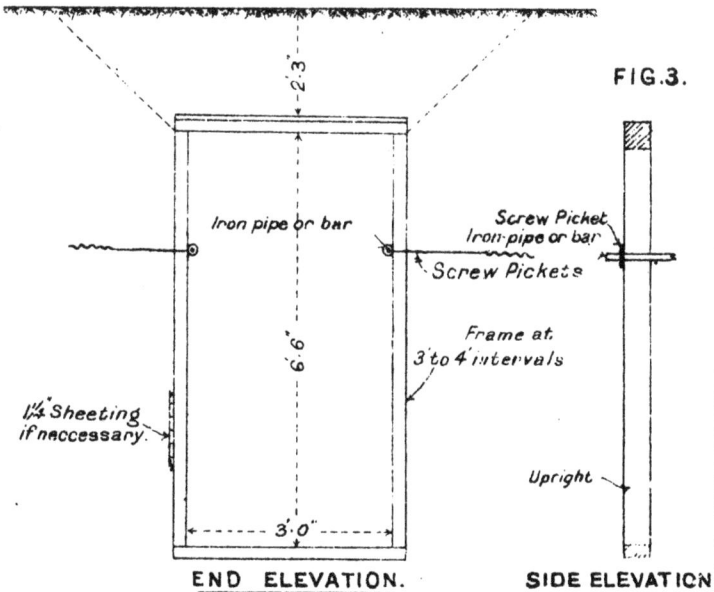

2'3"

FIG. 3.

Iron pipe or bar

Screw Picket
Iron-pipe or bar

Screw Pickets

6'6"

Frame at
3 to 4 intervals

1¼" Sheeting
if neccessary.

Upright

3'0"

END ELEVATION.

SIDE ELEVATION.

Plate 48.

TYPE OF SUMP.

SECTION.

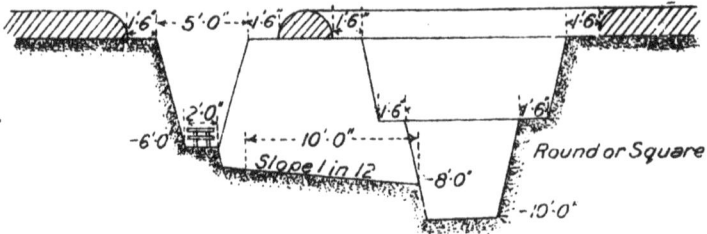

Round or Square

PLAN

Sump consists of a circular or square hole revetted if necessary

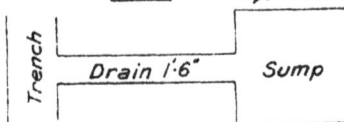

Trench

Drain 1·6"   Sump

Note:- Sump should never be put in where drainage out to lower ground can be arranged.

Drain formed of corrugated iron sheets (a).
Or boxed-lined with wooden box (b).

WRONG

Drains of the Types (a) & (b) can be used when a trench is sited on sloping ground sufficient to ensure the flow of water as below.

Box Drain slope 1 in 12

Plate 49.

DEFENCE OF FIRE TRENCH AGAINST BOMBING.

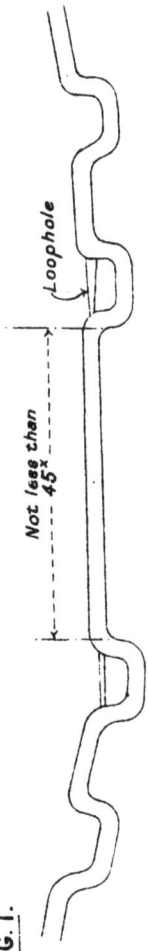

Not less than 45ˣ

Loophole

FIG. I.

TRAVERSING AN UNTRAVERSED TRENCH.

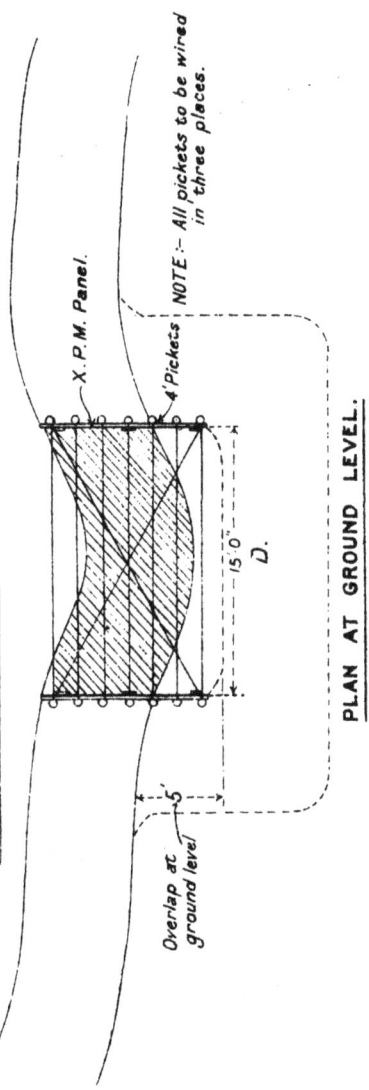

X.P.M. Panel.

4' Pickets

NOTE :- All pickets to be wired in three places.

15'0"

D.

PLAN AT GROUND LEVEL.

Overlap at ground level

5'

FIG. 2.

Malby & Sons Lith

8921 11084 P•3115/387

Plate 50.

BRIDGE TRAVERSE.

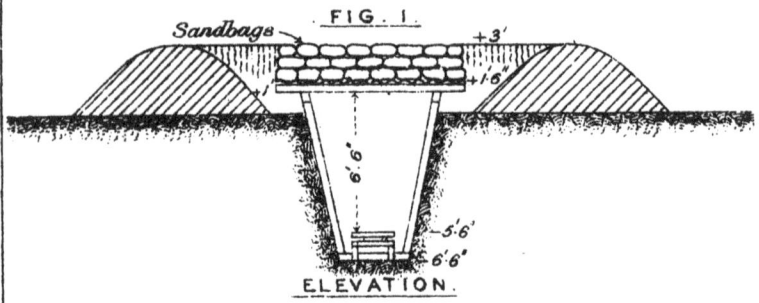

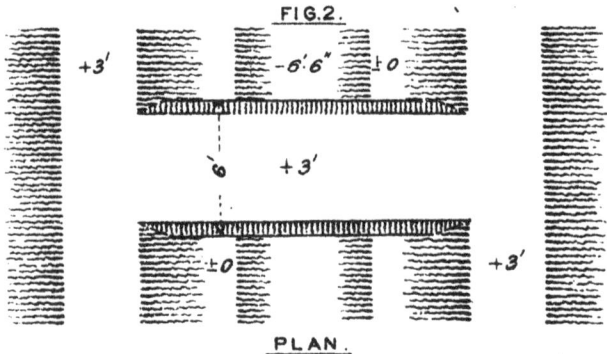

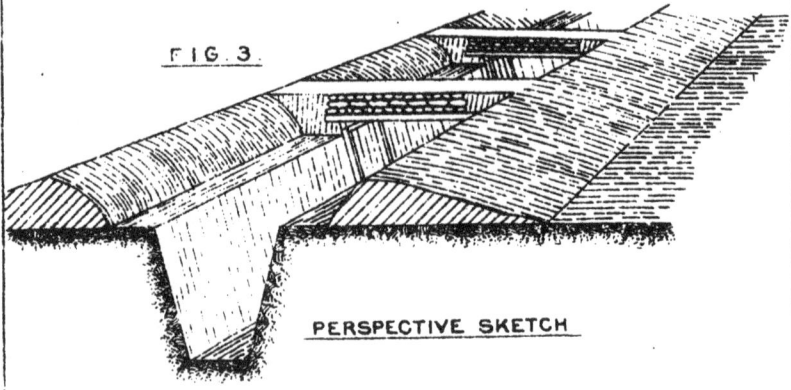

FIG. 1.

Sandbags

+3'

+1'6

6'6"

5'6'
6'6"

ELEVATION.

FIG. 2.

+3'

-6'6" ±0

6'

+3'

±0

+3'

PLAN.

FIG. 3.

PERSPECTIVE SKETCH.

Plate 51.

LOOPHOLES.

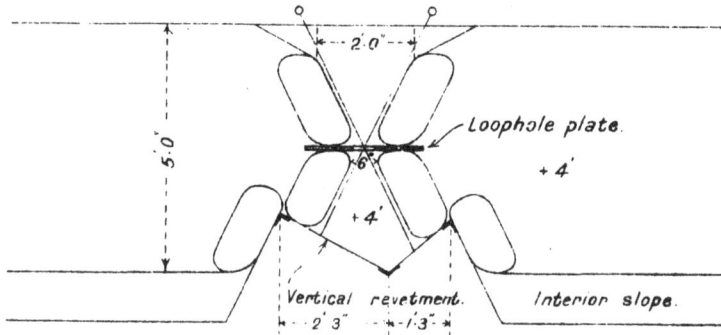

Exterior slope.

2' 0"

Loophole plate.

+ 4'

5' 0'

·6'

+ 4'

Vertical revetment.

Interior slope.

2' 3"   1' 3"

PLAN AT LEVEL 4 FEET SHOWING
FIRST LAYER OF SANDBAGS FORMING LOOPHOLE
FIG. I.

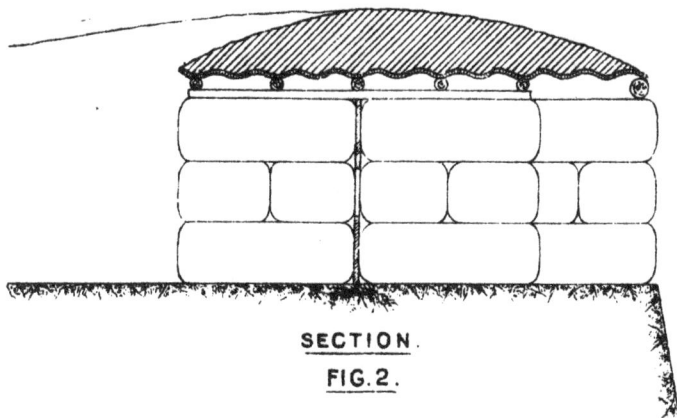

SECTION.
FIG. 2.

B.921.11084. P.R. 3115/387

Malby & Sons Lith.

Plate 52.

LOOPHOLE PLATE.

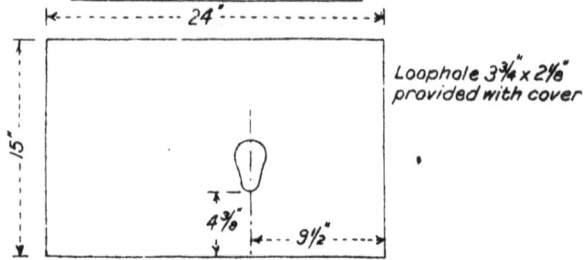

24"

15"

4⅜"   9½"

Loophole 3¾" x 2⅛"
provided with cover

LOOPHOLE BOX.

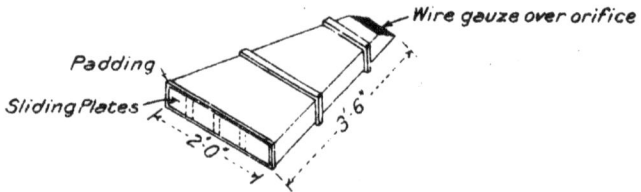

Wire gauze over orifice

Padding

Sliding Plates

2'0"   3'6"

MOCK SANDBAG CEMENT LOOPHOLE.
FOR OBSERVATION

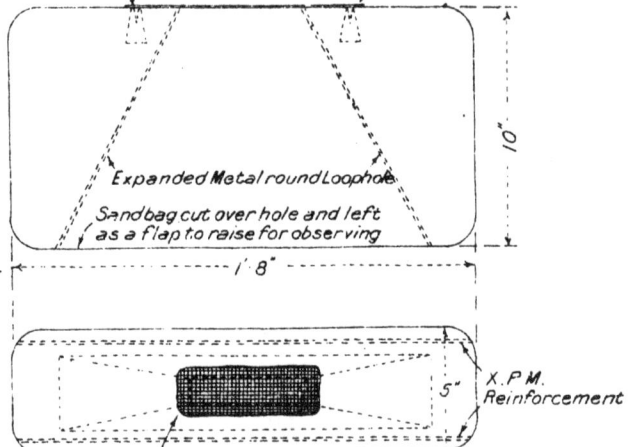

Gauze tacked over and sandbag to
be cut out over hole

Expanded Metal round Loophole

Sandbag cut over hole and left
as a flap to raise for observing

10"

1'8"

5"   X.P.M.
Reinforcement

Wire gauze tacked on under sandbag and painted same colour as sandbag
FINISHED ARTICLE

Plate 53.

SNIPER'S POST.

SECTION.

VIEW FROM FRONT.

When front parapet is covered with tins of all kinds, the tin used to disguise the loophole is very difficult to identify even at 10 yds. range.

OBLIQUE LOOPHOLE.

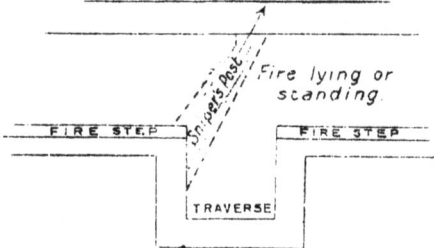

Plenty of Dummy Loopholes should be provided.

8921. 1084 PP9115/387   Maloy & Sons Lith

Plate 54

SPLINTER AND BULLET-PROOF
OBSERVATION POST.

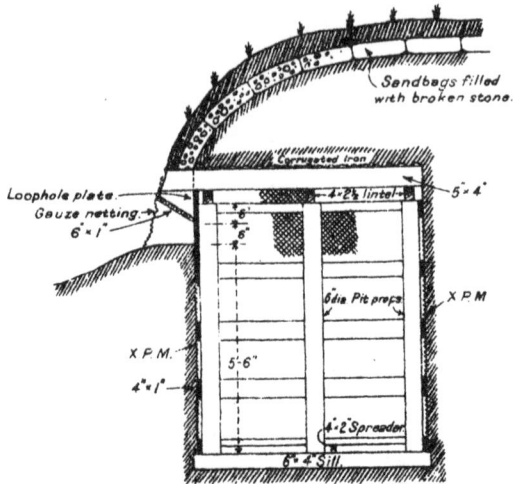

Sandbags filled with broken stone.

Corrugated Iron

Loophole plate

Gauze netting 6"×1"

4·2½ lintel — 5"×4"

6dia Pit props — X P M

X P M.

4"×1"

5'-6"

4·2 Spreader

6"×4" Sill

SECTION.

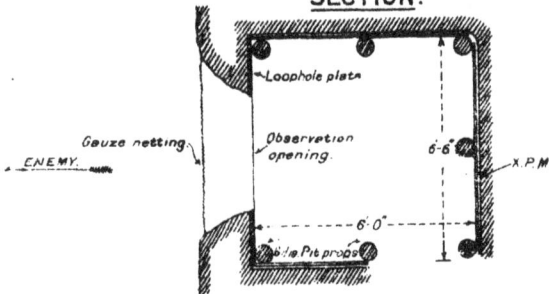

Loophole plate

Gauze netting

ENEMY.

Observation opening.

6'-6"

X. P. M.

6'-0"

6 dia Pit props

PLAN.

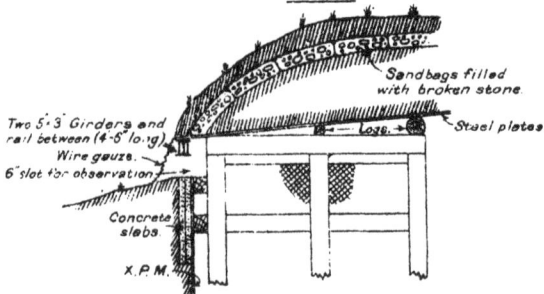

Sandbags filled with broken stone.

Two 5·3 Girders and rail between (4·5 long)

Wire gauze.

6" slot for observation.

Loss. — Steel plates

Concrete slabs

X. P. M.

ALTERNATIVE HEAD PIECE.

8421 1084 PP.2115/287

Malby & Sons. Lith

Plate 55.

SPLINTER-PROOF OBSERVATION POST.

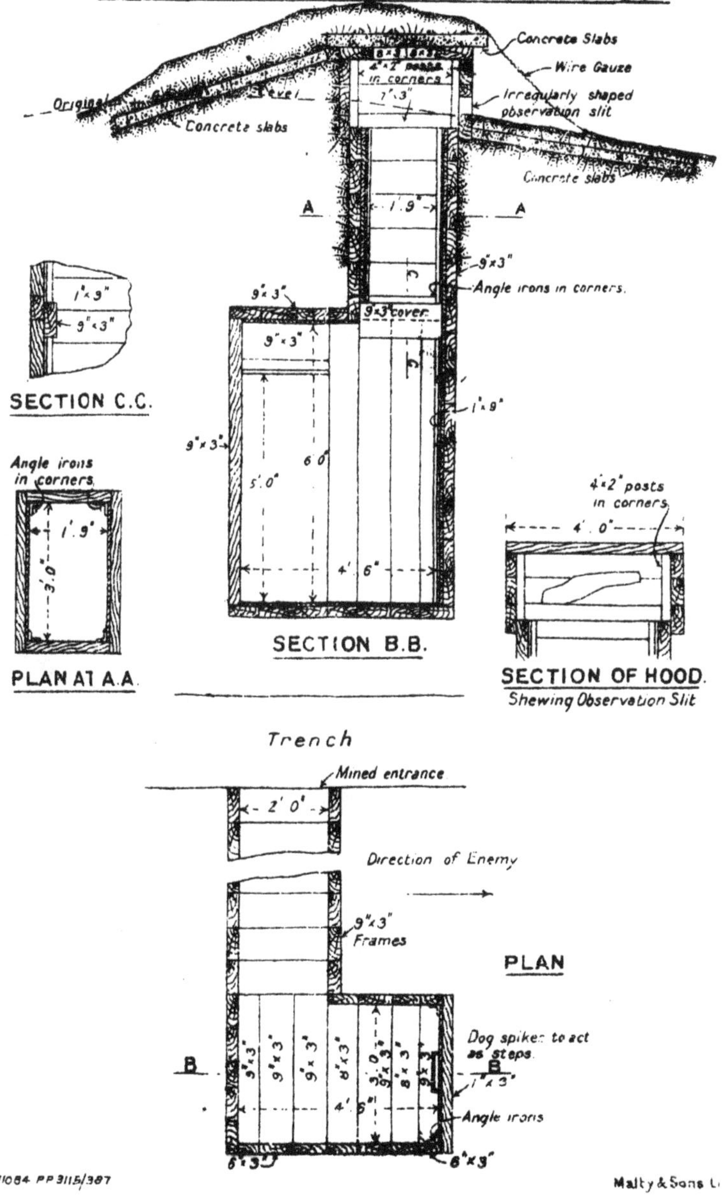

Concrete Slabs

Wire Gauze

4"×2" posts in corners

Irregularly shaped observation slit

Original

Concrete slabs

1'.3"

Concrete slabs

1'.9"

9"×3"

Angle irons in corners.

A — A

SECTION C.C.

1'×9"

9"×3"

PLAN AT A.A.

Angle irons in corners.

1'.9"

3'.0"

9"×3"

9"×3"

9"×3" cover.

5"

6'.0"

5'.0"

4'.6"

1'×9"

SECTION B.B.

SECTION OF HOOD.

Shewing Observation Slit

4"×2" posts in corners

4'.0"

Trench

Mined entrance

2'.0"

Direction of Enemy

9"×3" Frames

PLAN

Dog spikes to act as steps.

1'×3"

Angle irons

B — B

9"×3"   9"×3"   9"×3"   9"×3"   9"×3"   9"×3"   9"×3"

3'.0"

4'.6"

6'.3"

6'.3"

6"×3"

8921 11084 PP 3115/387

Malby & Sons L.

Plate 56.

METHOD OF SLOPING THE SIDES OF A TRENCH.

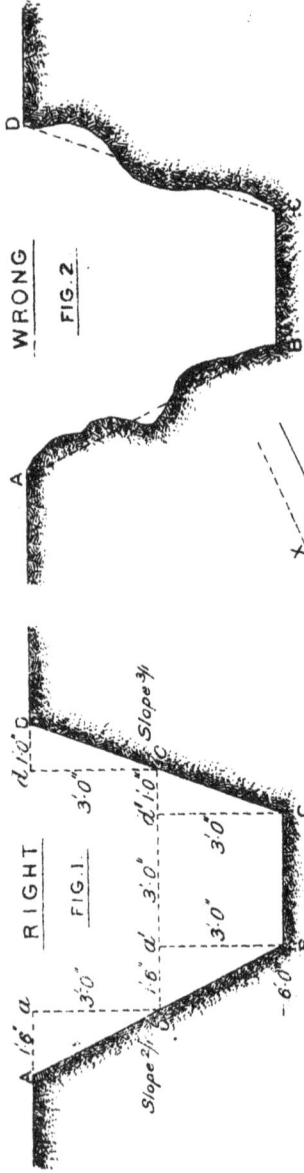

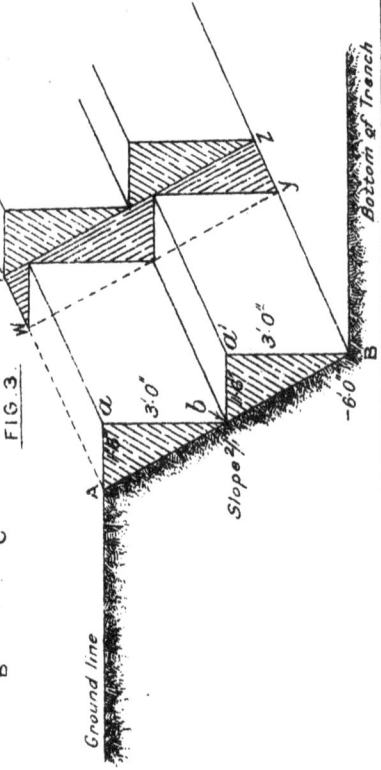

RIGHT   FIG. 1.

WRONG   FIG. 2.

FIG. 3.

Ground line

Bottom of Trench

Plate 57.

ANCHORAGES.

FIG. 1
BREASTWORK.

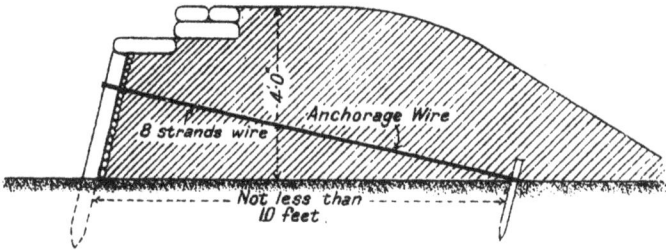

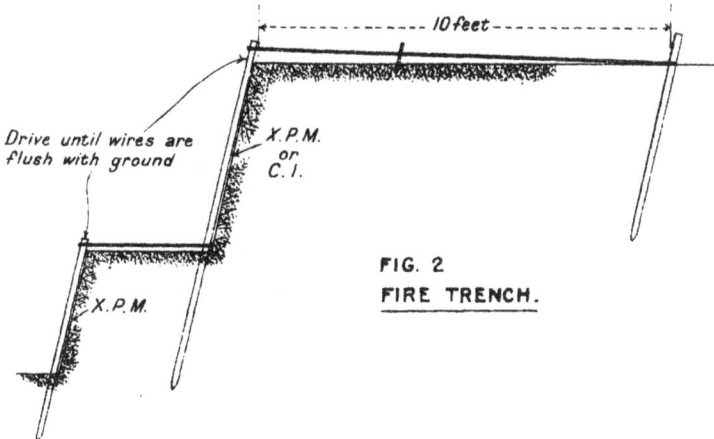

4.0"

Anchorage Wire

8 strands wire

Not less than
10 feet

10 feet

Drive until wires are
flush with ground

X.P.M.
or
C.I.

FIG. 2
FIRE TRENCH.

X.P.M.

11084. PP.3115/387

Plate 58.

SANDBAG REVETMENT.

ELEVATION

SECTION

Correct
English Bond

Correct Section

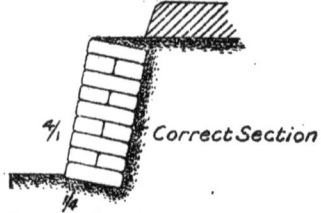

Foundation should be cut at
right angles to slope and always
brought to a solid bottom.

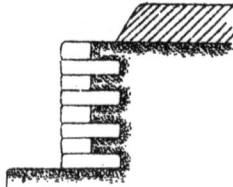

Parapet

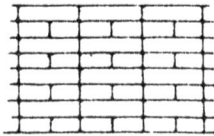

Wrong (Joints not Broken)

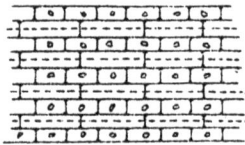

Wrong (Vertical)

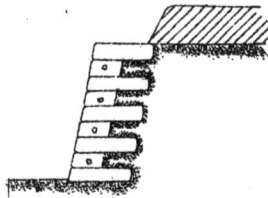

Wrong (Seams and Choked
Ends of Bags outward).

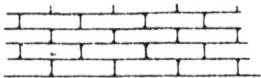

Wrong (Bags not at Right Angles
to Slope).

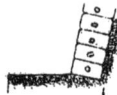

Wrong (All Stretchers and no Headers).

Maltby & Sons. Lith

Plate 59.

ADAPTATION OF HEDGES AND WALLS.

FIG. I.

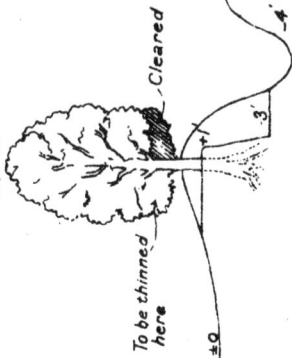

To be thinned here

Cleared

±0

+

−3′

FIG. 2.

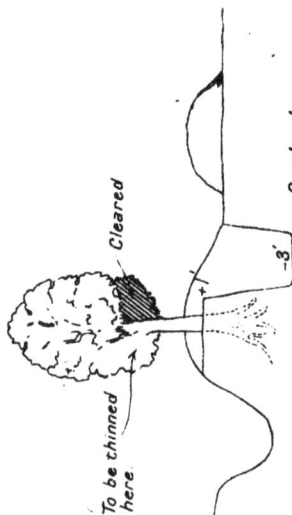

To be thinned here

Cleared

+

−3′

FIG. 3.

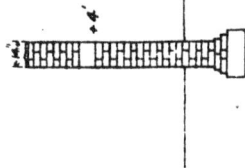

+4′

FIG. 4.

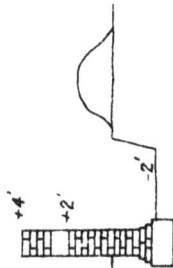

+4′

+2′

−2′

FIG. 5.

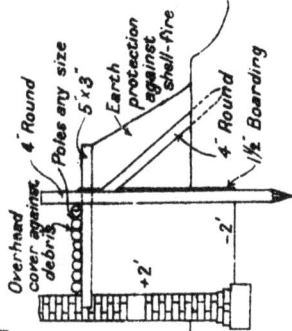

Overhead cover against debris

4″ Round

Poles any size

5″×3″

Earth protection against shell-fire

4″ Round

½″ Boarding

+2′

−2′

Plate 60.

ADAPTATION OF CUTTINGS AND EMBANKMENTS.

EMBANKMENT.

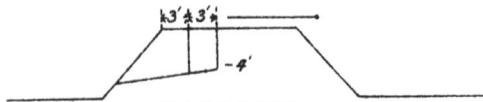

SECTION.

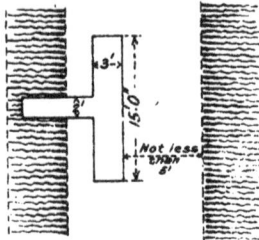

PLAN.

CUTTING.

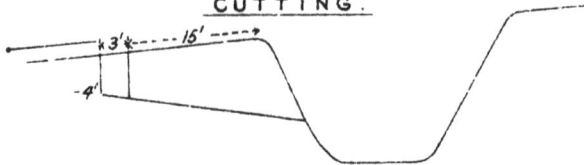

SECTION.

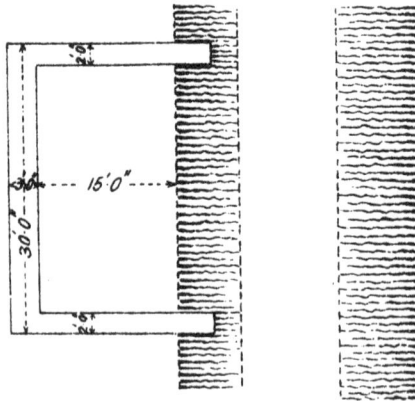

PLAN.

Plate 61.

CORRUGATED IRON BLOCK HOUSE.
(CIRCULAR)

FIG. 1. PERSPECTIVE VIEW.

FIG. 2. SECTION.

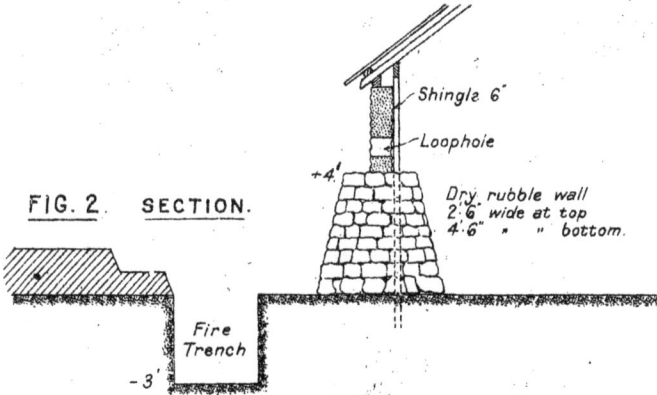

Shingle 6″

Loophole

+4′

Dry rubble wall
2′6″ wide at top
4′6″ ″ ″ bottom.

Fire Trench

-3′

FIG. 3. PLAN.

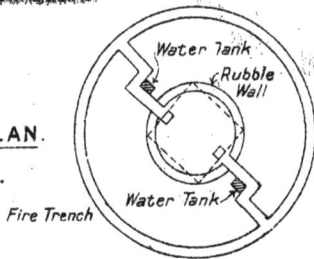

Water Tank

Rubble Wall

Water Tank

Fire Trench

8921.11084. PP 3115/987.

Malby & Sons. Lith.

Plate 62.

LOG BLOCK HOUSE.

Plate 63.

FIG. I.

Bamboo Stockade.
(bayonet pattern) Lushai '89.

FIG. 2

Afghan sangar.

8921 11084. PP.3115/387

Malby & Sons, Lith

Plate 64.

IMPROVEMENTS TO SHELL-HOLES.

FIG. I.

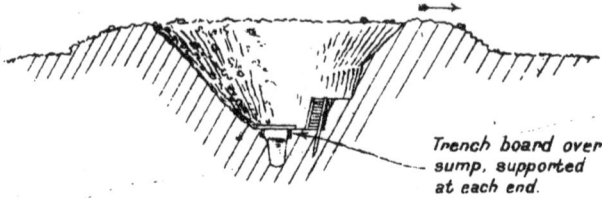

Trench board over sump, supported at each end.

FIG. 2. **SMALL SHELL-HOLES OCCUPIED AND LARGE ONE USED AS SUMP.**

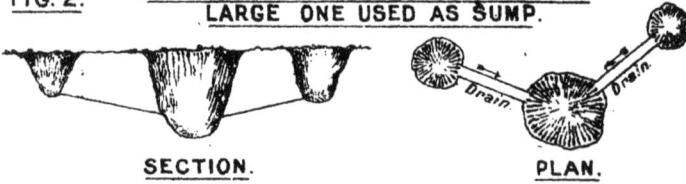

SECTION.   PLAN.

FIG. 3. **NO EXCAVATED EARTH TO SHOW ABOVE GROUND LEVEL.**

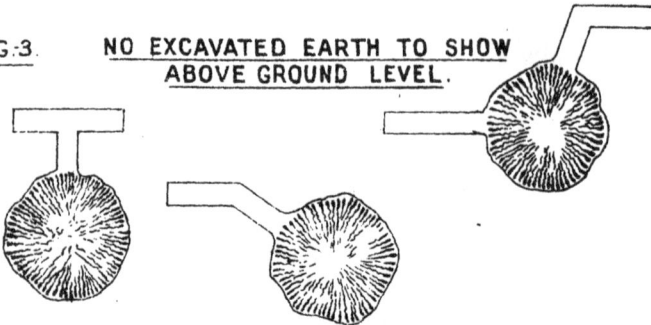

FIG. 4. **FRONT FACE CUT AWAY TO FORM FIRE STEP.**

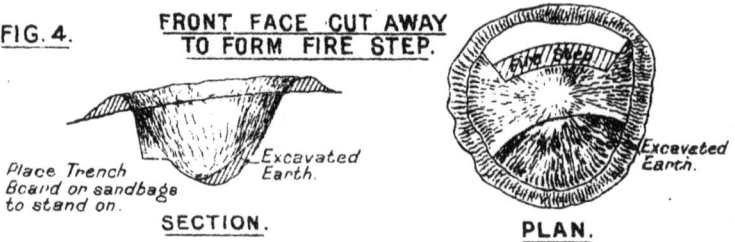

Place Trench Board or sandbags to stand on.

Excavated Earth.

Excavated Earth.

SECTION.   PLAN.

8921.11084. PP3115/987.   Malby & Sons. Lith.

Plate 65.

TWO SHELL-HOLES WITH CAMOUFLAGED CONNECTION.

Fig. 1.

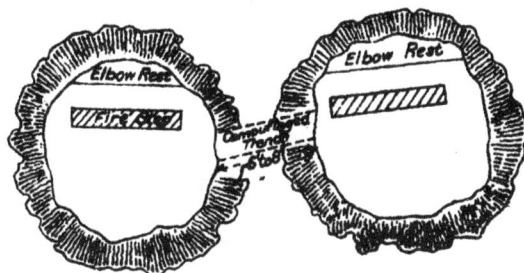

Elbow Rest

Elbow Rest

Camouflaged Stop

PLAN

Fig. 2. SHELL-HOLE POSITION

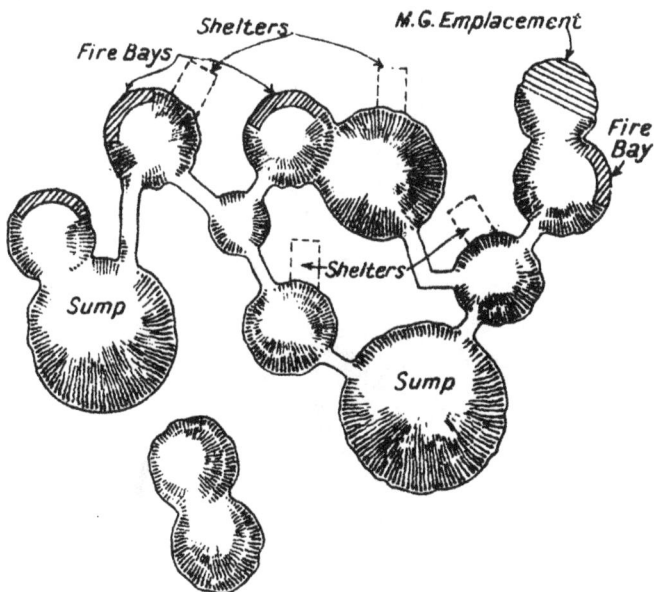

Fire Bays

Shelters

M.G. Emplacement

Fire Bay

Sump

Shelters

Sump

Malby & Sons. Lith.

Plate 66.

IMPROVEMENTS TO SHELL-HOLES.

FIG.I.

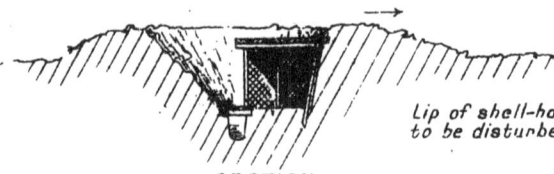

Lip of shell-hole not
to be disturbed.

SECTION.

Weather-proof cover
for Lewis gun and crew.
Gun fires from top
of shelter.

All excavated earth
to be dumped in
neighbouring holes.

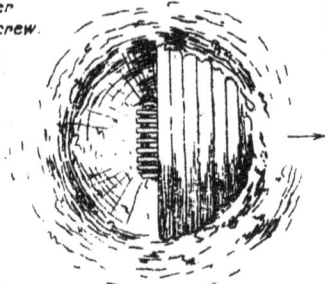

PLAN.

FIG.2.

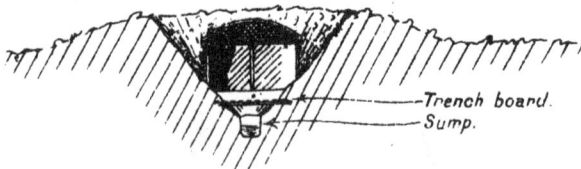

Trench board.
Sump.

Showing shelter let into side of
shell-hole. roof of curved sheets of C.I.

Malby & Sons. Lith.

Plate 67.

DEFENCE OF CRATERS.

NEAR LIP DEFENDED.

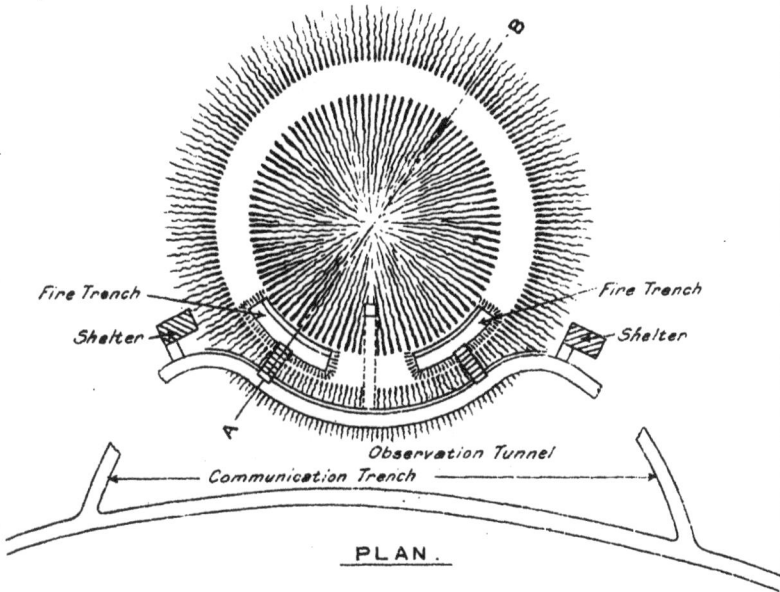

B

Fire Trench

Shelter

Fire Trench

Shelter

A

Observation Tunnel

Communication Trench

PLAN.

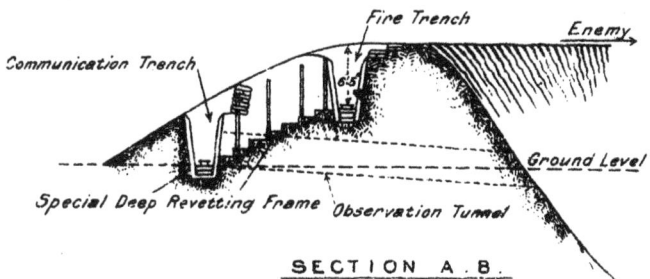

Fire Trench

Enemy

Communication Trench

Ground Level

Special Deep Revetting Frame

Observation Tunnel

SECTION A.B.

8921. 11084. PP. 3116/887

Malby & Sons. Lith.

Plate 68.

DEFENCE OF CRATERS.

BY BOMBING TRENCHES WHERE THERE IS NO FIELD OF FIRE.

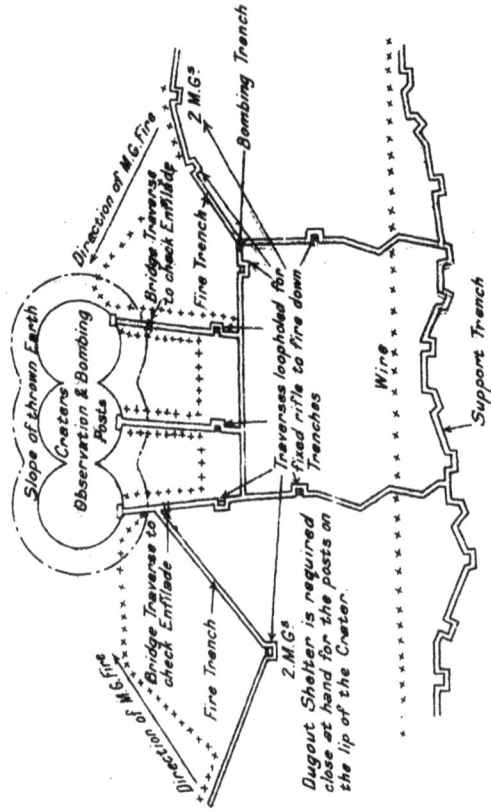

Plate 69.

SPECIAL DEEP REVETTING FRAMES.

ELEVATION.   SECTION.

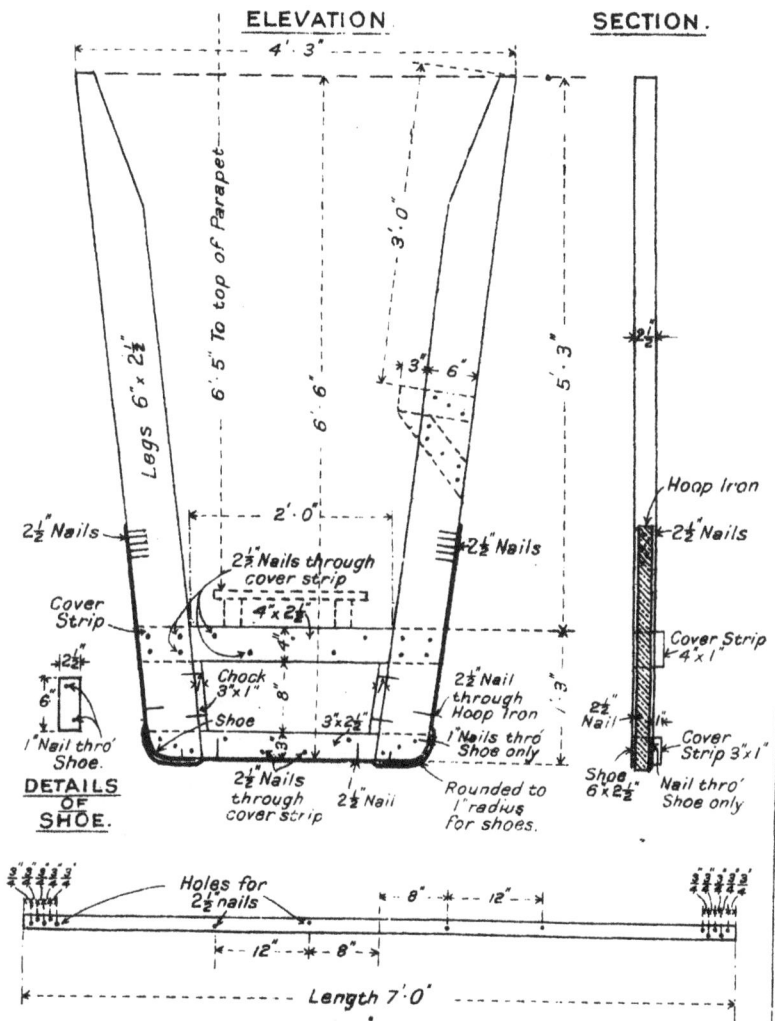

4' 3"

Legs 6" × 2½"

6' 5" To top of Parapet

6' 6"

3' 0"

3" 6"

5' 3"

2½" Nails

2½" Nails

Cover Strip

2½" Nails through cover strip

4" × 2½"

2' 0"

Chock 3" × 1"

Shoe

3" × 2½"

2½" Nail through Hoop Iron

1" Nails thro' Shoe only

Hoop Iron

2½" Nails

Cover Strip 4" × 1"

2½" Nail

Cover Strip 3" × 1"

Shoe 6" × 2½"

Nail thro' Shoe only

2½"

6"

1" Nail thro' Shoe.

DETAILS OF SHOE.

2½" Nails through cover strip

2½" Nail

Rounded to 1" radius for shoes.

Holes for 2½" nails

8"   12"

12"   8"

Length 7' 0"

DETAILS OF HOOP IRON.

Plate 70

GAS - PROOF CURTAINS.

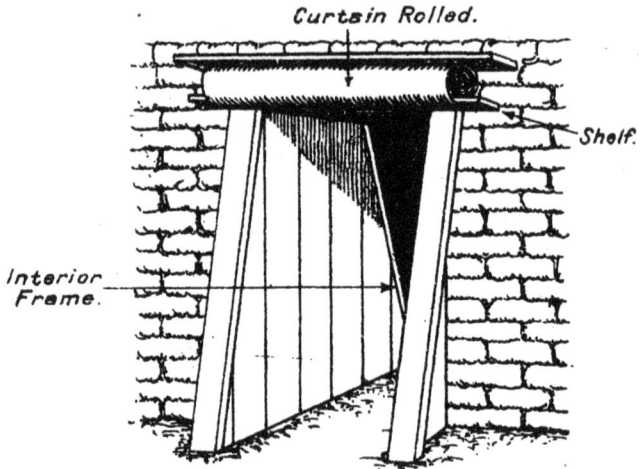

Curtain Rolled.

Shelf.

Interior Frame.

FIG.1. Open.

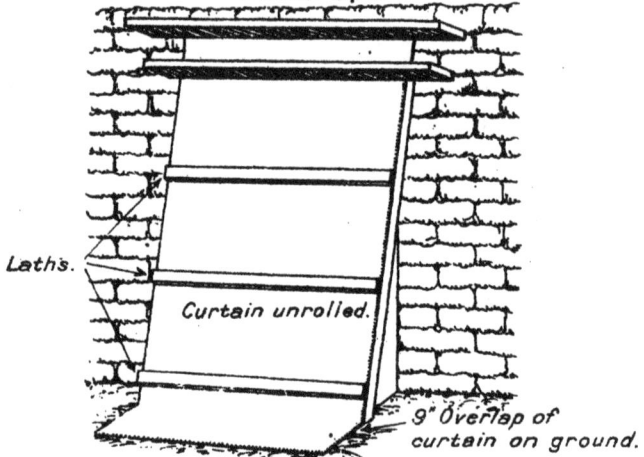

Laths.

Curtain unrolled.

9" Overlap of curtain on ground.

FIG.2. Closed.

2921.11084.PP3115/387

Malby & Sons. Lith.

Plate 71

GAS-PROOF CURTAINS.

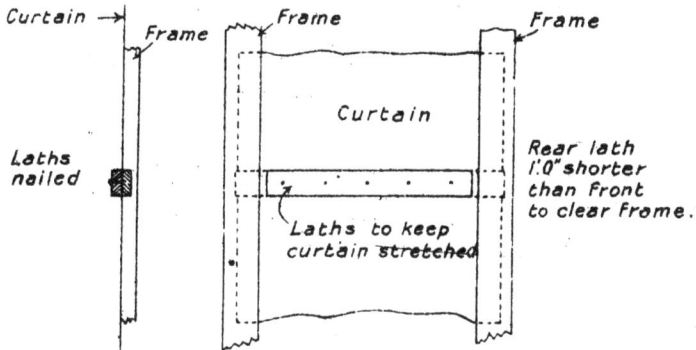

Curtain →

Frame

Frame

Frame

Laths
nailed

Curtain

Rear lath
1:0" shorter
than front
to clear frame.

Laths to keep
curtain stretched

FIG. I.
DETAILS OF LATHS KEEPING CURTAIN
STRETCHED.

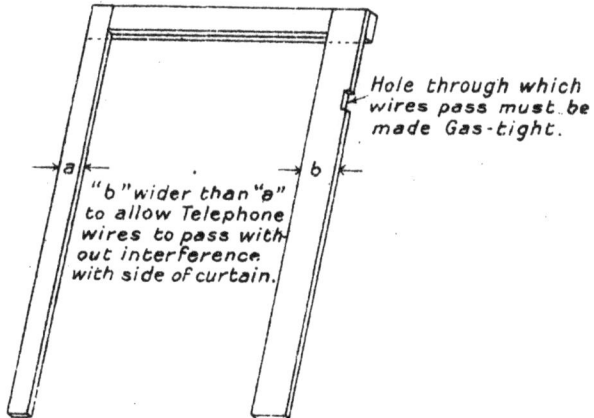

Hole through which
wires pass must be
made Gas-tight.

a

b

"b" wider than "a"
to allow Telephone
wires to pass with-
out interference
with side of curtain.

FIG. 2.
PERSPECTIVE OF FRAME WHEN TELEPHONE
WIRES PASS ALONG GALLERY.

8921.11084.PP3115/987

Malby & Sons, Lith.

Plate 72

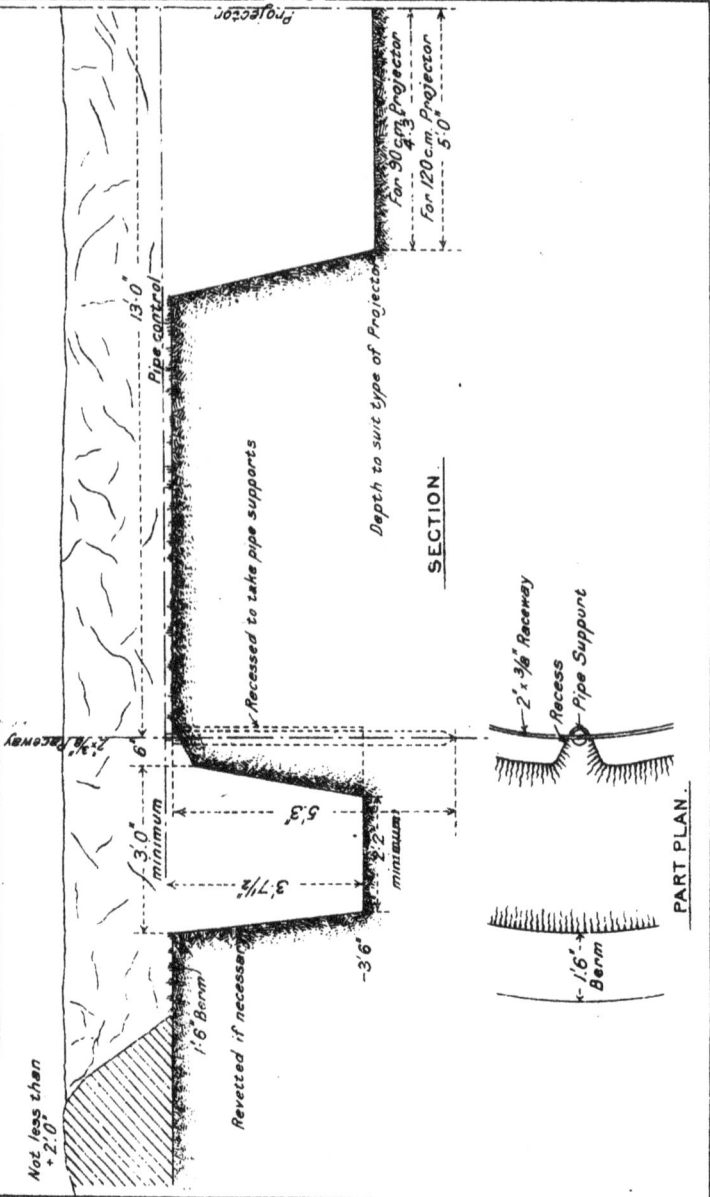

TYPE OF SEARCHLIGHT EMPLACEMENT.
SUITABLE FOR SKEW GEAR PIPE CONTROL WITH TELESCOPE.

Projector

Projector

For 90 c.m. Projector — 4'.3"
For 120 c.m. Projector — 5'.0"

Depth to suit type of Projector

SECTION

13'.0"

Pipe control

Recessed to take pipe supports

2'.6" Raceway

3'.0" minimum

5'.3"

3'.7½"

2'.2" minimum

3'.6"

1'.6" Berm

Revetted if necessary

Not less than
+ 2'.0"

2' x 3/8" Raceway
Recess
Pipe Support

1'.6" Berm

PART PLAN.

Malby & Sons. Lith

Plate 73

B.C. POST.

Speaking Tubes

Map Table

Bed

Telephonist's Table

PLAN.   SECTION.   FIG.1.

Speaking Tubes

Map Table

Bed

9'6"

10'3"

9'6"

3'6"

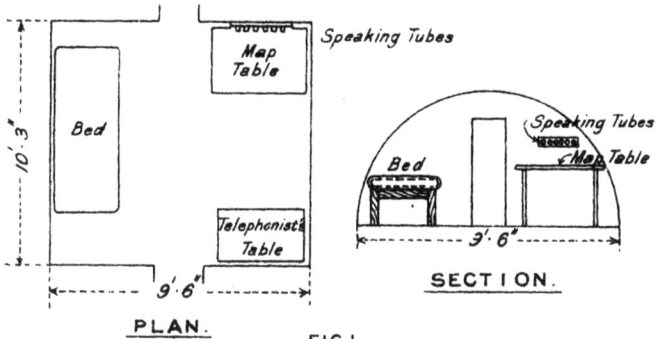

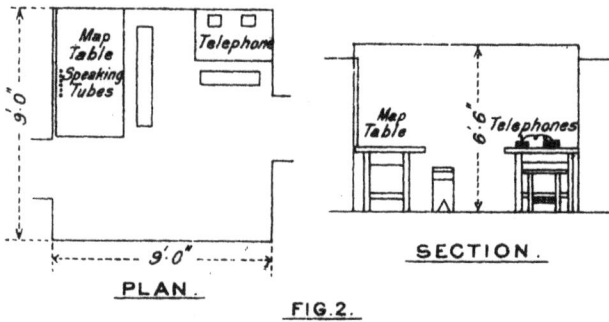

TELEPHONE DUG-OUT.

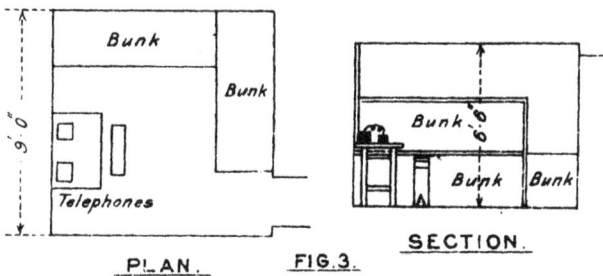

Map Table
Speaking Tubes
Telephone

PLAN.   SECTION.   FIG.2.

Map Table   Telephones

9'0"   9'0"   6'6"

Bunk

Bunk

Telephones

PLAN.   FIG.3.   SECTION.

Bunk   Bunk   Bunk

9'0"   6'6"

Plate 74

18-POUNDER AMMUNITION SHELTER.
(FOR DRY GROUND.)

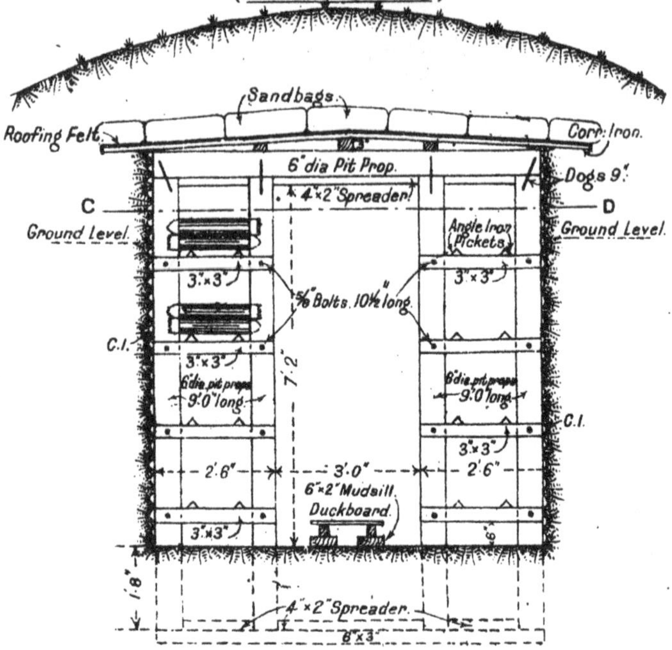

Sandbags.

Roofing Felt. Corr: Iron.

6" dia Pit Prop.

Dogs 9".

4"×2" Spreader.

C — — — — — — — — D

Ground Level. Angle Iron Pickets. Ground Level.

3"×3" 3"×3"

⅝ Bolts. 10½" long.

7'.2"

C.I. 6"dia pit props. 9'.0" long. 6"dia pit props. 9'.0" long. C.I.

3"×3" 3"×3"

— 2'.6" — — 3'.0" — — 2'.6" —

6"×2" Mudsill.

Duckboard.

3"×3"

1'.8"

4"×2" Spreader.

8'×3'

SECTION A.B.

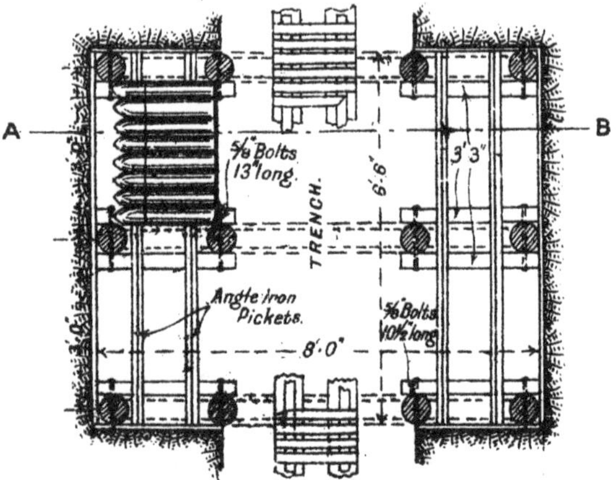

A — — — — — — — — B

⅝ Bolts 13" long.

6'.6" 3'.3"

TRENCH.

Angle Iron Pickets. ⅝ Bolts 10½" long.

— 8'.0" —

PLAN AT C.D.

Plate 75

18-POUNDER AMMUNITION SHELTER.
(FOR WET GROUND)

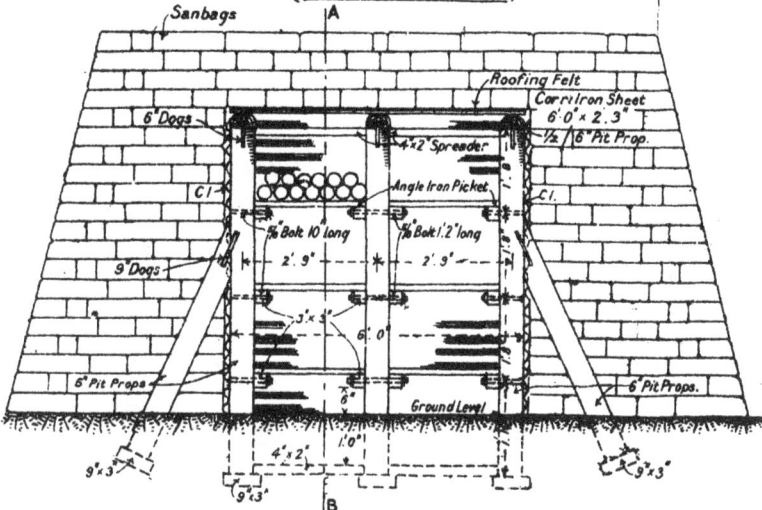

Sanbags

Roofing Felt
Corr: Iron Sheet
6'.0" × 2.3"

6" Dogs
4"×2" Spreader
½ 6" Pit Prop.

C.I.
Angle Iron Picket
C.I.

⅝ Bolt 10 long
⅝ Bolt 1.2' long

2'. 9"
2'. 9"

9" Dogs

3"× 3"
6'. 0"

6" Pit Props
6" Pit Props.

4"
6"

Ground Level

4'. 2"
1'.0"

9"× 3
9"×3"

9'.3'

FRONT ELEVATION.

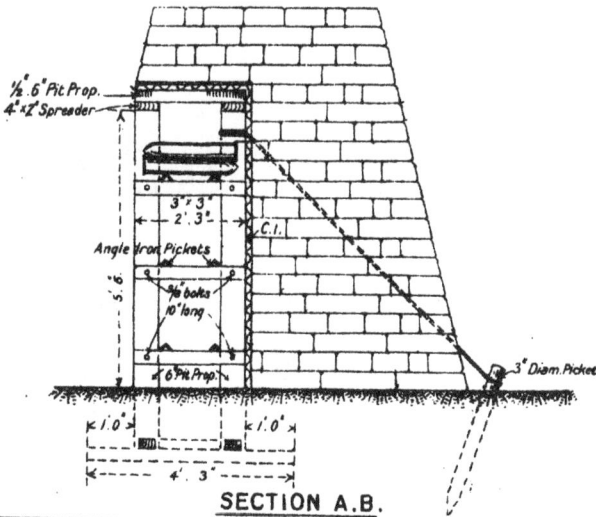

½ 6" Pit Prop.
4"×2" Spreader

3'. 3"
2'. 3"

Angle Iron Pickets
C.I.

⅝ bolts
10' long

5'. 6"

6" Pit Prop.
3" Diam. Picket

1'.0"
1'.0"

4'. 3"

SECTION A.B.

Plate 76

SPLINTER-PROOF SHELTER FOR AMMUNITION.

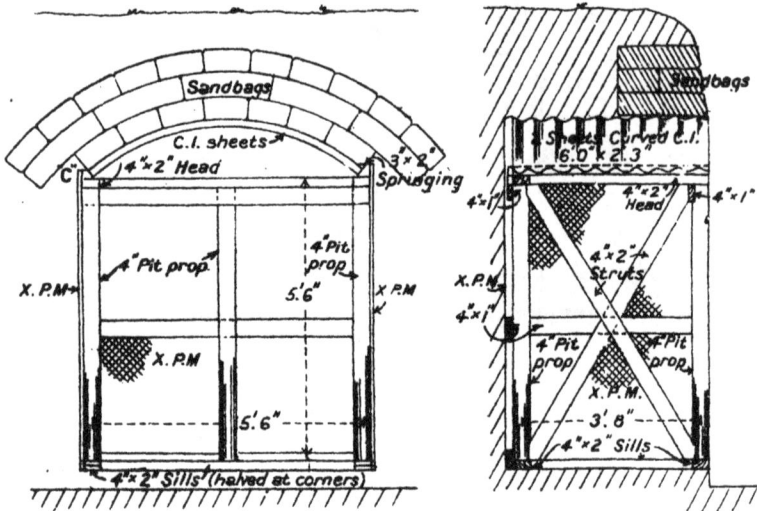

FRONT ELEVATION

SECTION

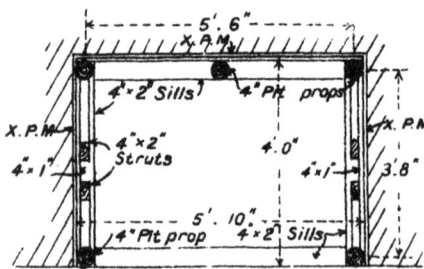

PLAN

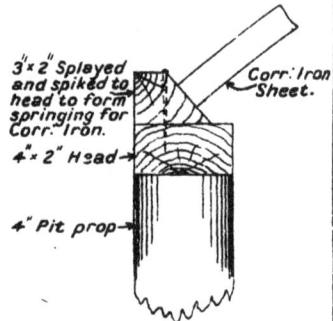

DETAIL "C"

Shelves for ammunition etc:
must not be supported on the framework of the shelter.

Malby & Sons. Lith.

EMPLAC

These Dimensions are sugg
the size of Emplacement
case will vary with the gr
Dimensions given are gove

FIG. I.

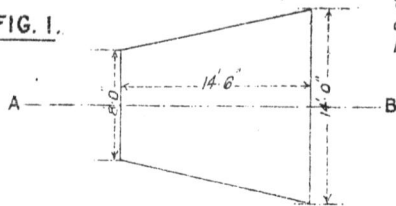

A ——— B

14' 6"

PLAN

Allows 80° switch.

FIG.2.

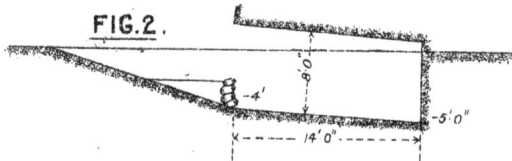

8'·0"

−4'

−5'·0"

14'·0"

SECTION AB

18 PᴰR

60 PᴰR MK. II.

FIG. 3.

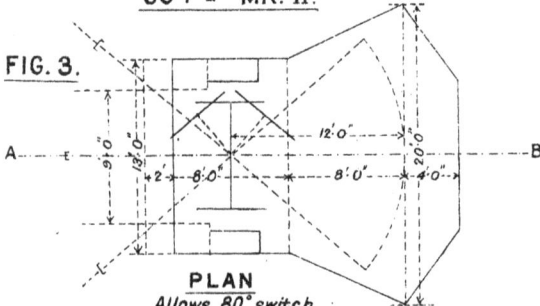

A ——— B

12'·0"

20'·0"

13'·0"

9'·0"

2' 8'·0" 8'·0" 4'·0"

PLAN

Allows 80° switch

FIG. 4.

Position of muzzle
at 35° Elevation.

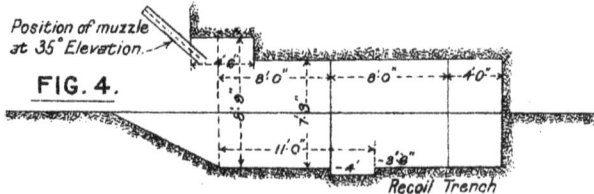

8'·0" 8'·0" 4'·0"

7'·3"

11'·0"

−4' 3'·6"

Recoil Trench

SECTION A.B.

FIG. 5.

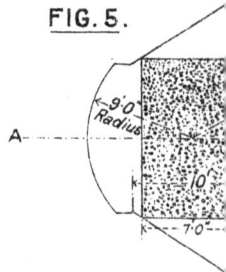

A ———

9'·0"
Radius

10'·0"

7'·0"

PLA

FIG. 6.

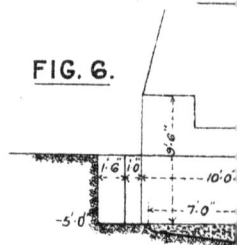

9'·6"

1'·6" 1'·0" 10'·0"

7'·0"

−5'·0"

S
6" H(

Plate 77.

EMENTS.

~ested as a general guide to
s. The details in any particular
ound and other factors. The
~rned by the angle of switch assumed.

FIG. 7.

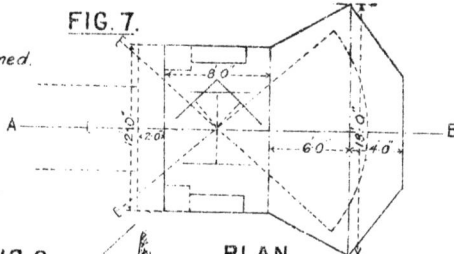

A — B

FIG. 8.

PLAN
Allows 80° switch

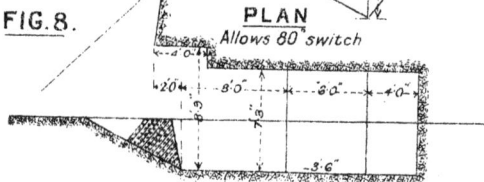

SECTION AB

60 P<u>DR</u> MK. I.

FIG. 9.

A — B

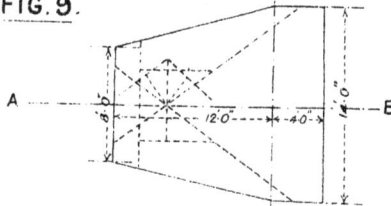

PLAN
Allows 70° switch

FIG. 10.

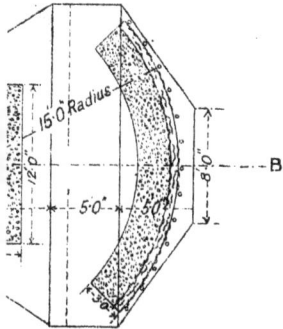

15·0 Radius

B

~N (Allows 70° switch)

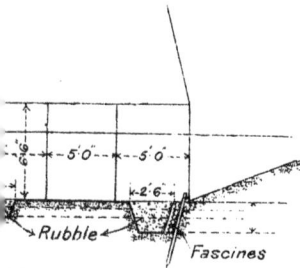

Rubble

Fascines

ECTION. A.B.

)W. (26 CWTS).

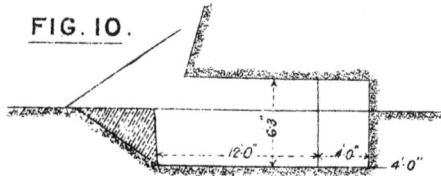

SECTION AB
4·5" HOW.

Malby & Sons, Ltd

Plate 78

SPLINTER-PROOF TRAVERSE.
PROTECTION TO ENTRANCES.

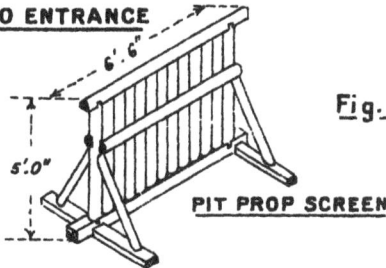

ELEVATION

Fig.1.

Hoop iron to clip over rail — Steel rail

9"×3"×4'.0" shield or ½ pit props.

Corr: Iron

Wheel

Muzzle

10" Pit prop.

Clip

3" — 4'.0"

Corr: Iron

SECTION OF SHIELD

Sandbags

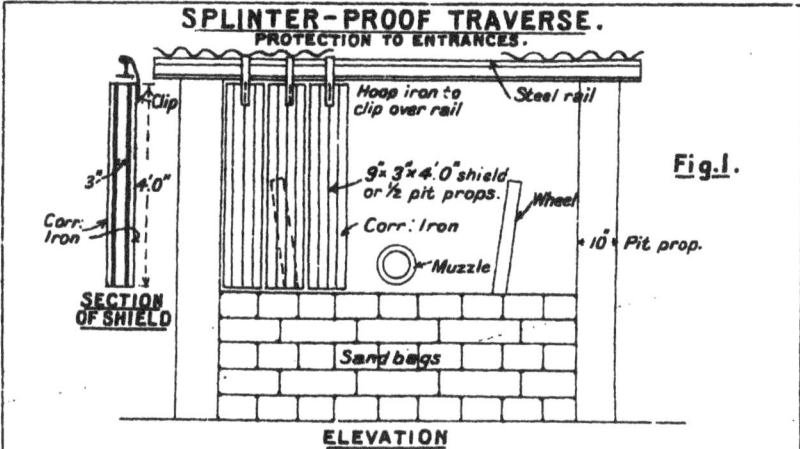

PROTECTION TO ENTRANCE

Fig.3.

6'.6"

5'.0"

PIT PROP SCREEN

SANDBAG SCREEN

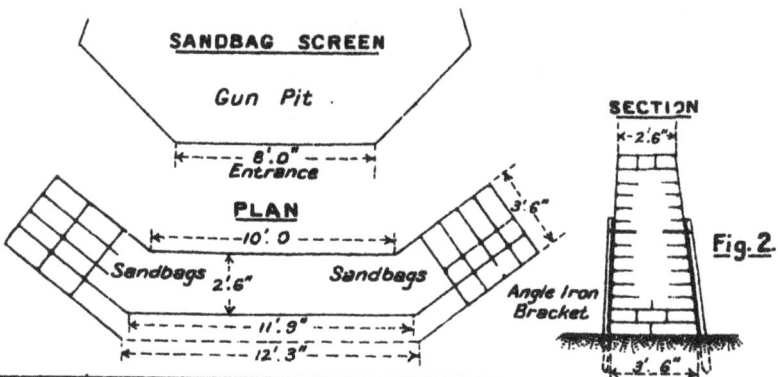

Gun Pit.

8'.0" Entrance

PLAN

10'.0"

Sandbags 2'.6" Sandbags

11'.9"

12'.3"

SECTION

2'.6"

3'.6"

Fig.2.

Angle Iron Bracket

3'.6"

Plate 79

TRAIL-SUPPORT SHEWING FIXED
SUPPORT AT "A" AND CUSHION AT "B".

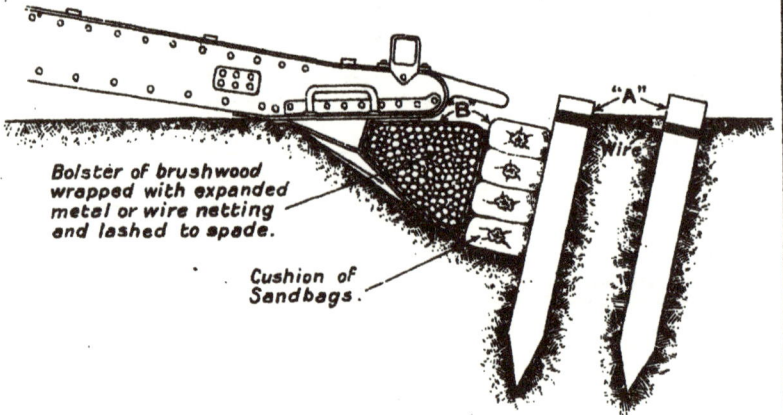

"A"

Wire

Bolster of brushwood
wrapped with expanded
metal or wire netting
and lashed to spade.

Cushion of
Sandbags.

ALTERNATIVE DESIGN WITHOUT BOLSTER.

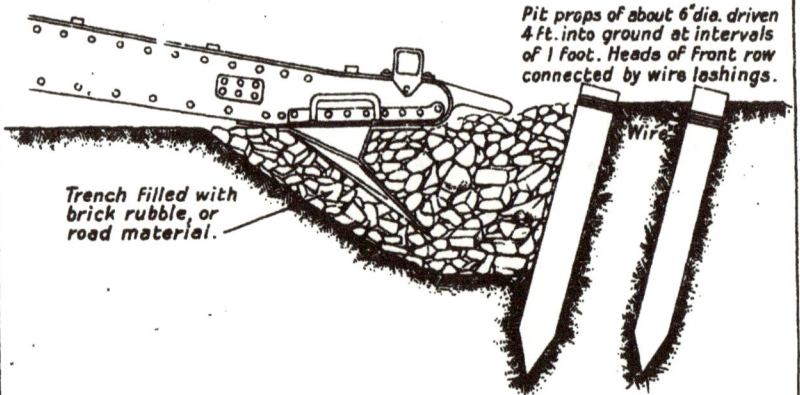

Pit props of about 6˝dia. driven
4 ft. into ground at intervals
of 1 foot. Heads of front row
connected by wire lashings.

Wire

Trench filled with
brick rubble, or
road material.

Malby & Sons. Lith.

Plate 80

PLATFORM FOR 18-POUNDER.

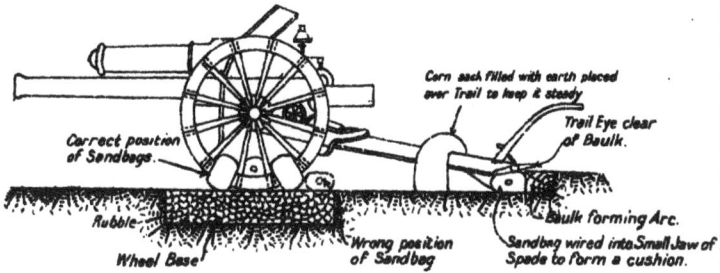

Corn sack filled with earth placed over Trail to keep it steady

Trail Eye clear of Baulk.

Correct position of Sandbags.

Rubble

Wheel Base

Wrong position of Sandbag

Baulk forming Arc.

Sandbag wired into Small Jaw of Spade to form a cushion.

DETAIL OF TRAIL SUPPORT
(ALTERNATIVE)

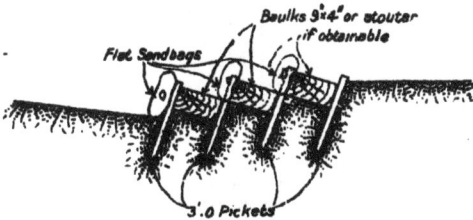

Baulks 9"x4" or stouter if obtainable

Flat Sandbags

3'.0 Pickets

FIG.2.   **SECTION.**

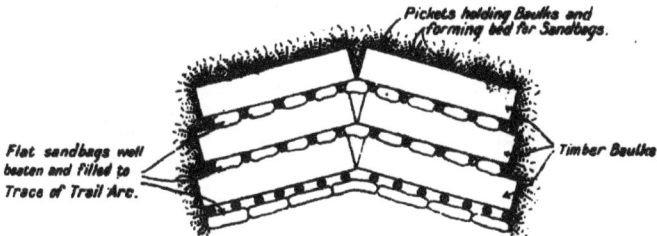

Pickets holding Baulks and forming bed for Sandbags.

Flat sandbags well beaten and filled to Trace of Trail Arc.

Timber Baulks

FIG.3.   **PLAN.**

8921.11084. PPS115/967   Malby & Sons, Lith.

Plate 81

PLATFORM FOR 4·5" HOWITZER.

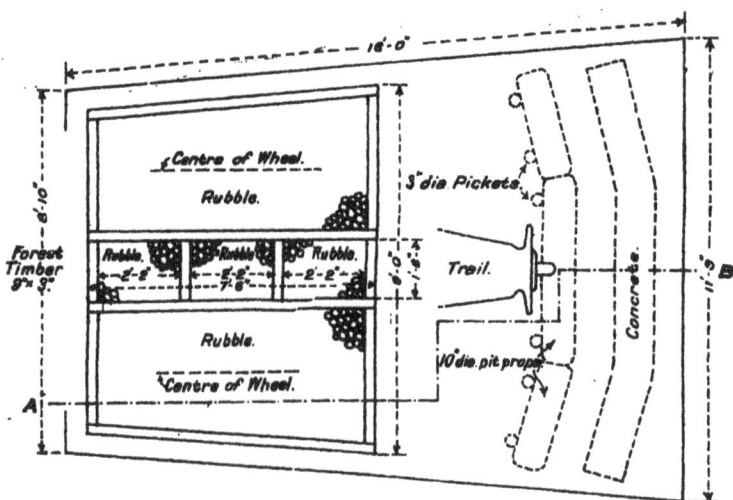

16'-0"

Centre of Wheel.

Rubble.

3" dia Pickets.

Forest Timber 9"× 9".

Rubble. Rubble. Rubble.
2'-2" 2'-2" 2'-2"
7'-6"

Trail.

Rubble.

Concrete.

Centre of Wheel.

10" dia. pit props.

A

B

6'-0"

6'-10"

11'-3"

PLAN.

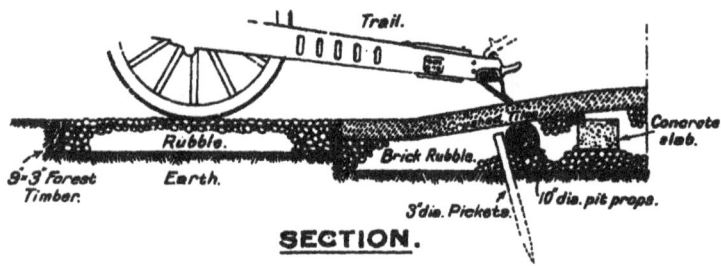

Trail.

Rubble.

Concrete slab.

9"×3" Forest Timber.

Earth.

Brick Rubble.

3" dia. Pickets.

10" dia. pit props.

SECTION.

6821.11064.PP.9115/387

Malby&Sons.Lith

Plate 82

WOODEN WHEEL-BED AND WHEEL-GUIDES, FOR 60 PR B.L. GUN MK IV/L CARRIAGE.

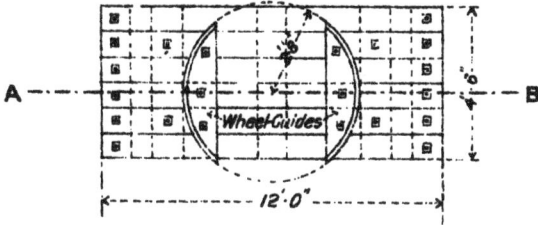

A - - - - - B

Wheel-Guides

4'6"

12'0"

Sleeper 10 x 4½   Earth and Stones.
(well rammed.)

SECTION THROUGH A.B.

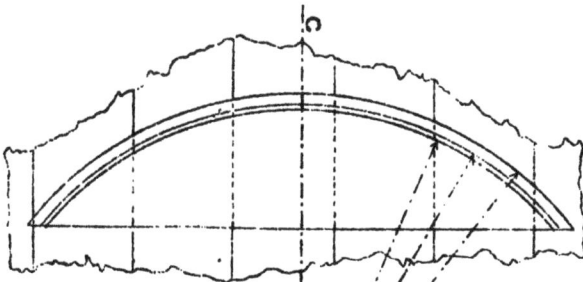

C

2'2" 2'6" 2'8"

SECTION THROUGH C.D.

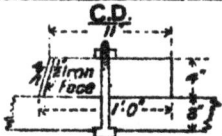

½ iron face

1'0"

ENLARGED PLAN AND SECTION OF WHEEL GUIDES.

Malby & Sons, Lith.

Plate 83

PLATFORM FOR 6" HOWITZER.

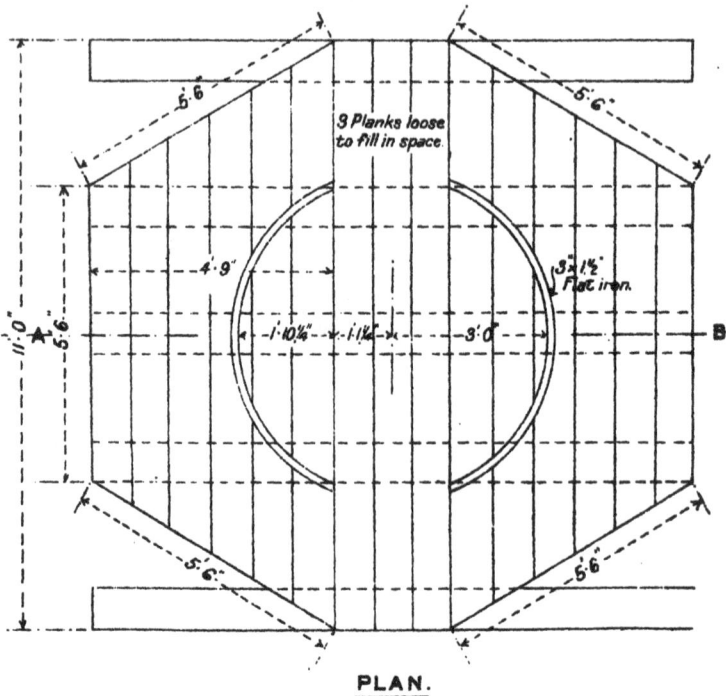

3 Planks loose
to fill in space.

5'·6"

5'·6"

4'·9"

3·1½
Flat iron.

11'·0"

5'·6"

1'·10¼"   1'·1¼"   3'·0"

A

B

5'·6"

5'·6"

PLAN.

11'·9"

6'·0"

2'·3"

CROSS SECTION
A.B.

8821 11084. P.F.3115/387

Malby & Sons. Lith

Plate 82

WOODEN WHEEL-BED AND WHEEL-GUIDES,
FOR 60 Pʳ B.L. GUN Mᴷ IV/L CARRIAGE.

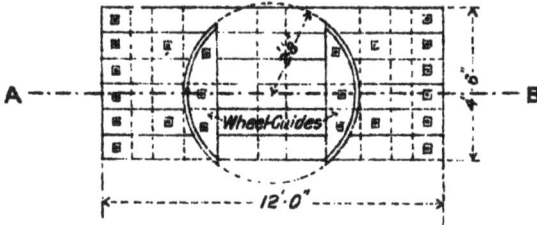

A —————————————————————————————— B

Wheel-Guides

12'·0"

SECTION THROUGH
A.B.

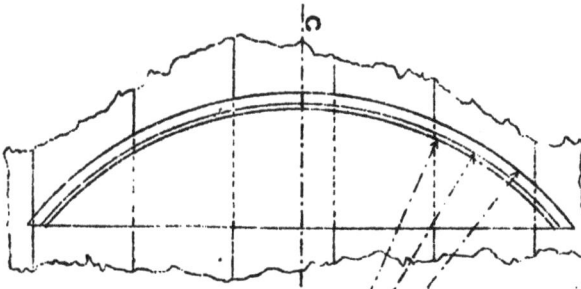

Sleeper 10 x 4½   Earth and Stones.
(well rammed.)

SECTION THROUGH
C.D.

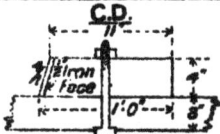

Iron
Face

1'·0"

ENLARGED PLAN AND SECTION
OF WHEEL GUIDES.

Plate 83

PLATFORM FOR 6" HOWITZER.

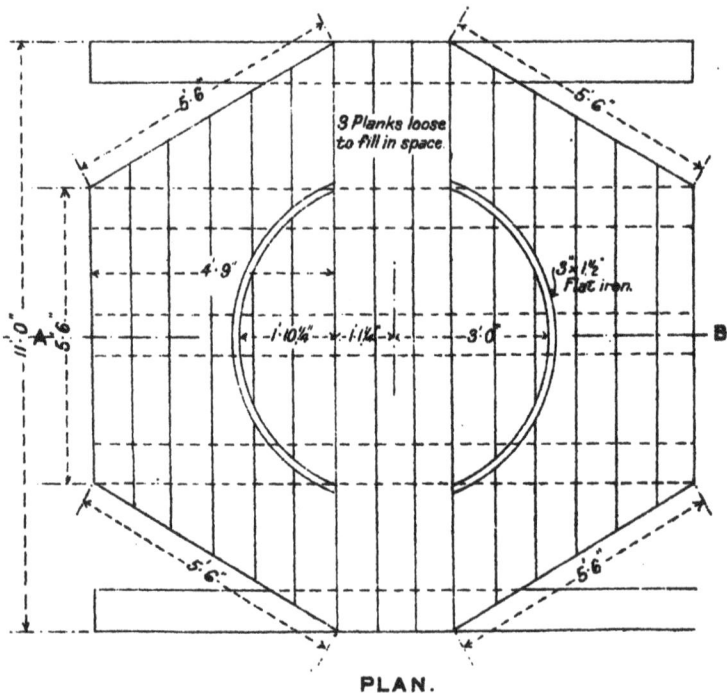

3 Planks loose
to fill in space.

5·6

5·6

5·6

5·6

4·9

3·1½
Flat iron.

11·0"

5·6

1·10¾" 1·1¼" 3·0"

A

B

PLAN.

11·9"

6·0"

2·3"

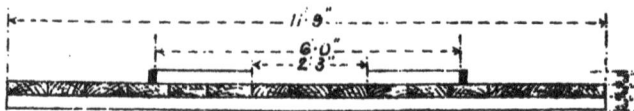

**CROSS SECTION
A.B.**

Malby & Sons. Lith

Plate 84

DOUBLE DECKED PLATFORM
FOR 8" HOWITZER.

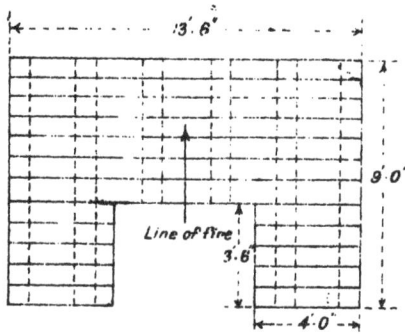

PLAN.

SECTION.

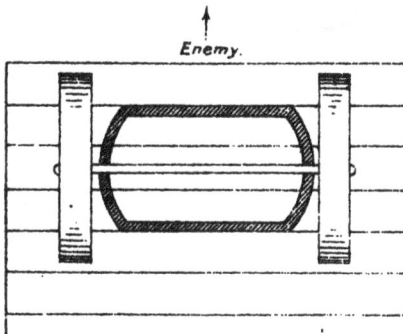

GUIDE ARC.

Malby & Sons, Lith.

Plate 85.

ORGANIZATION OF A DEFENSIVE SYSTEM
(DIAGRAMMATIC ONLY).

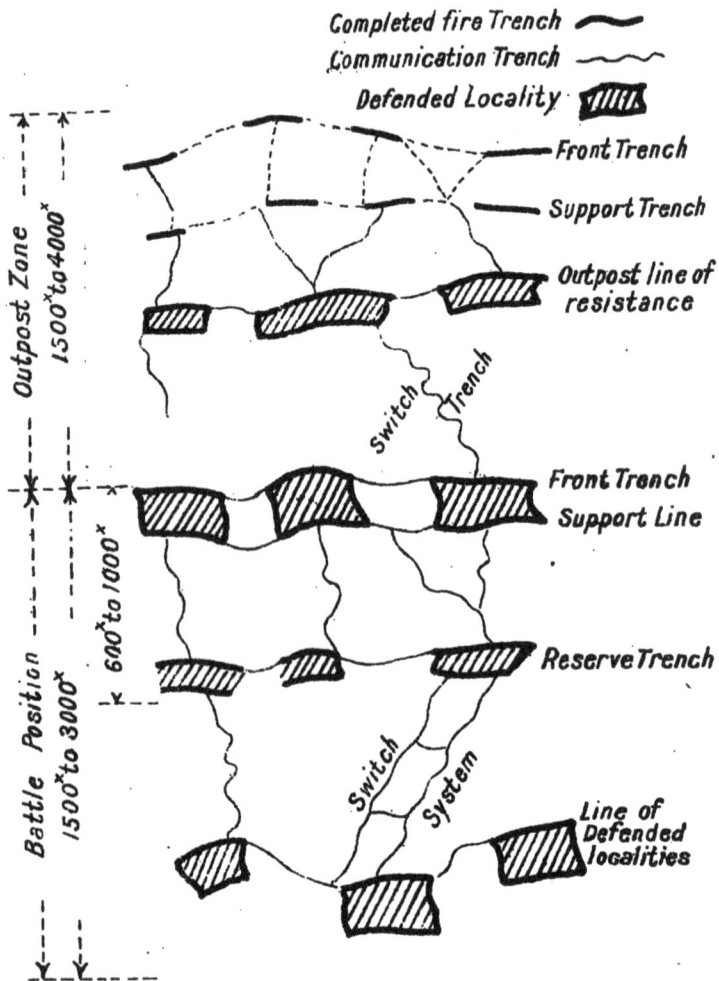

Completed fire Trench

Communication Trench

Defended Locality

Front Trench

Support Trench

Outpost line of resistance

Outpost Zone 1500ˣ to 4000ˣ

Switch Trench

Front Trench

Support Line

Battle Position 1500ˣ to 3000ˣ

600ˣ to 1000ˣ

Reserve Trench

Switch System

Line of Defended localities

Plate 86.

DEFENCE POST AT JUNCTION OF TRENCHES.

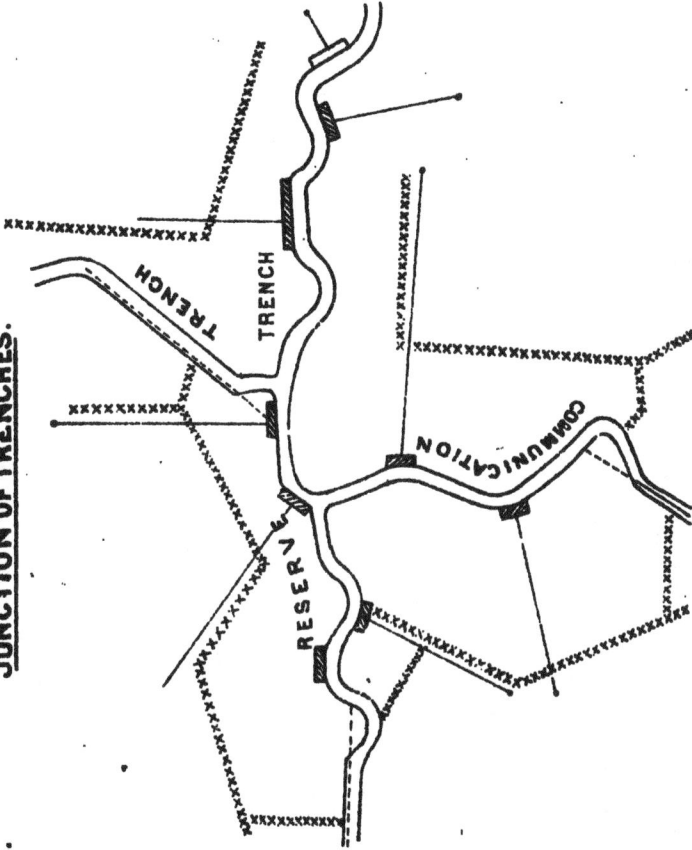

METHODS OF PRODUCING ALL ROUND FIRE.

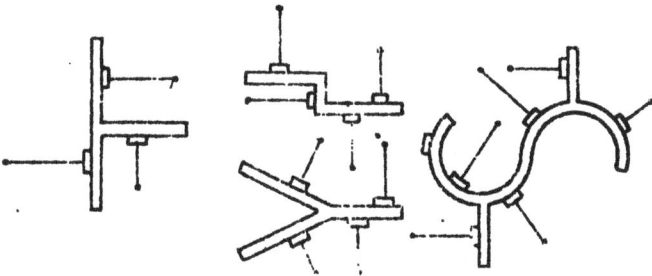

TRENCH

TRENCH

RESERVE

COMMUNICATION

Malby & Sons, Lith.

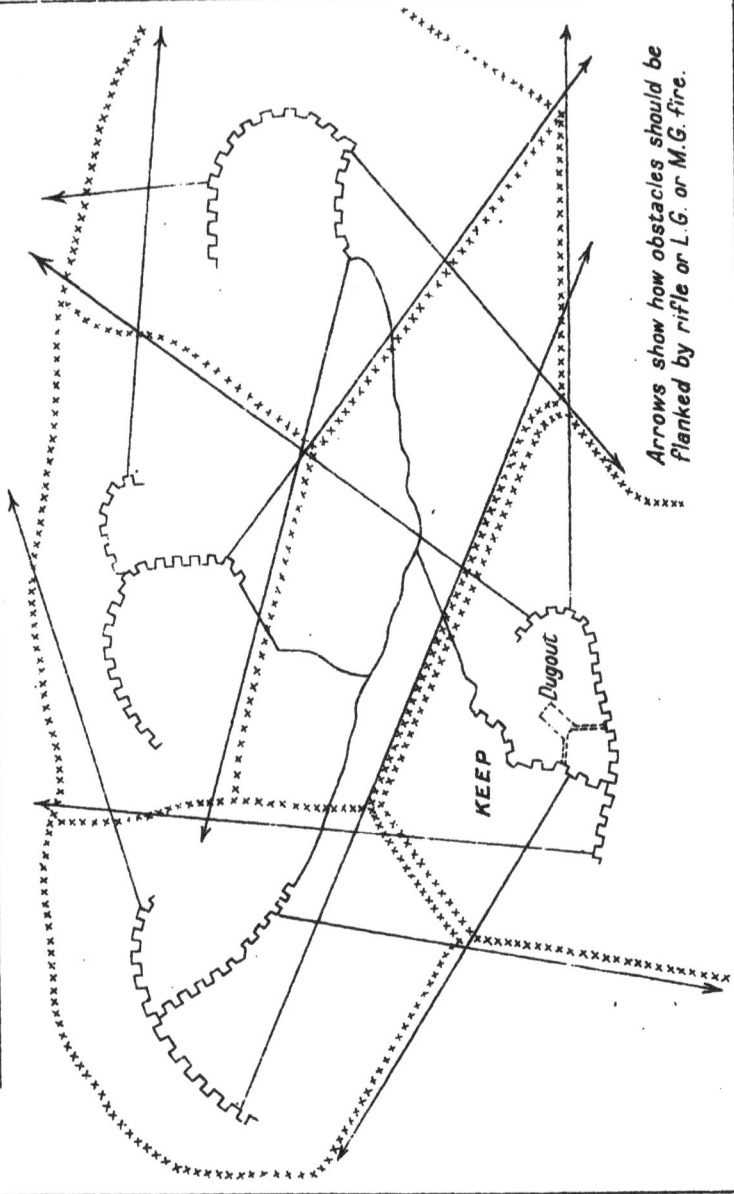

Plate 37.

DEFENDED LOCALITY IN ISOLATED POSITION FOR 4 PLATOONS.

KEEP

Dugout

Arrows show how obstacles should be flanked by rifle or L.G. or M.G. fire.

Malby & Sons. Lith.

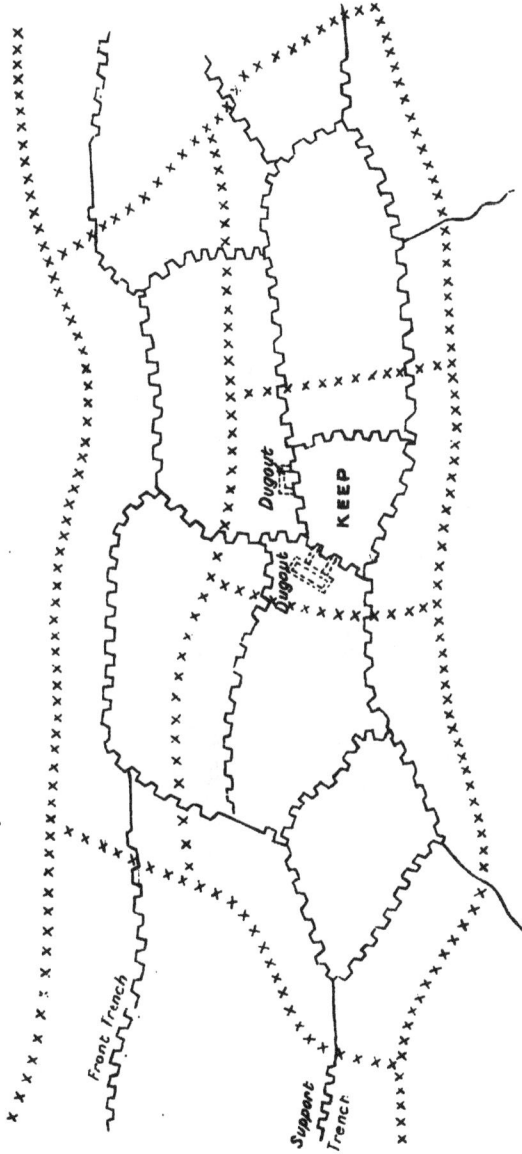

Plate 88.

DEFENDED LOCALITY AS PART OF A TRENCH SYSTEM.

KEEP

Dugout

Dugout

Front Trench

Support Trench

Malby & Sons. Lith.

(422) 1084 PP 3115/387

Plate 89.

DEFENCE OF VILLAGE.

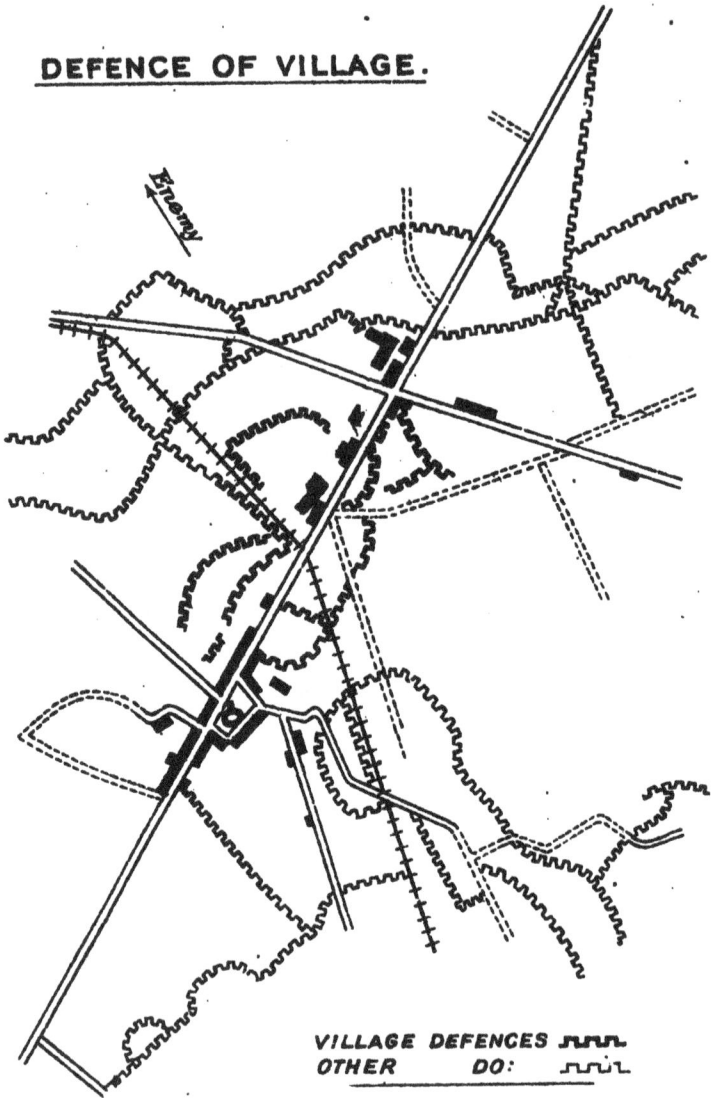

Enemy

VILLAGE DEFENCES ⊓⊔⊓⊔

OTHER DO: ⊓⊔⊓⊔

Plate 30.

DEFENCE OF VILLAGE.
DETAIL OF INTERIOR DEFENCES.

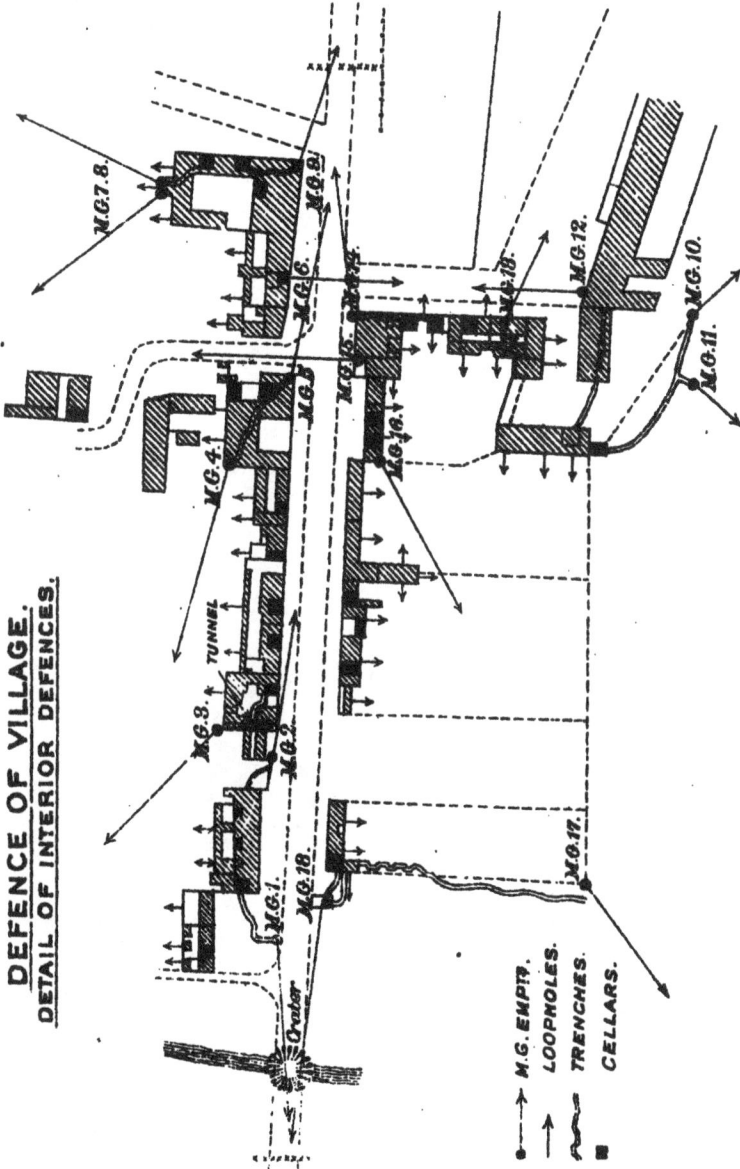

M.G.7.8.

M.G.9.

M.G.6.

M.G.12.

M.G.10.

M.G.11.

M.G.4.

M.G.8.

M.G.2.

M.G.18.

M.G.1.

M.G.17.

TUNNEL

Outer

●——➤ M.G. EMPTS.

↑ LOOPHOLES.

TRENCHES.

■ CELLARS.

Malby & Sons, Lith.

8821./1069f. PR.5H5/387.

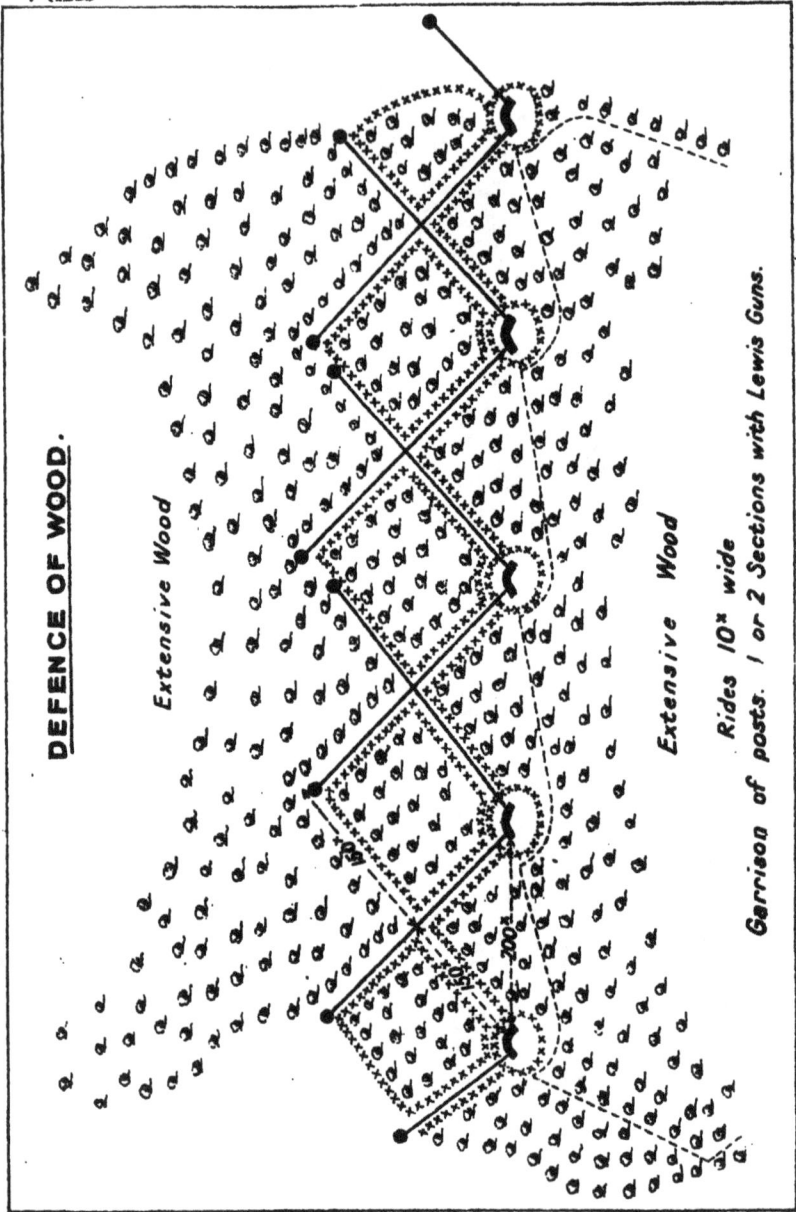

Plate 91.

DEFENCE OF WOOD.

Extensive Wood

Extensive Wood

Rides 10ˣ wide

Garrison of posts. 1 or 2 Sections with Lewis Guns.

Malby & Sons. Lith

Plate 92.

OBSTACLES AND CLEARING IN WOOD.

Malby & Sons, Lith.

8921. 11084. PP 3115/367.

Plate 43.

DEFENSIBLE POST.

PLAN.

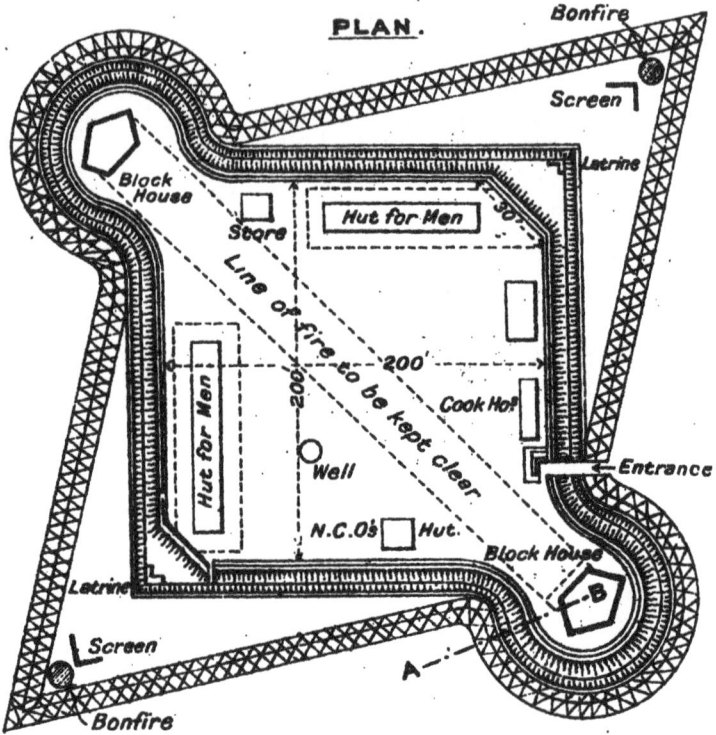

Bonfire

Screen

Latrine

Block House

Store

Hut for Men

30'

Line of fire to be kept clear

200'

200'

Hut for Men

Cook Ho!

Entrance

Well

N.C.Os   Hut.

Block House

B

A

Latrine

Screen

Bonfire

SECTION ON A.B.

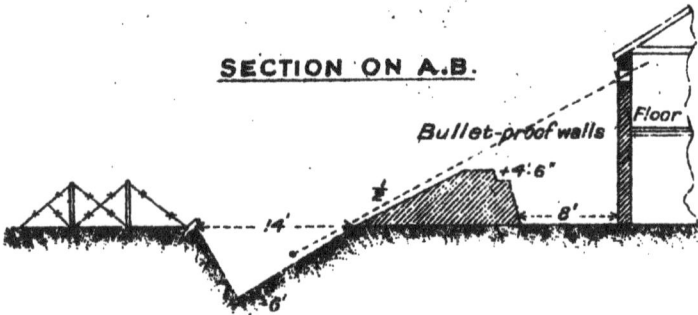

Bullet-proof walls

Floor

+4.6'

1'

14'

8'

6'

Malby & Sons.Lith.

Plate 94.

USE OF NATURAL CAMOUFLAGE. .

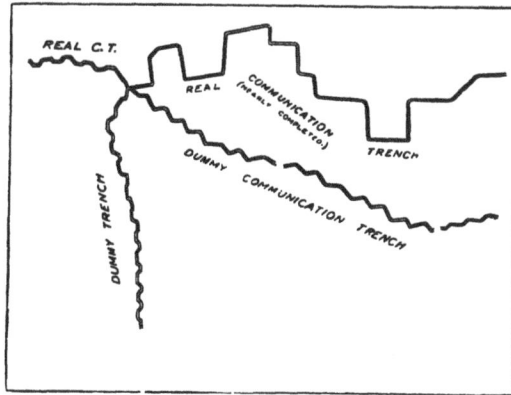

Plate 95

TRACK DISCIPLINE.

Fig. 1.

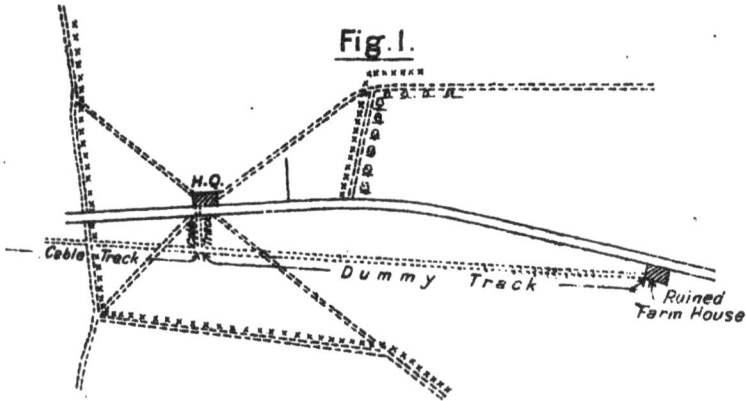

SHOWING TRACKS LIKELY TO BE MADE LEADING TO A H.Qs. AND METHOD OF PLACING WIRE TO CONFINE TRAFFIC TO APPROVED ROUTES.

Fig. 2.

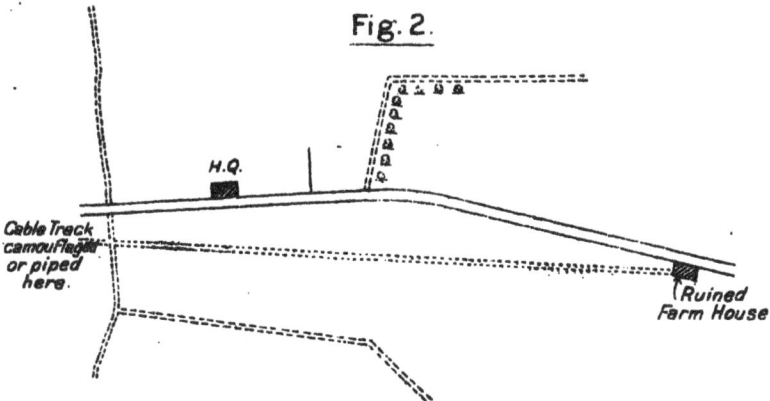

TRACKS AS THEY APPEAR ON AIR PHOTOGRAPHS.

Malby & Sons. Lith.

Plate 96

O. P. IN BRICK WALL.
painted gauze between A and B

Malby & Sons. Lith.

Plate 97.

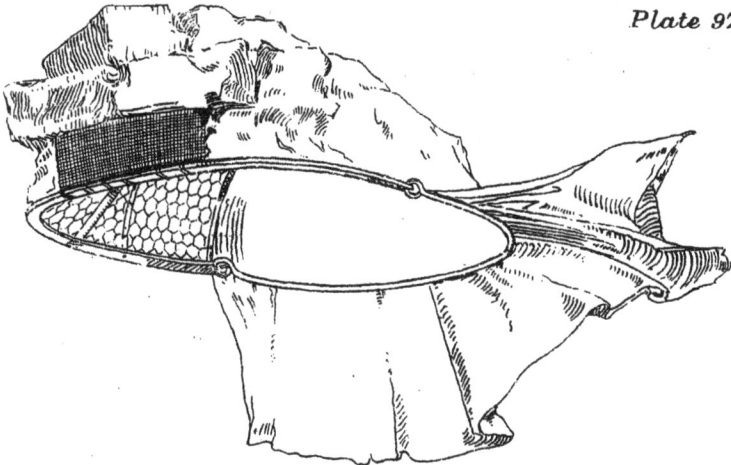

PORTABLE BEEHIVE O.P. UNARMOURED.
Weight about 10 lbs.
Showing "brick" Camouflage. Any surroundings can be similarly imitated.

PORTABLE BEEHIVE O.P. IN USE.

8521.11084-PP3115/387

Maloy & Sons.L:th

Fig. 1. Concealment of Embrasures   *Plate* 98.

Splinter and bullet-proof M.G. emplacement in open field. Covered approach connects it to heavily strutted cellar. Position of loophole indicated by dotted line.

Fig. 2.

M.G. emplacement guarding main approach. Loophole of painted canvas x.........x Emplacement connected to strengthened cellar.

Plate 99.

CONCEALMENT OF EMBRASURES.

Splinter-proof M.G. emplacement inside a house. Concrete front, rail and sandbag roof. Loophole x.........x is covered with wire gauze painted to look like the bottom 6″ of the shutter. Emplacement is made strong enough to support the debris of the house when it collapses under shell fire, and is connected to a cellar with strengthened roof.

Plate 100.

WIRE NETTING AND CANVAS SCREEN.

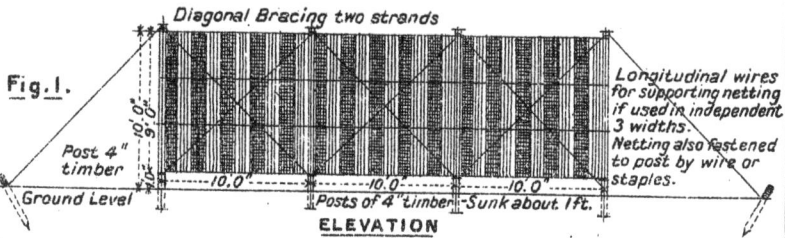

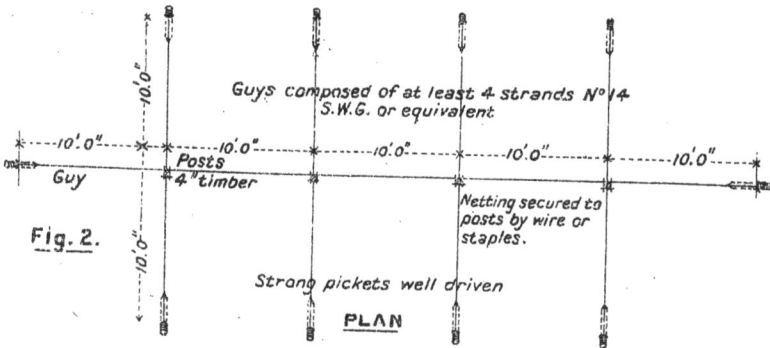

Fig.1.

Diagonal Bracing two strands

Longitudinal wires for supporting netting if used in independent 3 widths. Netting also fastened to post by wire or staples.

Post 4" timber

Ground Level

10.0"

10.0"

9.0"

10.0"

Posts of 4" timber Sunk about 1 ft.

ELEVATION

Fig. 2.

Guys composed of at least 4 strands Nº 14 S.W.G. or equivalent

10.0"

Guy

10.0"

Posts 4" timber

10.0"

10.0"

10.0"

10.0"

Netting secured to posts by wire or staples.

Strong pickets well driven

PLAN

IMITATION BRICK WALL

Enemy

Fig. 3.

ROAD

Dummy Wall

Enemy

Real wall removed

Fig. 4.

ROAD

9921.11084. PP3115/387. 62500. 12.21.

Malby & Sons, Lith.

Plate 101.

SCREENING ROADS.

FIG.1. PERPENDICULAR TO FRONT.

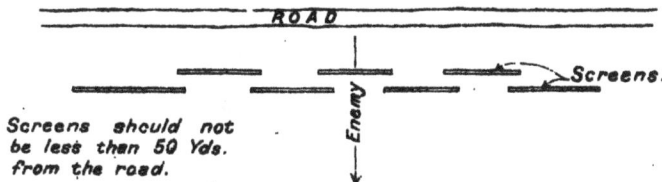

ROAD

Screens.

Screens should not
be less than 50 Yds.
from the road.

FIG.2. PARALLEL TO FRONT.

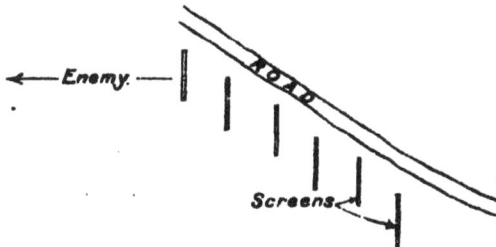

Enemy

ROAD

Screens.

FIG.3. OBLIQUE TO FRONT.

8821.11084 PP 3115/387. 62500·12·21   Malby & Sons. Lith

Plate 102.

TAPE TEMPLET.

Fig. I.

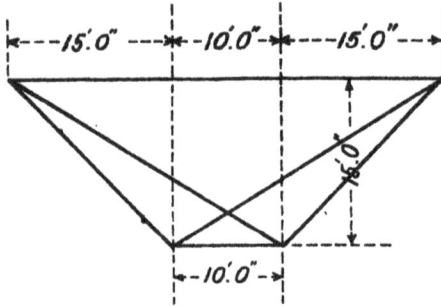

USE OF HAIRPINS FOR RAPID TRACING.

Fig. 2.

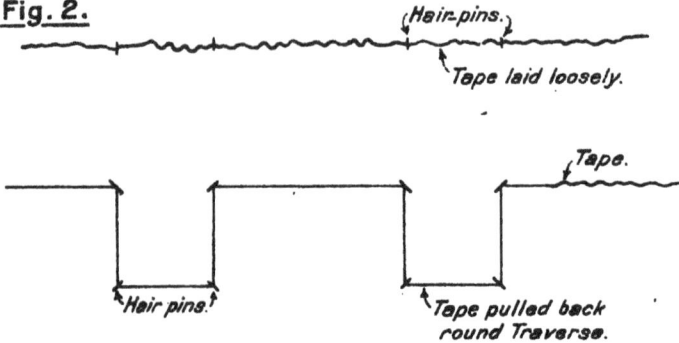

Hair-pins.

Tape laid loosely.

Tape.

Hair pins.

Tape pulled back round Traverse.

Plate 103.

KNOTS.

FIG. 1.
Thumb

FIG. 2.
Figure of 8

FIG. 3.
Reef

FIG. 4.
Single Sheet
Bend

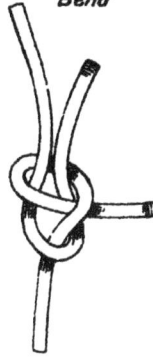

FIG. 5.
Double Sheet
Bend

FIG. 6.
Hawser Bend

Seizing

FIG. 7.
Commencement
of Bowline

FIG. 8.
Bowline
Completed

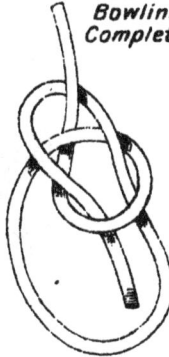

FIG. 9.
Timber Hitch

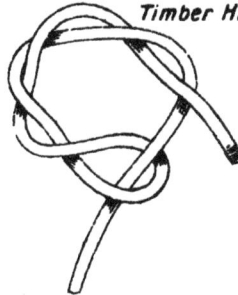

Malby & Sons. Lith.

Plate 104

KNOTS.

Fig.1

Clove Hitch

Fig.2

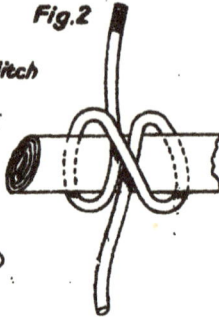

Fig.3

2 Half Hitches

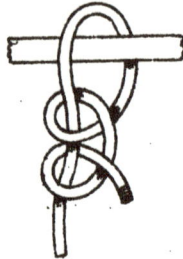

Fig. 4

Round Turn &
2 Half Hitches

Fig.5

Fisherman's
Bend

Fig. 6

Running Knot

Fig. 7

Lever Hitch

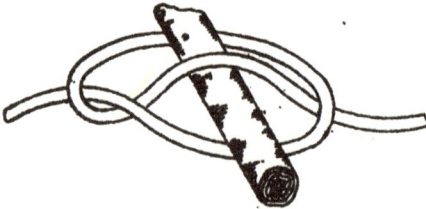

Fig.8.

Man Harness Hitch

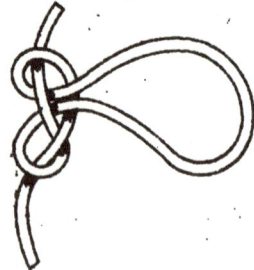

Malby & Sons, Lith.

Plate 105.

KNOTS.

Fig. 1.

Fig. 2.

Fig. 3.

Cat's Paw on Centre of Rope

a   *b*

Double Blackwall Hitch

Single Blackwell Hitch

Draw Hitch.

Fig. 4.   **Fig. 5.**   **Fig. 6.**

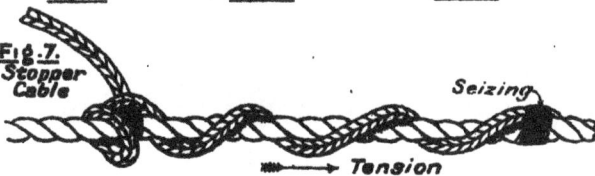

Fig. 7.
Stopper Cable

Seizing

→ *Tension*

Malby & Sons. Lith.

Plate 106.

LASHINGS.

FIG. I.

FIG. 2.

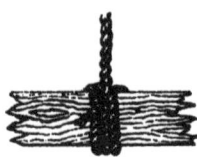

FIG. 3.

FIG. 4.
SQUARE LASHING.

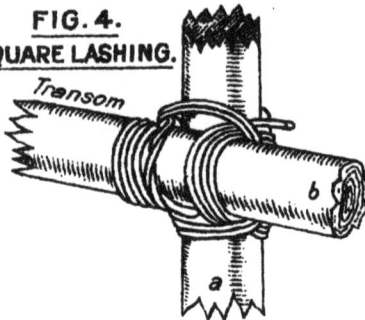

FIG. 5.
DIAGONAL LASHING.

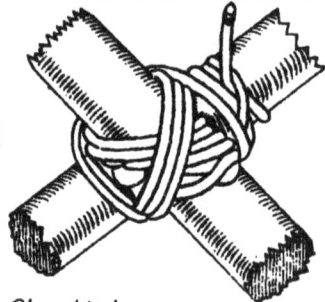

FIG. 6.

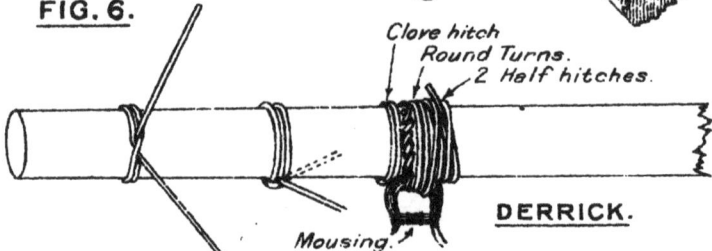

Clove hitch
Round Turns.
2 Half hitches.

Mousing.

DERRICK.

FIG. 7.

USE OF SELVAGEE.

Malby &Sons. Lith.

Plate 107

HOLDFASTS & ANCHORAGES.

Fig. I.
3.2.1. HOLDFAST.

Fig. 2.
BAULK HOLDFAST.

Fig. 3.
LOG ANCHORAGE.

Filled in and rammed after completion of bridge.

Revetted if necessary

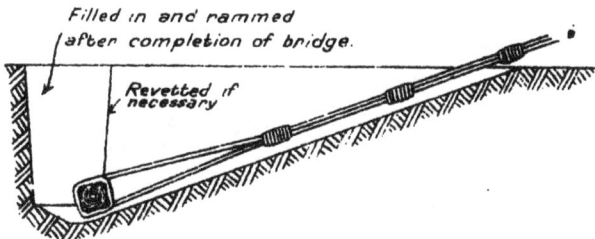

C92. 11084 P P 3115/387.   Malby & Sons. Lith.

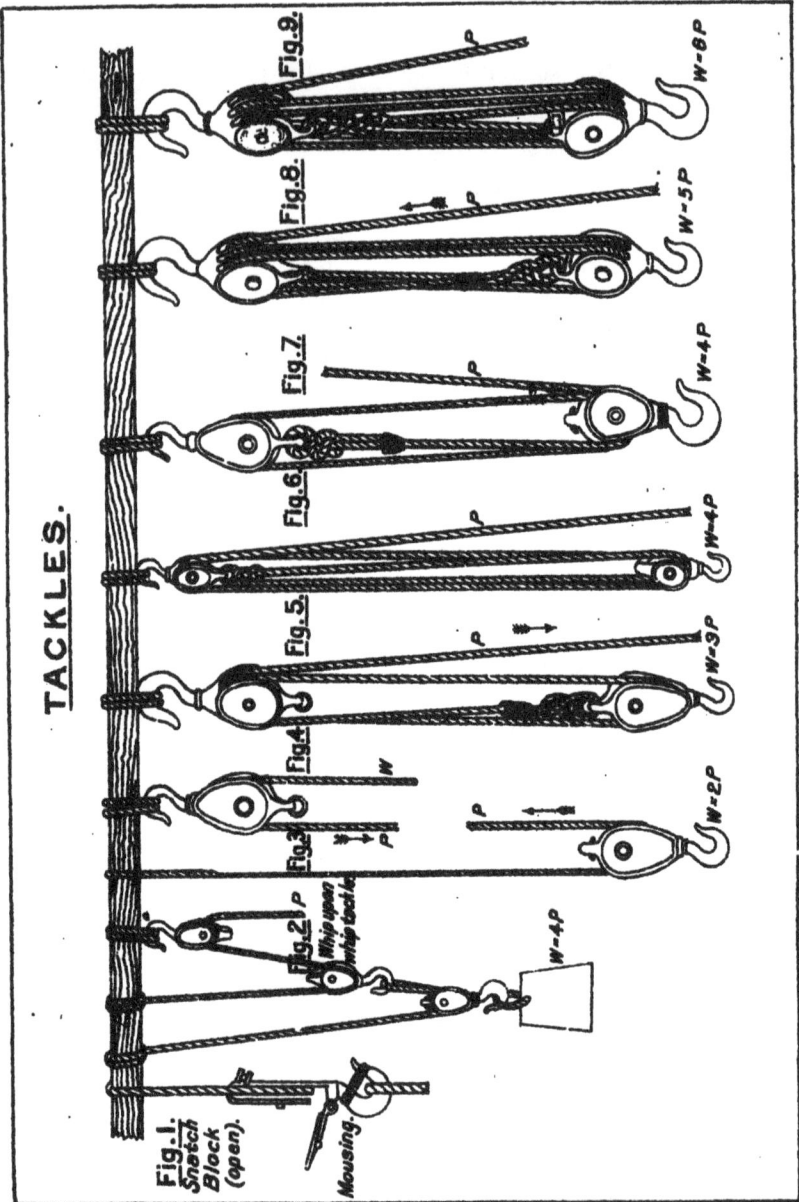

Plate 108.

TACKLES.

Fig. 1.
Snatch Block
(open).

Mousing.

Fig. 2.
Whip upon
whip tackle.

Fig. 3.

Fig. 4.

Fig. 5.

Fig. 6.

Fig. 7.

Fig. 8.

Fig. 9.

W=2P

W=3P

W=4P

W=4P

W=4P

W=5P

W=6P

P

W

Malby & Sons, Lith.

Plate 109.

USE OF SPARS.

Fig. 1.
SHEER
LASHING.

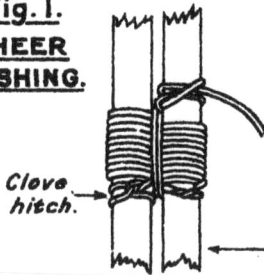

Clove
hitch.

← Sheers. →

Fig. 2.

Guy.

Guy.

Sling.

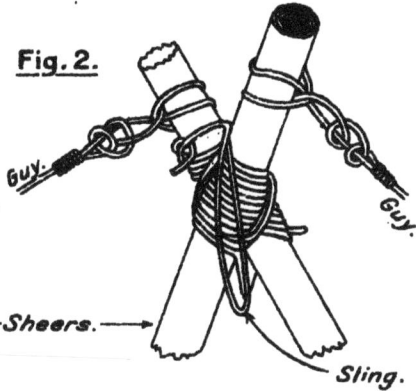

Fig. 3.
GYN
LASHING.

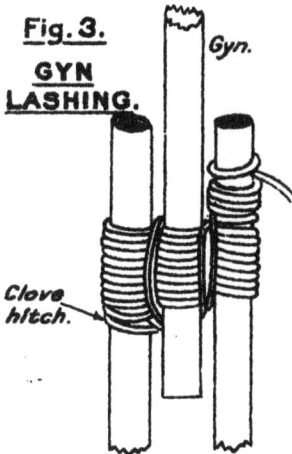

Gyn.

Clove
hitch.

Fig. 4.
LEADING BLOCK
FOR TACKLE.

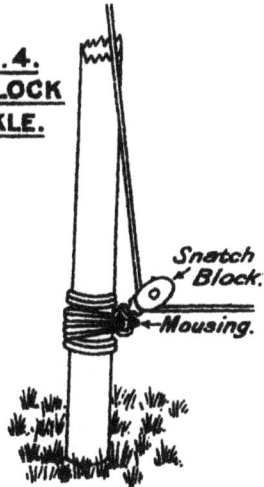

Snatch
Block.

← Mousing.

Plate 110.

SWINGING DERRICK.

Fig.1.

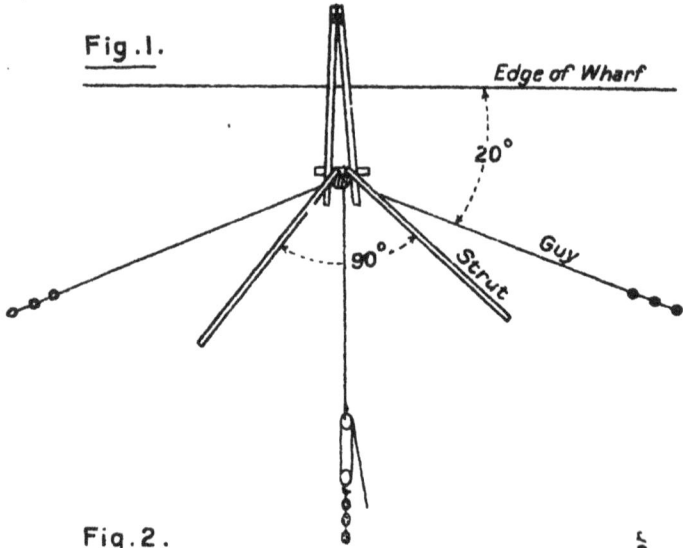

Edge of Wharf

20°

90°

Guy

Strut

Fig.2.

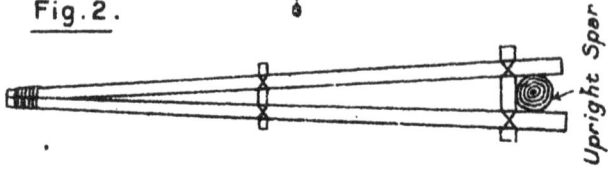

Upright Spar

Fig.3.

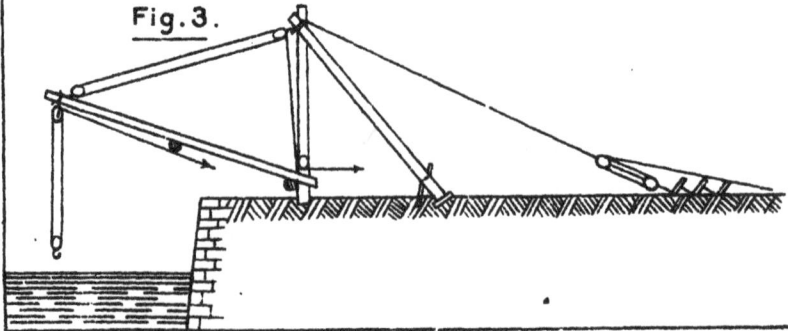

Plate III.

SECTION OF A GAP.

FIG. 1.

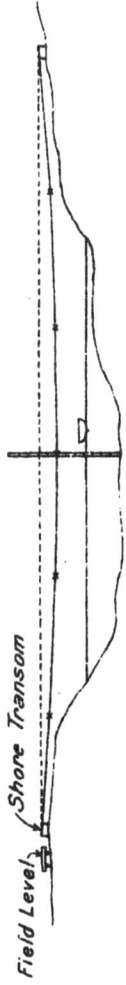

Field Level

Shore Transom

SHORE END.

FIG. 3.

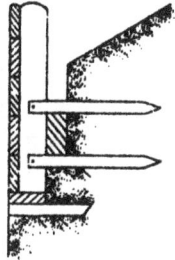

PARTS OF A BRIDGE.

FIG. 2.

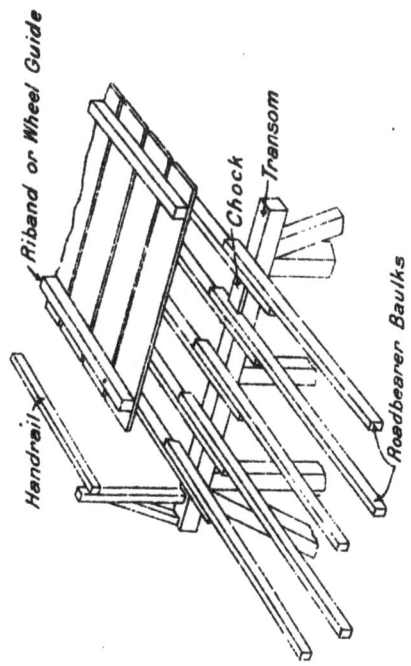

Riband or Wheel Guide

Chock

Transom

Handrail

Roadbearer Baulks

Plate 112.

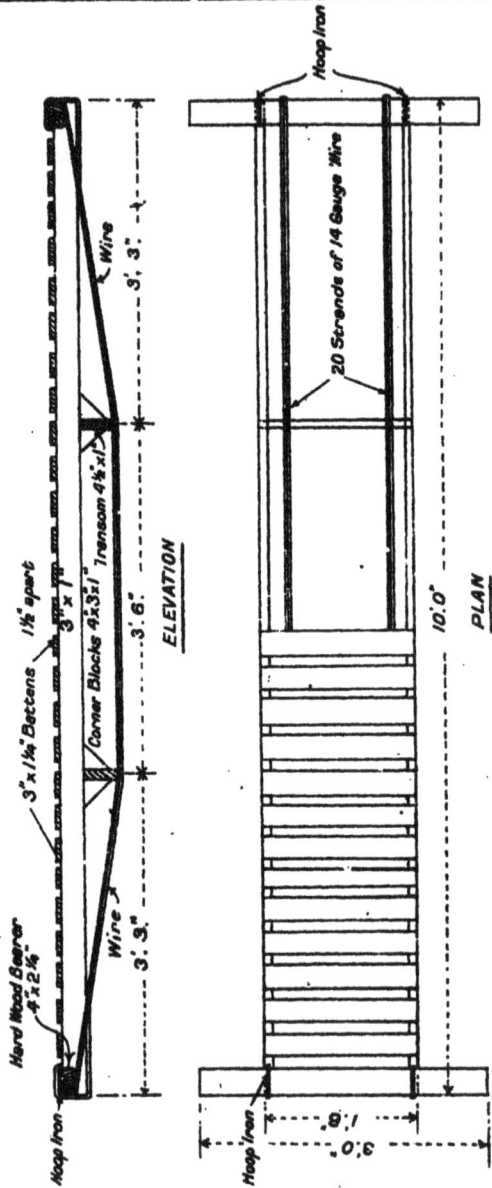

INFANTRY FOOTBRIDGE.

TRUSSED BEAM BRIDGE.

ELEVATION

Hard Wood Bearer 4" x 2½"

3" x 1½" Battens 1½" apart

3" x 1"

Corner Blocks 4½" x 3" x 1" Transom 4½" x 1"

Wire

Wire

Hoop Iron

3'. 3".

3'. 6".

3'. 3".

PLAN

Hoop Iron

20 Strands of 14 Gauge Wire

Hoop Iron

10'. 0".

1'. 8".

3'. 0".

Plate 113.

INFANTRY FOOTBRIDGE.
FLOAT FOOTBRIDGE.

Showing alternative floats of PETROL TINS, CORK & METAL CYLINDERS.

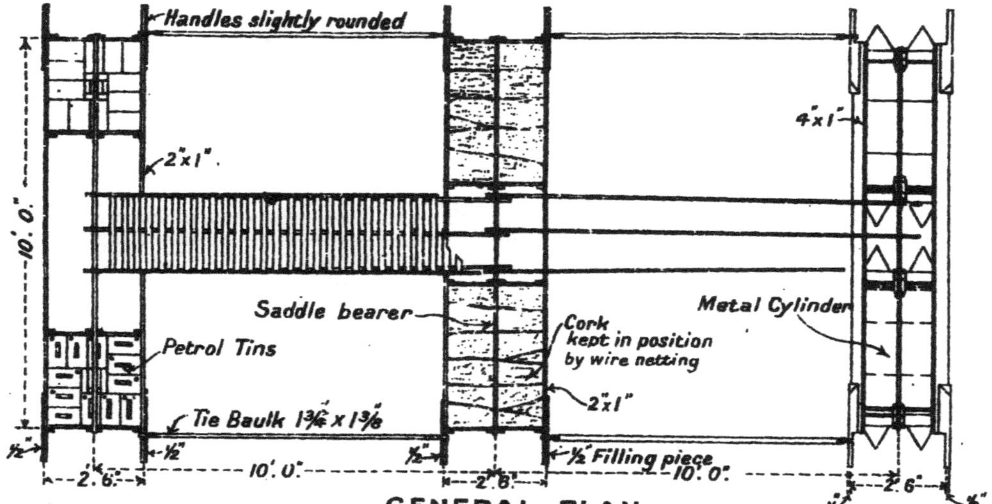

Handles slightly rounded

4"x1"

2"x1"

Metal Cylinder

Cork kept in position by wire netting

2"x1"

½" Filling piece

Saddle bearer

Petrol Tins

Tie Baulk 1¾" x 1¾"

2"x1"

10' 0"

½"

2' 6"

2' 6"

2' 6"

10' 0"

10' 0"

½"

½"

½"

GENERAL PLAN.

4"x1"

SIDE VIEW.

Plate 114.

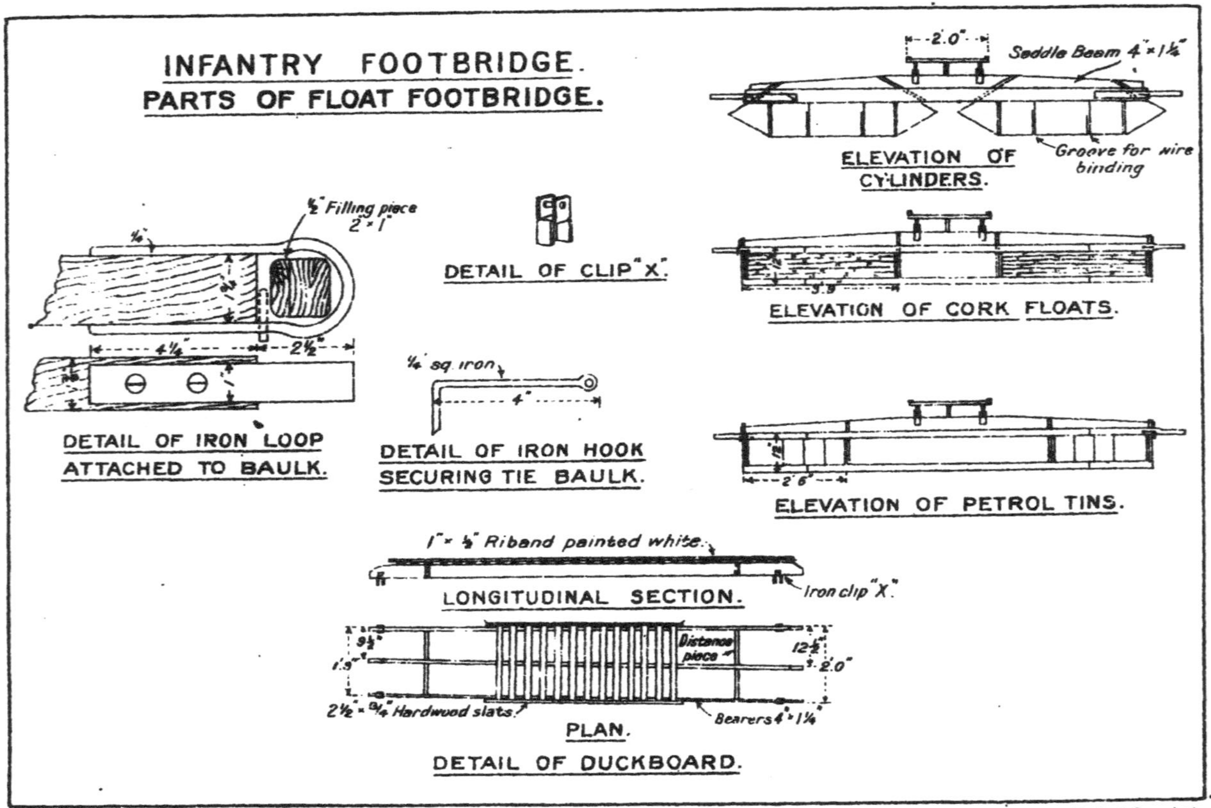

INFANTRY FOOTBRIDGE.
PARTS OF FLOAT FOOTBRIDGE.

Saddle Beam 4″×1¼″

Groove for wire binding

2′·0″

ELEVATION OF CYLINDERS.

ELEVATION OF CORK FLOATS.

2′·6″

ELEVATION OF PETROL TINS.

DETAIL OF CLIP "X".

¼″ sq iron

1″

4″

DETAIL OF IRON HOOK
SECURING TIE BAULK.

¼″ Filling piece
2″×1″

2½″

2″

1″

¼″

DETAIL OF IRON LOOP
ATTACHED TO BAULK.

Iron clip "X".

1″× ½″ Riband painted white.

LONGITUDINAL SECTION.

Distance piece

12½″

2′·0″

Bearers 4″×1¼″

9¼″

1′·9″

2½″× ¾″ Hardwood slats.

PLAN.

DETAIL OF DUCKBOARD.

Malby & Sons, Lith.

Plate 115.

INFANTRY FOOTBRIDGE.
CASK FOOTBRIDGE.

For 54 gallon casks

PLAN
Fig.1.

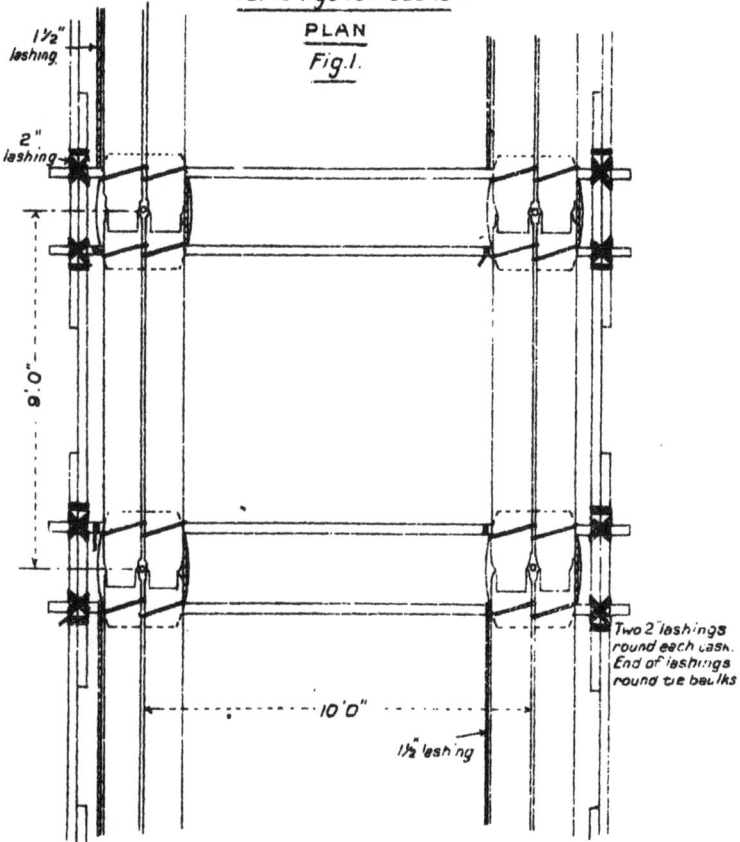

1½" lashing

2" lashing

9'0"

Two 2" lashings round each cask. End of lashings round tie baulks

10'0"

1½" lashing

When time presses or materials are lacking, planks one side only will suffice for infantry in file.

ELEVATION

2" lashing round casks, gunwales & tie baulks.

1½" lashing round chesses

Fig 2.

8321 11084. PP3115/397   Malby & Sons Lith

Plate 116.

ARTILLERY TRENCH BRIDGE.

Fig.1.

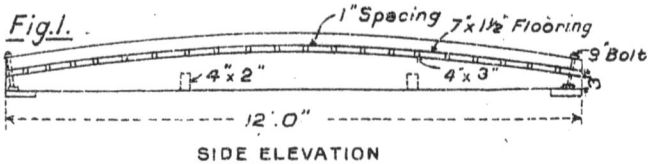

1" Spacing 7'x 1½" Flooring

4"x 2" 4"x 3" 9"Bolt

12'. 0"

SIDE ELEVATION

Fig.2.

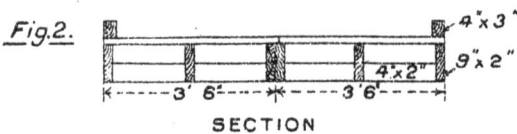

4"x 3"

4"x 2" 9"x 2"

3' 6" 3' 6"

SECTION

CRIB PIER IN SHALLOW WATER.

Fig.3.

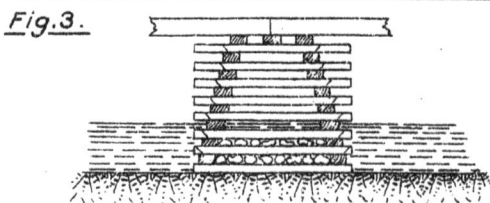

CRIB CAUSEWAY FOR TANKS,
BUILT OF RAILWAY SLEEPERS.

Fig.4.

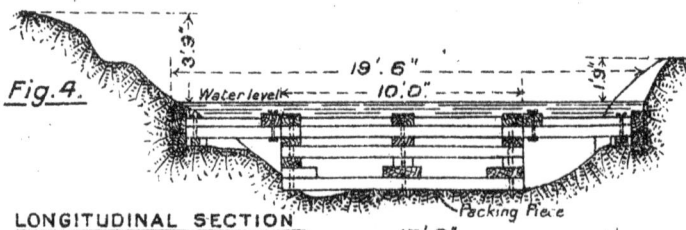

3'.3" 1'.9"

19'. 6"

Water level 10'. 0"

LONGITUDINAL SECTION

Packing Piece

Fig.5.

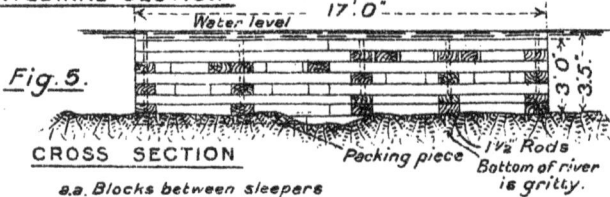

Water level 17'. 0"

3'.0" 3'.5"

CROSS SECTION

Packing piece 1½" Rods

Bottom of river is gritty.

a.a. Blocks between sleepers

Malby & Sons. Lith.

Plate 117.

FRAMED TRESTLE.

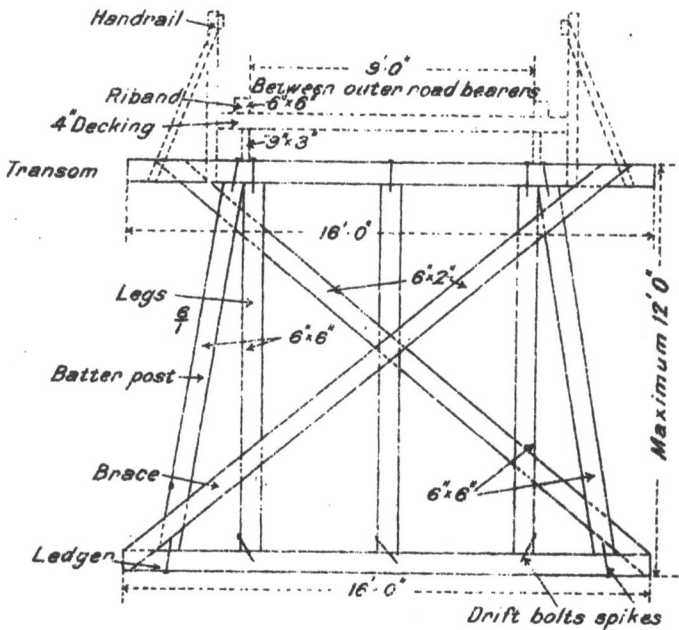

Handrail

Riband

4" Decking

Transom

9'0"
Between outer road bearers
6"x6"
9"x3"

16'0"

Legs

6"x2"

6"x6"

Batter post

Brace

6"x6"

Ledger

16'0"

Maximum 12'0"

Drift bolts spikes

All scantling minimum 6"x6" except braces 6"x2".
For heights 12' to 15' use 4 legs 6"x6" evenly spaced or 3 legs
minimum 7"x7".

Malby & Sons, Lith

Plate 118.

PLANK TRESTLE.

Out of 6"x 2" Timber

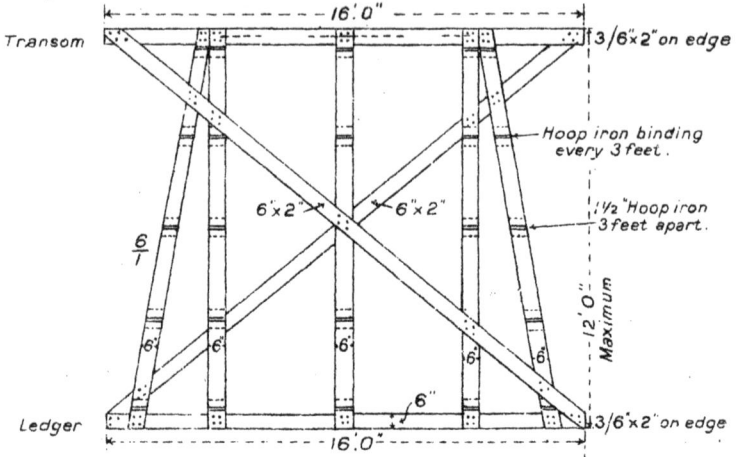

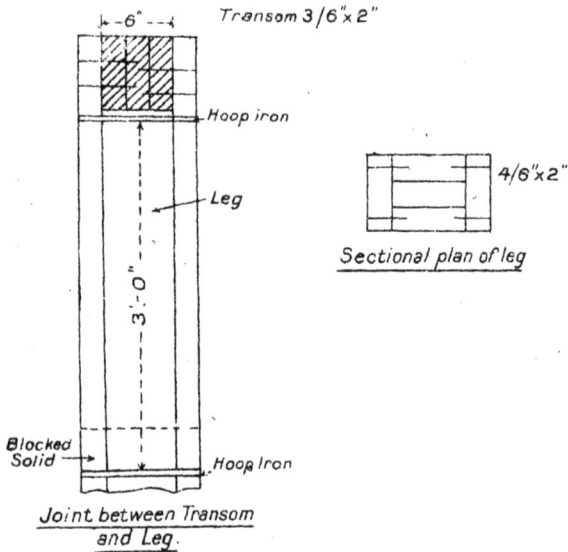

Transom

16'. 0"

3/6"x 2" on edge

Hoop iron binding
every 3 feet.

6"x 2"   6"x 2"

1½"Hoop iron
3 feet apart.

$\frac{6}{1}$

12' 0"
Maximum

6"   6"   6"   6"   6"

6"

Ledger

16'. 0"

3/6"x 2" on edge

Transom 3/6"x 2"

6"

Hoop iron

Leg

3'. 0"

4/6"x 2"

Sectional plan of leg

Blocked
Solid

Hoop Iron

Joint between Transom
and Leg.

Plate 119.

LASHED SPAR TRESTLE.

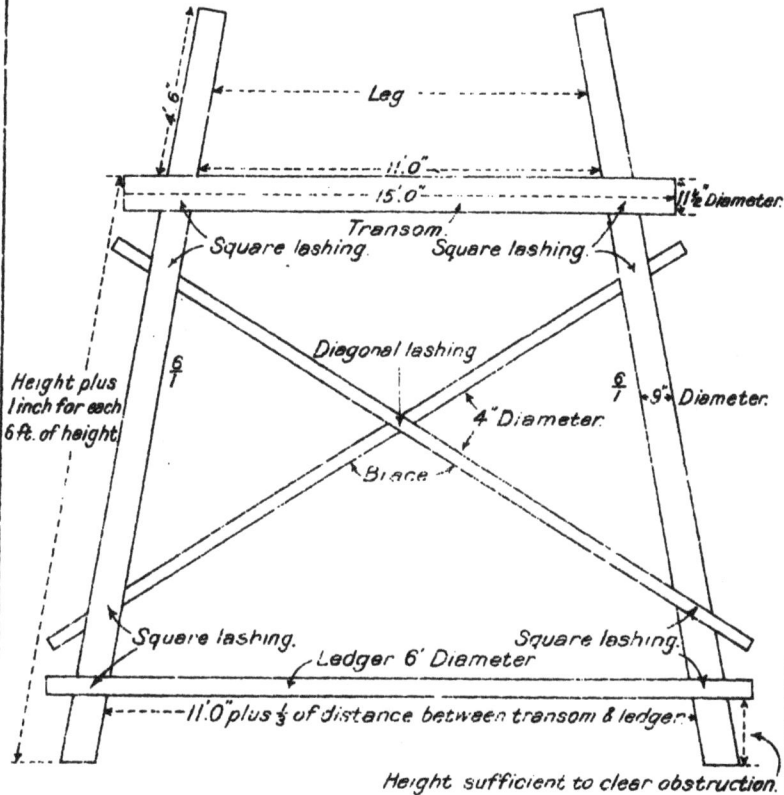

Leg

6'

11'0"

15'0"

11½" Diameter.

Transom.

Square lashing.   Square lashing.

Diagonal lashing

6/1

6/1   9" Diameter.

Height plus
1 inch for each
6 ft. of height.

4" Diameter.

Brace

Square lashing.   Square lashing.

Ledger 6' Diameter

11'0" plus ⅓ of distance between transom & ledger.

Height sufficient to clear obstruction.

Diameter of spars given are suitable for a 15 foot bay and a
15 feet height of trestle.

For other Spans & Heights:

Transom: 10" diameter for 10 feet bay increase ½ inch diameter
for every 2 feet increase in span of bay.

Legs: Reduce ½ inch for every foot decrease in height with 6 inch
minimum. With span of 10 feet legs can be ½ an inch lighter than
with 15 feet spans.

Plate 120.

LAUNCHING TRESTLES.

Fig. 1.

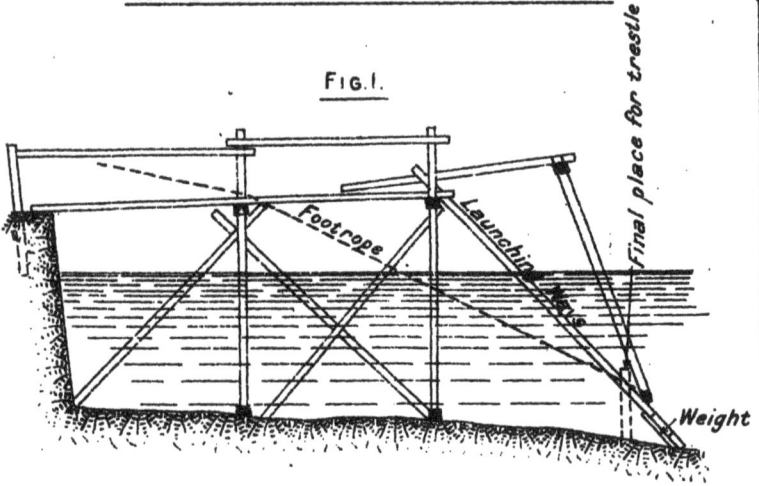

Final place for trestle

Footrope

Launching

Weight

SINGLE BARREL RAFT.

FIG. 2.

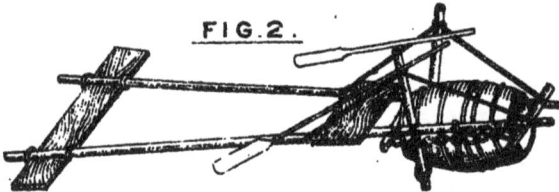

Plate 121.

BOAT PIER.

Fig.1. PLAN

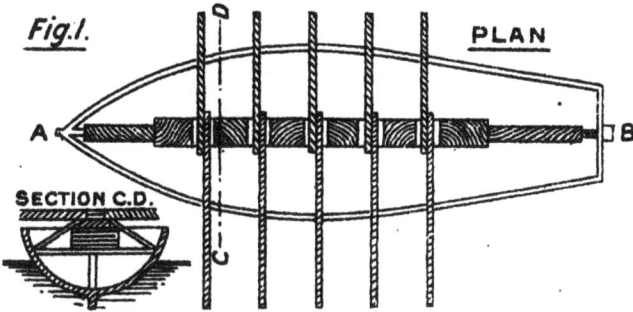

SECTION C.D.

Fig.2. SECTION ON A.B.

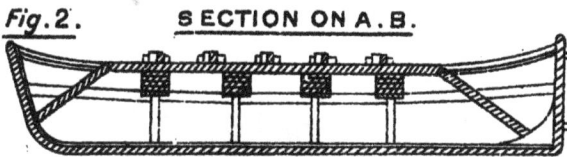

BARREL PIER.

Fig.3.

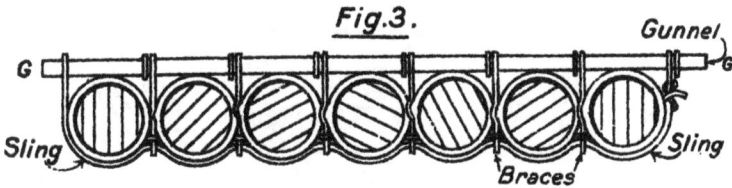

Gunnel

G

Sling Braces Sling

METHOD OF LASHING BARRELS TO GUNNELS.

Fig.4

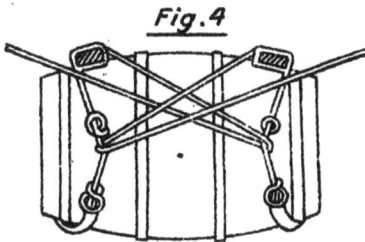

Fig.5.

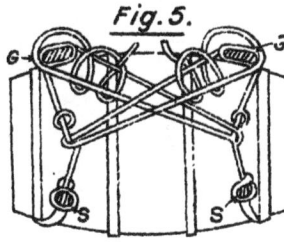

Plate 122.

BARREL BRIDGES.

Fig. 1.
FORMING BARREL BRIDGE FOR SWINGING.

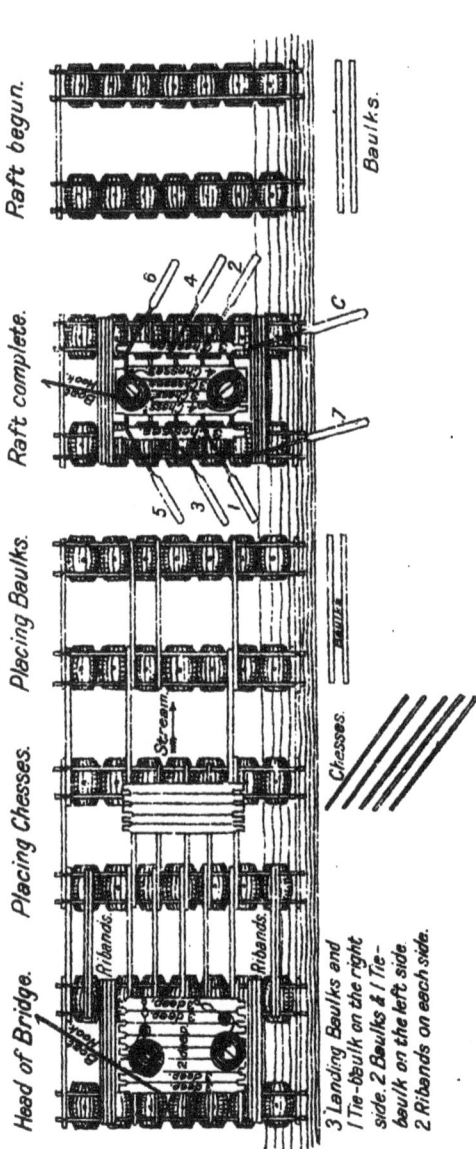

Head of Bridge.   Placing Chesses.   Placing Baulks.

Ribands.

Ribands.

Stream.

Chesses.

3 Landing Baulks and
1 Tie-baulk on the right
side. 2 Baulks & 1 Tie-
baulk on the left side.
2 Ribands on each side.

Fig. 2.
FORMING BARREL RAFTS FOR BRIDGING.

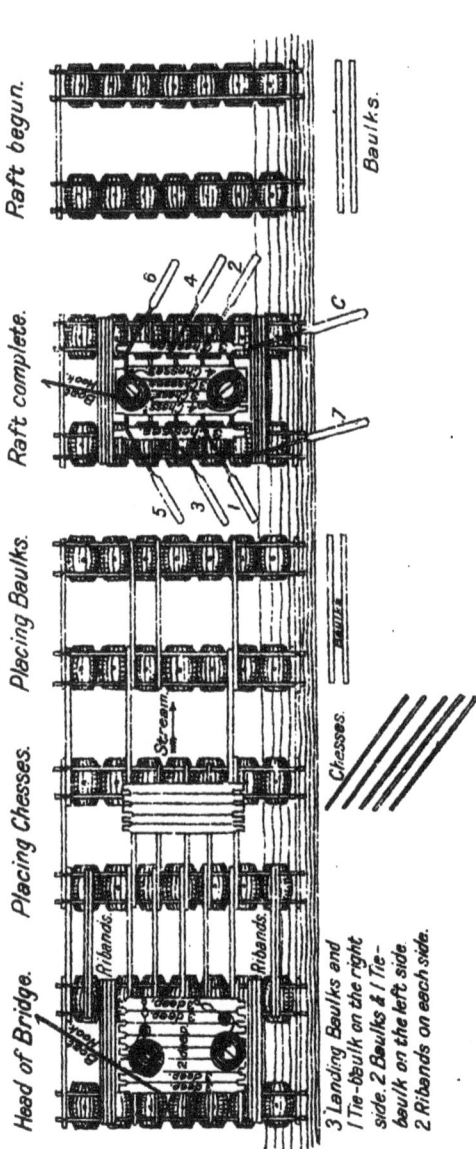

Raft complete.   Raft begun.

Baulks.

6
4
2
C
7
5
3
1

Malby & Sons, Lit.

Plate 123.

FLYING BRIDGE.

Fig. 1.

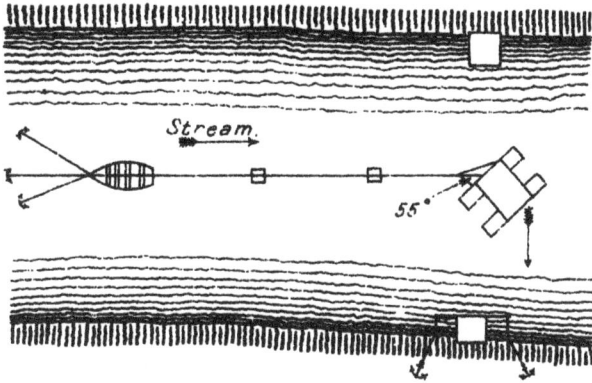

Stream.

55°

POSITION FOR FORD.

Fig. 2.

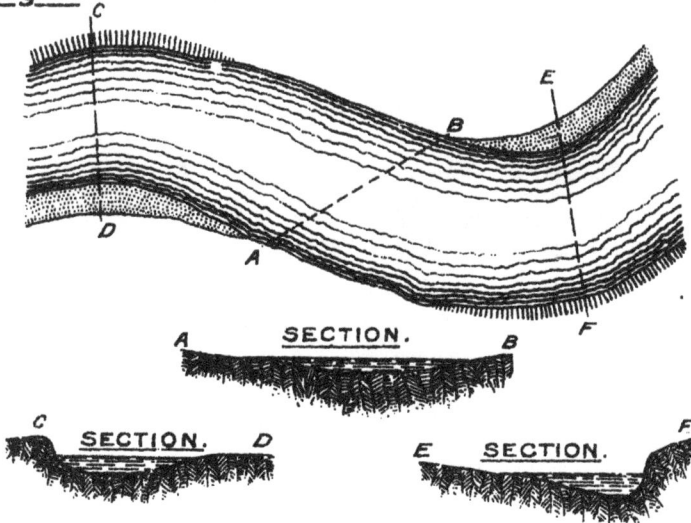

C

E

B

D

A

F

SECTION.

A — B

SECTION.

C — D

SECTION.

E — F

Mallo & Sons Lith

Plate 124.

BRIDGING EXPEDIENTS.

Tarpaulin 18x15' stuffed with straw, &c.

FIG.1.

Raft of four tarpaulins as Fig.1.

FIG.2.

10' to 12'

3'.6"

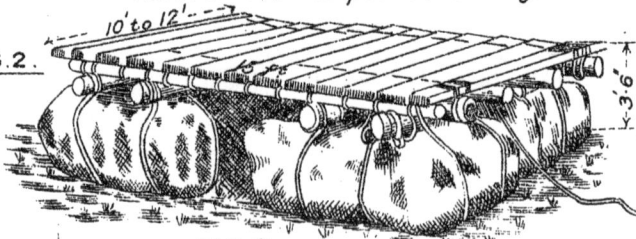

FIG.3.

9"

1'.0"

FIG.4.

4'.3"

2'.6"

4'.3"

2'.6"

12'.0

8'.0"

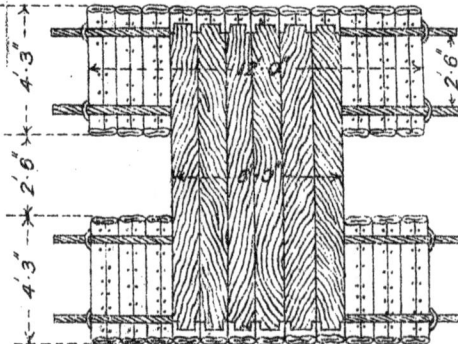

Raft of 24 ground sheets as Fig.3.

Malby & Sons, Lith.

Plate 125.

BIVOUAC SHEET BOAT.

Note:- Framing about 5"x¾". Bottom same
form as "B" but wider.

All frames to be lashed together, not nailed.
4 Carpenters 2½ hours per raft.

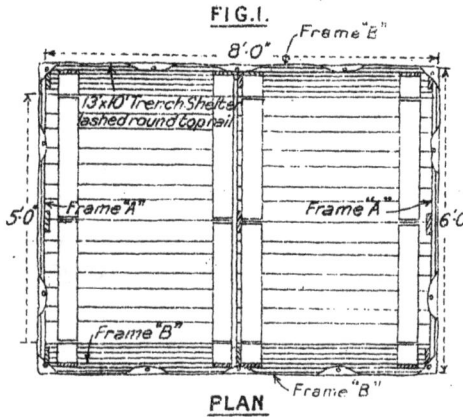

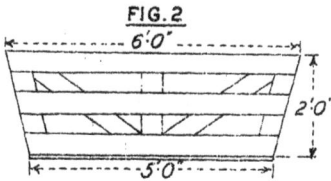

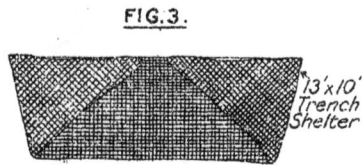

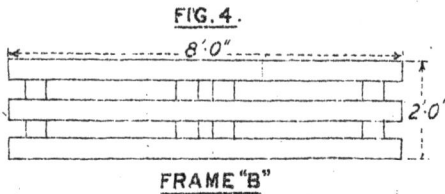

FIG.1.

8'·0"

Frame "B"

13'x10' Trench Shelter
lashed round top rail

Frame "A"

Frame "A"

5'·0"

6'·0"

Frame "B"

Frame "B"

PLAN

FIG.2

6'·0"

2'·0"

5'·0"

FRAME "A"

FIG.3.

13'x10'
Trench
Shelter

ELEVATION

FIG.4.

8'·0"

2'·0"

FRAME "B"

Malby & Sons. Lith.

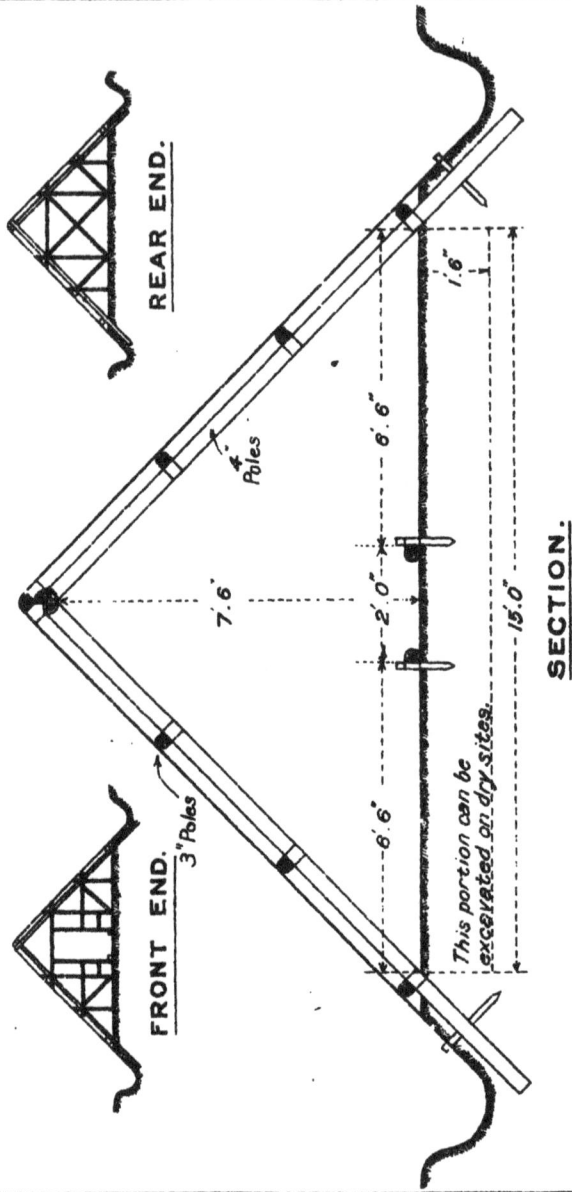

Plate 126.

PORTABLE TARPAULIN HUT.

REAR END.

FRONT END.

3" Poles

4" Poles

7.6"

6.6"

8.6"

2.0"

1.6"

15.0"

This portion can be excavated on dry sites.

SECTION.

Malby & Sons. Lith.

Plate 127.

DISPOSAL OF RAINWATER.

WRONG

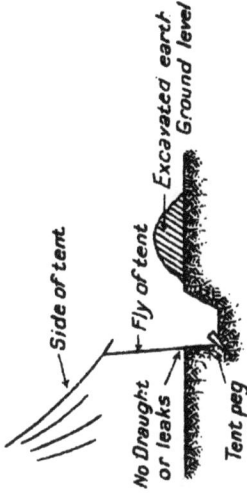

Side of tent

Fly of tent

Excavated earth

Draught & leaks

Tent peg

RIGHT

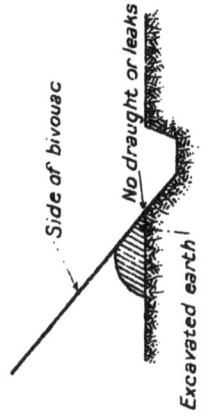

Side of tent

Fly of tent

Excavated earth

Ground level

No Draught or leaks

Tent peg

WRONG

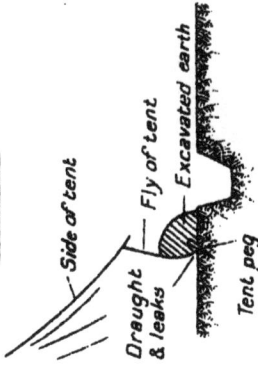

Side of bivouac

Excavated earth

Draught & leaks

RIGHT

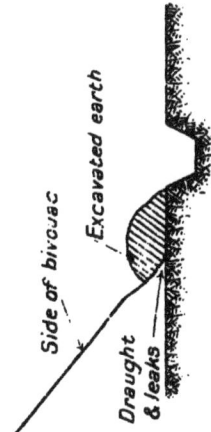

Side of bivouac

No draught or leaks

Excavated earth

Malby & Sons, Lith

Plate 128.

SPLINTER-PROOF PROTECTION FOR TENT.

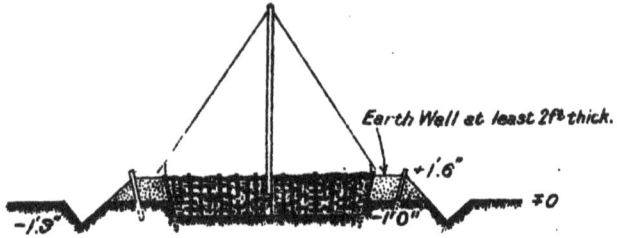

Earth Wall at least 2ft thick.

+1'.6"

-1'.3"   -1'.0"   ∓0

SECTIONAL ELEVATION.

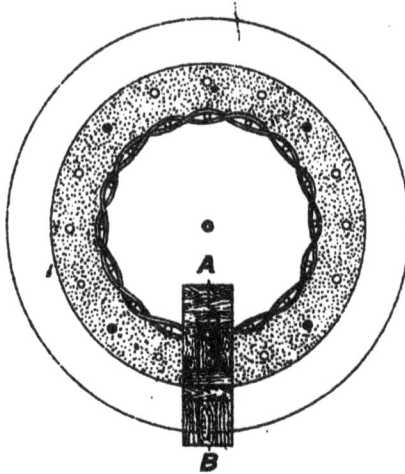

A

B

PLAN.

SECTION A-B.

Plate 129.

STABLES.

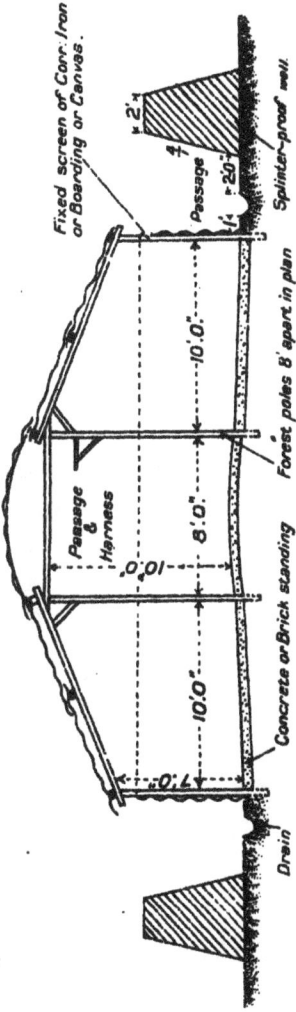

TYPICAL SECTION.

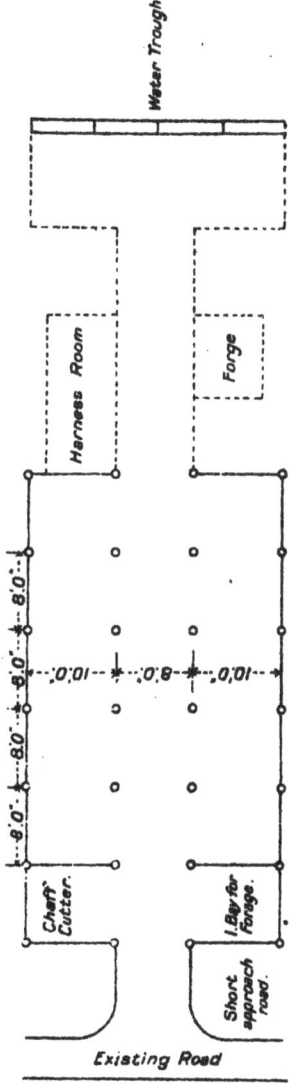

Fixed screen of Corr. Iron or Boarding or Canvas.

Passage

2'

2'0"

Splinter-proof wall.

10'0"

Passage & Harness

8'0"

10'0"

10'0"

7'0"

Drain

Concrete or Brick standing

Forest poles 8' apart in plan

SITE PLAN.

Water Trough

Harness Room

Forge

8'0"

8'0"

8'0"

10'0" 8'0" 10'0"

8'0"

Chaff Cutter.

1 Bay for Forage.

Short approach road.

Existing Road

Malby & Sons, Lith.

Plate 130.

WATER SUPPLY FROM WATER COURSES | WATER SUPPLY FROM MARSH.

OR HOLES.

Fig. 2.

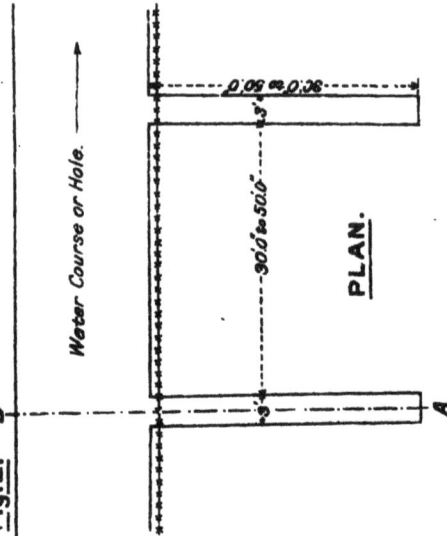

PLAN.

SECTION THRO' A-B.

Fig. 1.

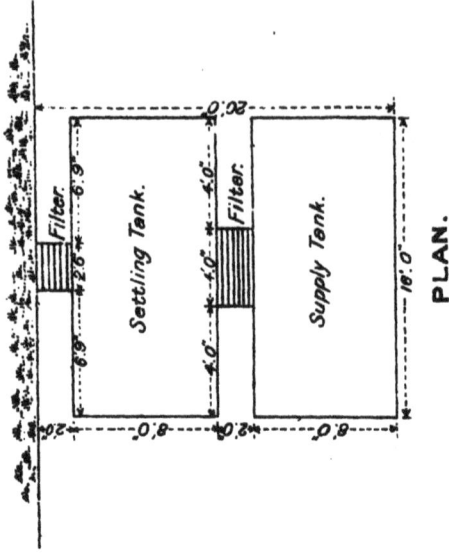

PLAN.

SECTION.

Mintern & Sons, Lith.

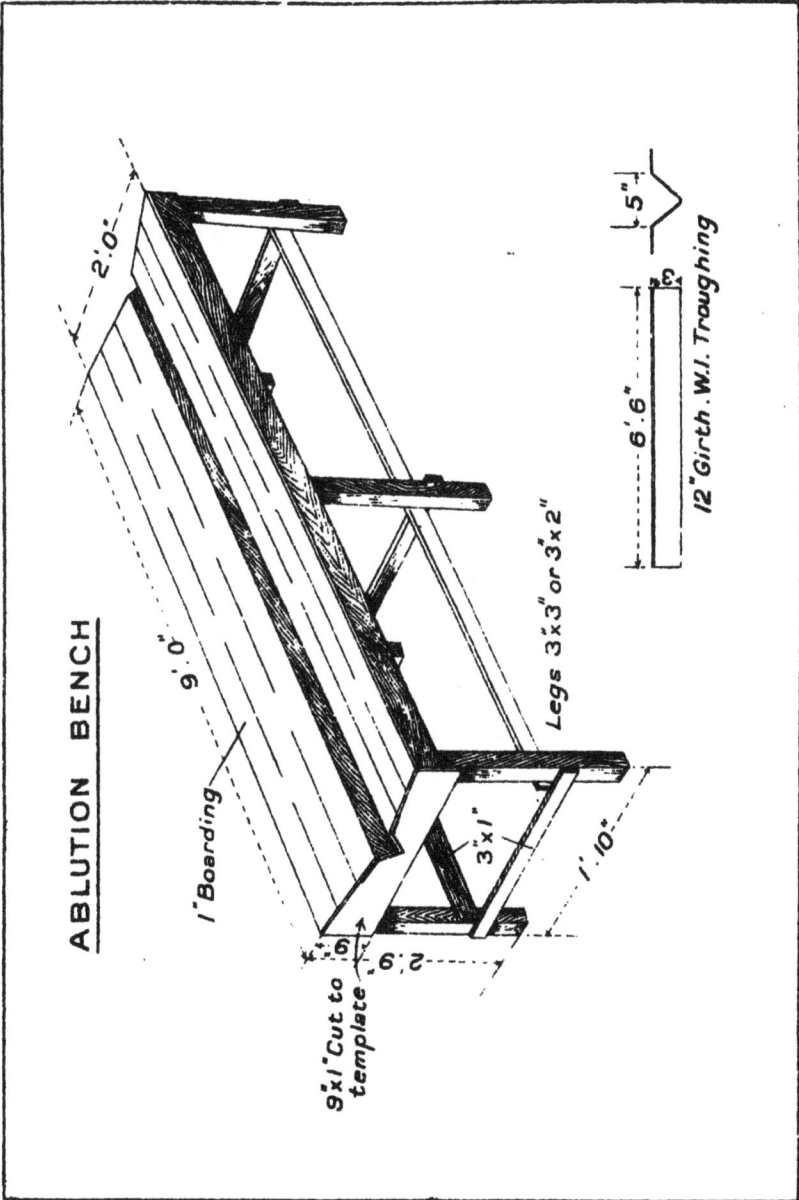

ABLUTION BENCH

Plate 131.

2'0"

9'0"

1" Boarding

Legs 3"x3" or 3"x2"

3"x1"

9"x1" Cut to template

2'9"

6"

1'10"

6'6"

5"

3"

12" Girth. W.I. Troughing

Plate 132.

KETTLE TRENCH.

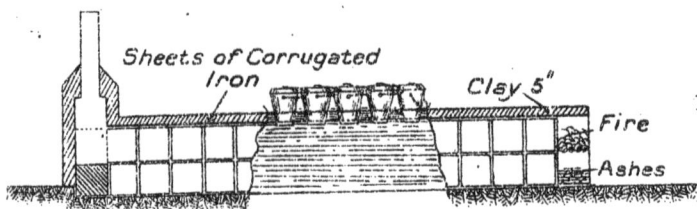

Sheets of Corrugated Iron

Clay 5"

Fire

Ashes

SECTIONAL ELEVATION

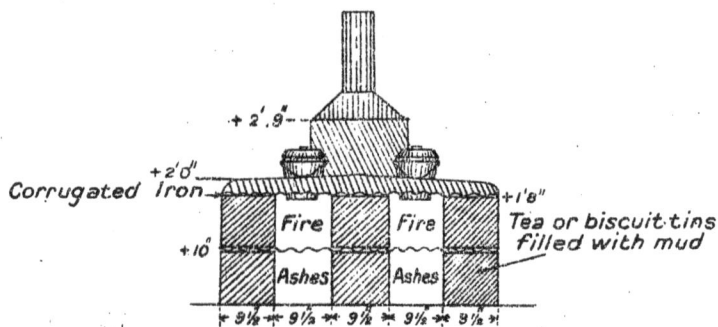

+2'.9"

+2'0"

Corrugated Iron

+1'8"

+10"

Fire

Fire

Ashes

Ashes

Tea or biscuit tins filled with mud

9½   9½   9½   9½   9½

ELEVATION

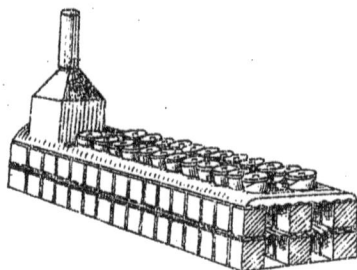

PERSPECTIVE VIEW.

Malby & Sons. Lith.

Plate 133.

CHAMBER OVEN.

Corrugated iron Top bend

Clay (Mud)

Top flue

Oven

+ 3' 6"

2' 3"
Oven

Corrugated iron shelf

Corrugated iron bottom bend.

Tray filled with mud detachable.

Fire

Ash

3' 9"

SECTION THROUGH CHIMNEY

Oven

2'-10"

Tray

Fire

3' 9"

SIDE VIEW.

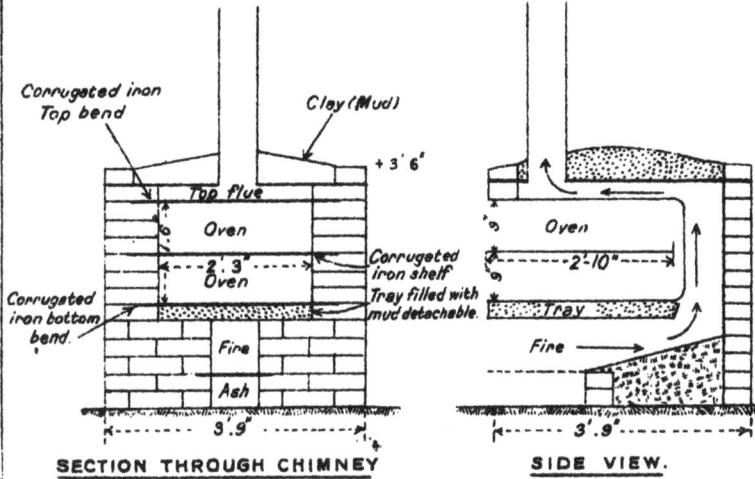

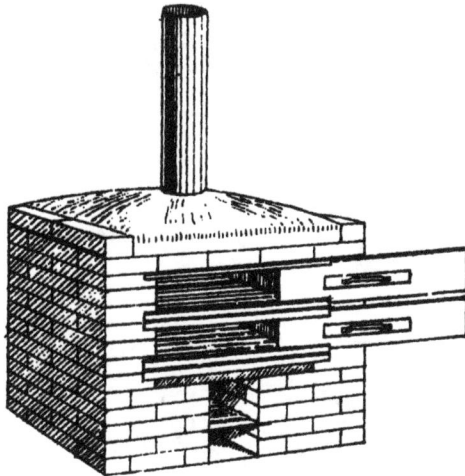

PERSPECTIVE VIEW.

8821.11884. PR3115/387. 62500-12·21   Malby & Sons. Lith

Plate 134.

COOKER FOR CAMPS AND BILLETS.

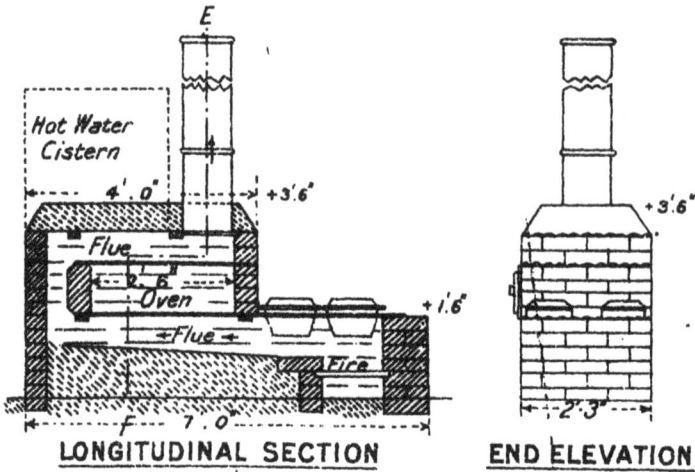

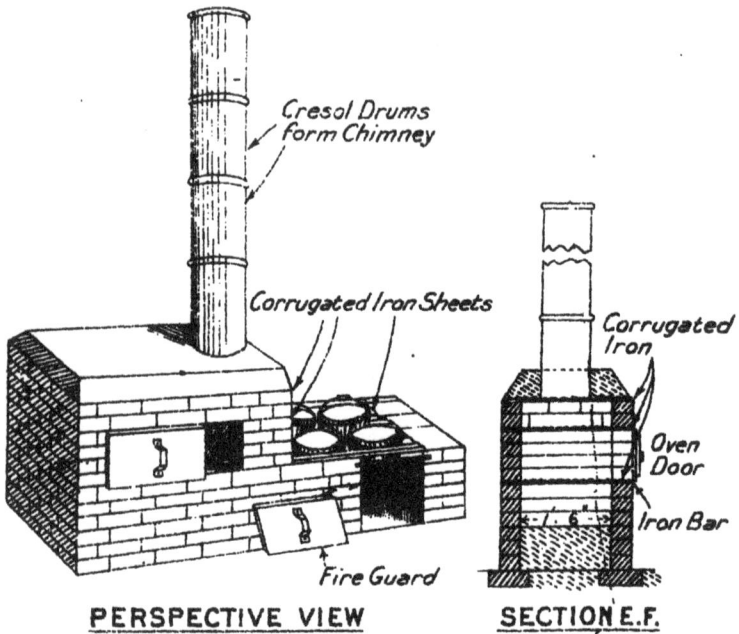

E

Hot Water
Cistern

4'. 0"

+3'.6"

Flue

2' 6"

Oven

+1'.6"

Flue

Fire

F 7. 0"

LONGITUDINAL SECTION

+3'.6"

2'.3"

END ELEVATION

Cresol Drums
form Chimney

Corrugated Iron Sheets

Corrugated
Iron

Oven
Door

1'. 6"

Iron Bar

Fire Guard

PERSPECTIVE VIEW

SECTION E.F.

Plate 135

PORTABLE FIELD KITCHEN.

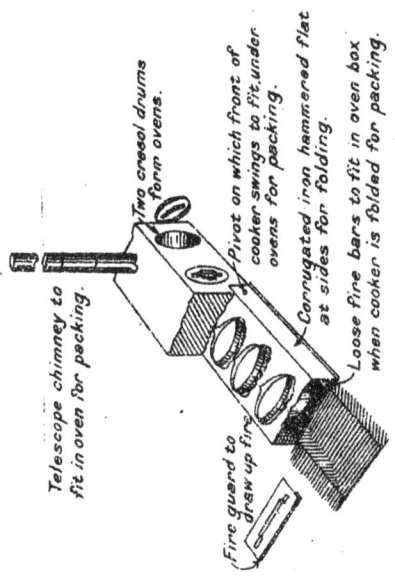

Telescope chimney to fit in oven for packing.

Two cresol drums form ovens.

Pivot on which front of cooker swings to fit under ovens for packing.

Corrugated iron hammered flat at sides for folding.

Loose fire bars to fit in oven box when cooker is folded for packing.

Fire guard to draw up fire.

PERSPECTIVE VIEW.

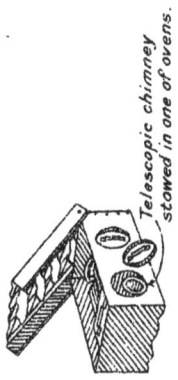

Telescopic chimney stowed in one of ovens.

Method of folding and packing.

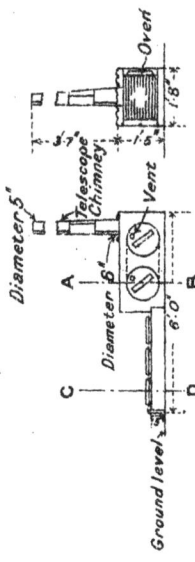

Diameter 5"

Oven

Telescope Chimney

Vent

A

Diameter 6"

B

6'0"

C

D

Ground level

SIDE ELEVATION. SECTION A.B.

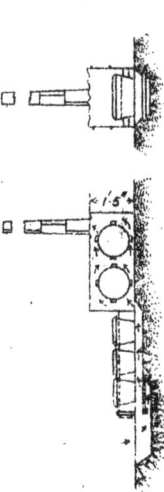

3'7"

1'8"

1'5"

1'6"

LONGITUDINAL SECTION. SECTION C.D.

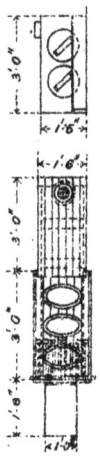

3'0"

1'6"

1'8"

3'0"

3'0"

1'6"

PLAN.

1'0"

Cooker folded complete

Malby & Sons, Lith.

B&2/11084 PP.3115/387.

Plate 136.

PORTABLE FLY-PROOF MEAT SAFE.

All Framing made from 3" x ¾" Timber.

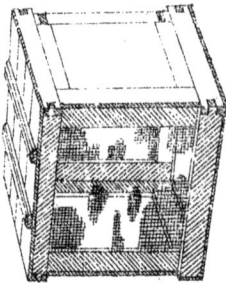

Meat Safe Closed.

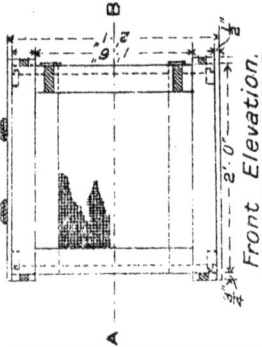

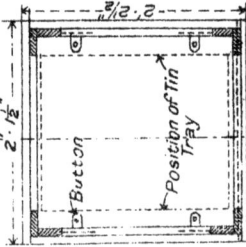

Detail of Button.

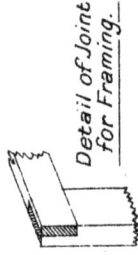

Detail of Joint for Framing.

Detail of Joint for Trays.

Perspective View.

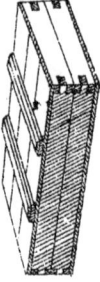

Elevation of Section C.D.

Front Elevation.

Button

Position of Tin Tray

Plan of Section at A.B.

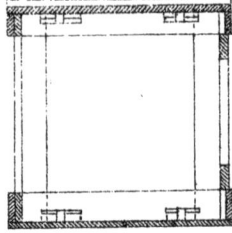

Malby & Sons Lith.

Plate 137.

TYPES OF URINALS.

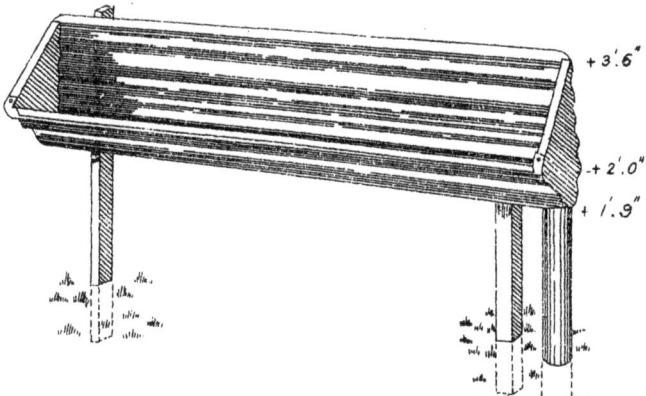

+ 3'.6"

+ 2'.0"

+ 1'.9"

PERSPECTIVE VIEW

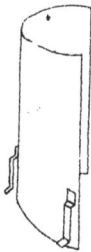

SIDE VIEW

+ 2'.10"

+ 1'.9"

Night
Urinal

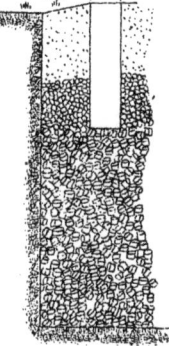

SECTION THRO'
SOAK PIT.

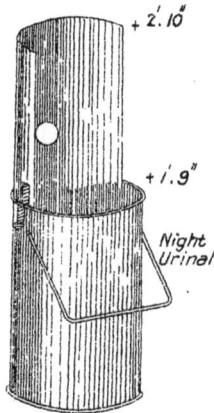

NIGHT URINAL.

8321.11084. P P 3115/387.

Malby & Sons. Lith.

Plate 138.

FLY-PROOF DEEP TRENCH LATRINE.

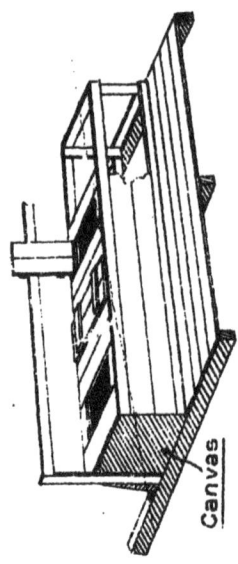

Canvas

PERSPECTIVE VIEW

Detail of Hinge

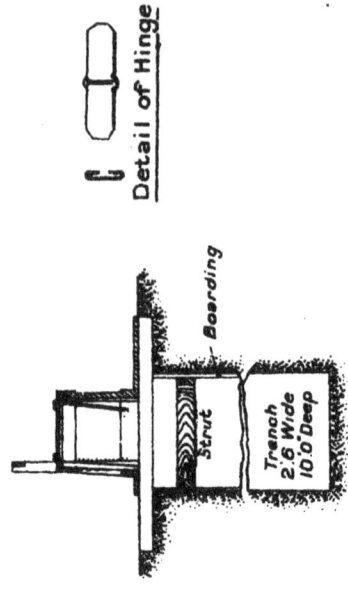

Boarding

Strut

Trench
2'6" Wide
10'0" Deep

SECTION A.B.

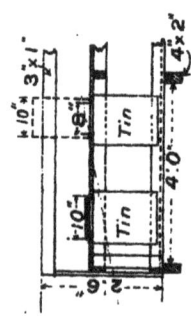

1'10" 3"×1"

6"

Tin

Tin

4'0"

4"×2"

1'10"

Tin

2'6"

SECTION C.D.

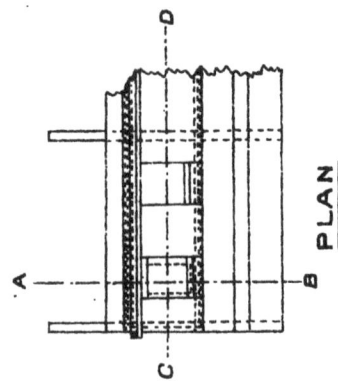

A

C

D

B   PLAN

Malby & Sons. Li:

8981 11064. P.P.9115/387.

Plate 139.

GREASE TRAPS.

Fig. 1.

Filled with brushwood

Outlet

SECTION

Water line
Bottom of inner
tin perforated

Fig. 2.

Tin anti-splash plate

Tin turned over and
nailed to wood framing

Petrol tin bottom
perforated

Tin turned over and
nailed to wood framing.

Baffle plates

SKETCH OF INTERCEPTOR

COMPLETED GREASE TRAP

B

Tin baffle plates

1'6"

1'6"

PLAN

Outlet

A

Joints of woodwork should be
filled with pitch or tar

Tin anti-splash plate

Ground level

Biscuit tin outlet at
bottom.

Petrol tins filled with
bracken or straw,
bottom perforated.

Outlet

Trench drain filled
with broken bricks.

SECTION A.B.

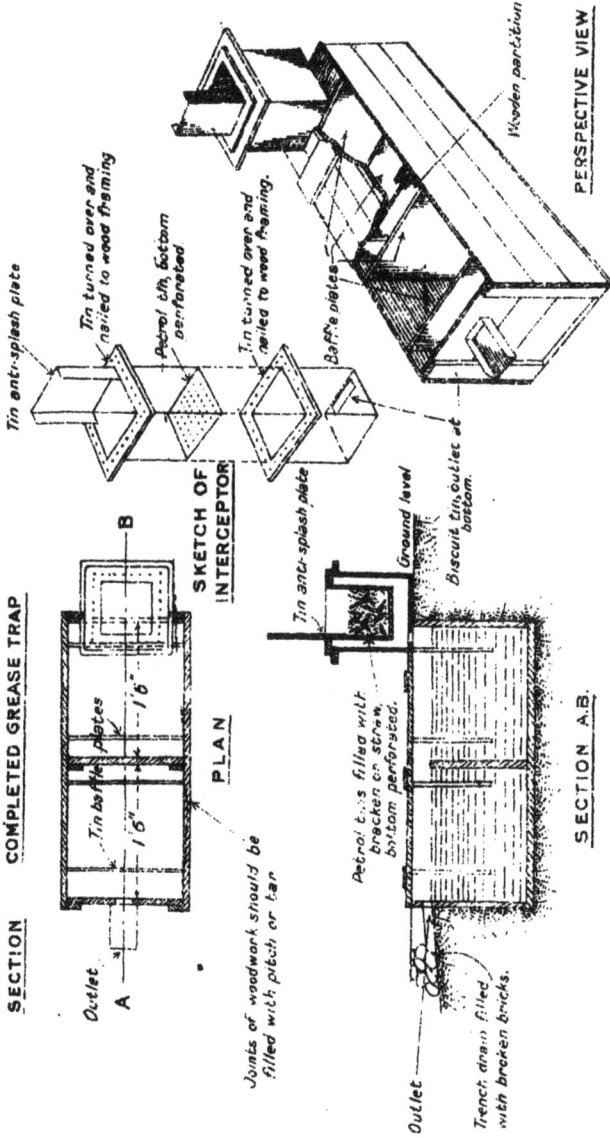

Malby & Sons, Lith

PERSPECTIVE VIEW

Wooden partition

Plate 140.

TEMPORARY INCINERATOR.

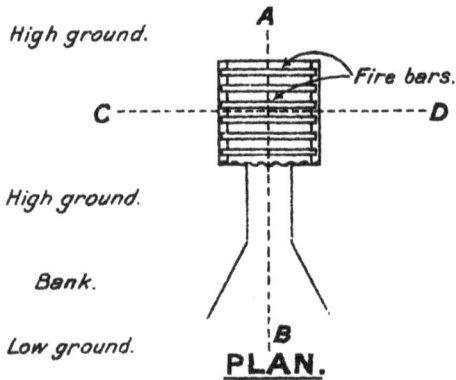

High ground.

Fire bars.

C ----- D

High ground.

Bank.

Low ground.

PLAN.

Incinerator dug out of side of bank.
Refuse fed at top of incinerator from high ground
and drawn from passage cut out of bank side.

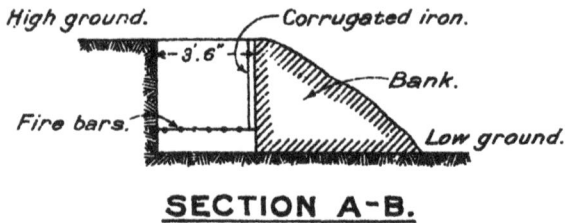

3′.6″

3.0″

3′.0″

Fire bars.

SECTION C-D.

High ground.

Corrugated iron.

3′.6″

Bank.

Fire bars.

Low ground.

SECTION A-B.

Malby & Sons, Lit'

Plate 141.

SMALL CORRUGATED STEEL SHELTER.

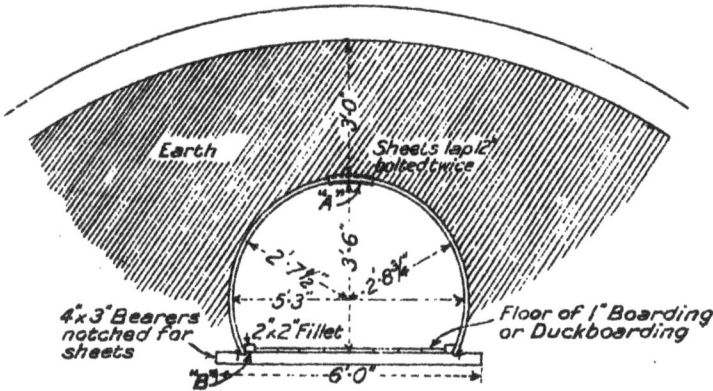

Earth

Sheets lap 12"
bolted twice

3'·0"

3'·6"

2'·7½"

2'·8¼"

5'·3"

4"x3" Bearers
notched for
sheets

2"x2" Fillet

Floor of 1" Boarding
or Duckboarding

"B"

6'·0"

SECTION

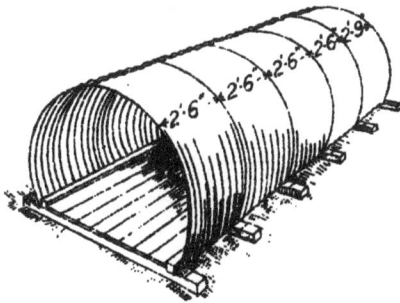

2'·6" 2'·6" 2'·6" 2'·6" 2'·9"

PERSPECTIVE

Note :—
Holes for bolts should be carefully punched
to ensure both bolts taking a bearing in
both sheets without distortion.

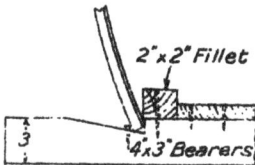

2"x2" Fillet

3

4"x3" Bearers

DETAIL AT "B"

12" Lap

Bolts ½" dia. 1¼" long,
screwed full length.

DETAIL AT "A"

Malby & Sons. Lith.

Plate 142.

SPLINTER-PROOF SHELTER.

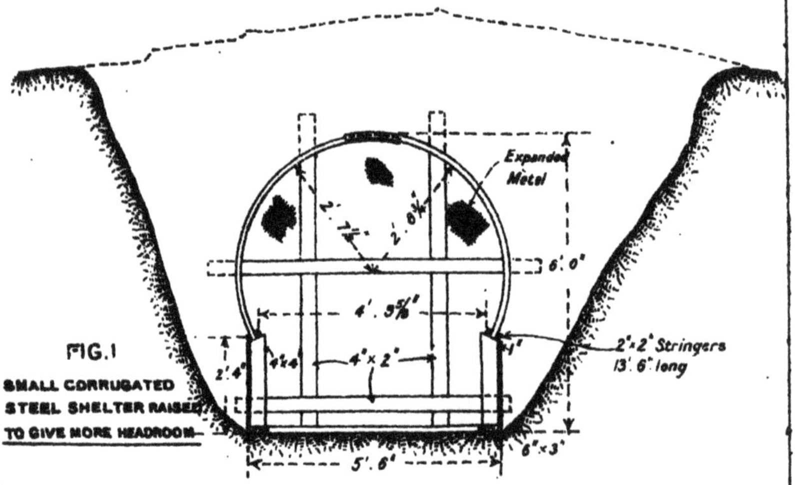

FIG.1

SMALL CORRUGATED
STEEL SHELTER RAISED
TO GIVE MORE HEADROOM

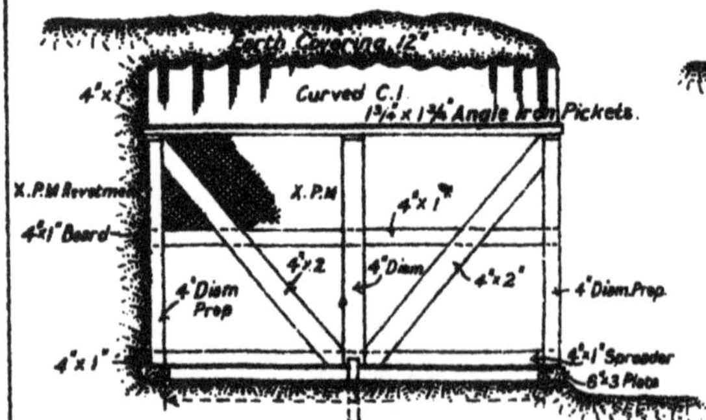

Expanded Metal

2'x2' Stringers 13' 6" long

FIG.2. **SECTION.**

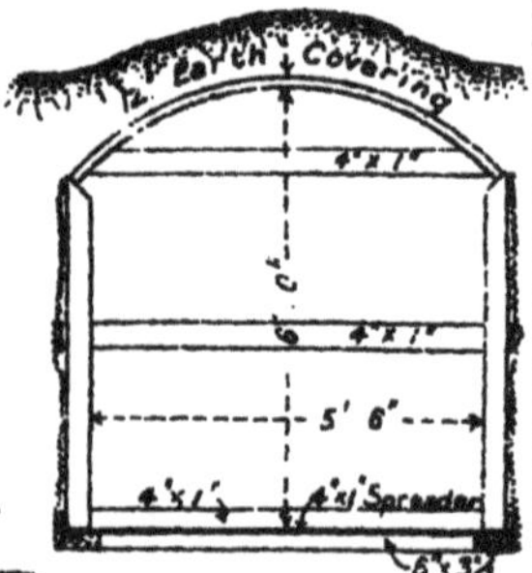

FIG.3.
FRONT ELEVATION.

USING SHEETS OF CURVED C.I.

LARGE CORRUGATED STEEL SHELTER.

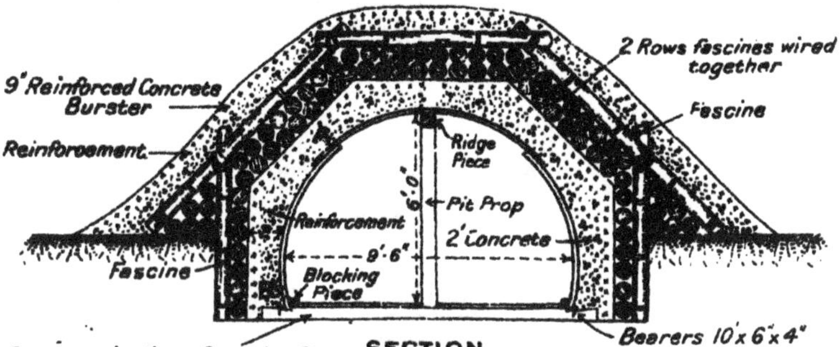
2 Rows fascines wired together
9" Reinforced Concrete Burster
Fascine
Reinforcement
Ridge Piece
6'·0"
Pit Prop
Reinforcement
2' Concrete
9'·6"
Fascine
Blocking Piece
Bearers 10"x 6"x 4"
Concrete in lieu of wooden floor in damp sites.

SECTION.

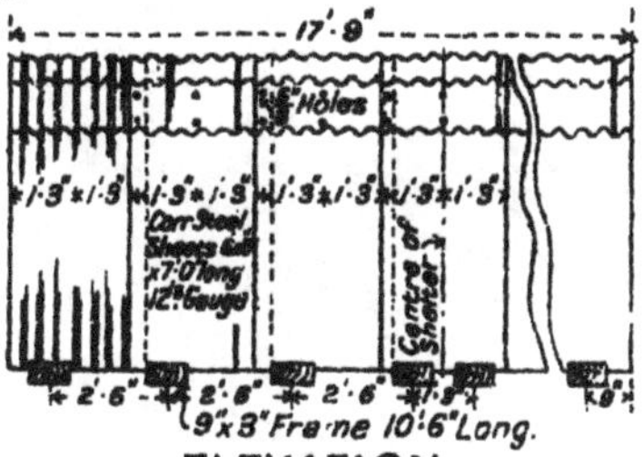
17'·9"
Holes
1'·3"
Corr. Steel Sheets 2'·9" x 7'·0" long 12 Gauge
Centre of Shelter
2'·6"
2'·6"
2'·6"
1'·3"
9"
9"x 3" Frame 10'·6" Long.

ELEVATION

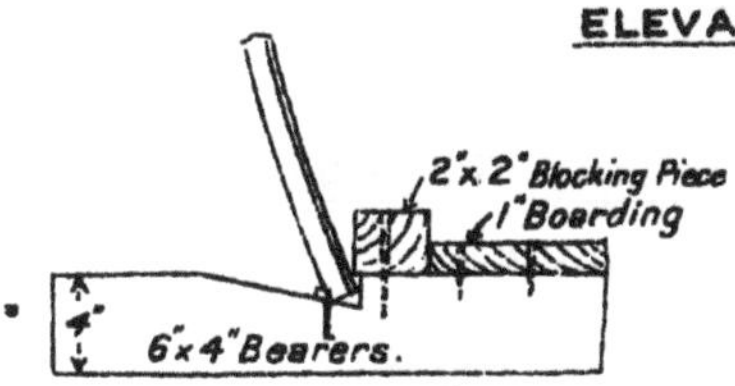
2"x 2" Blocking Piece
1" Boarding
4"
6"x 4" Bearers.

DETAIL AT "B"

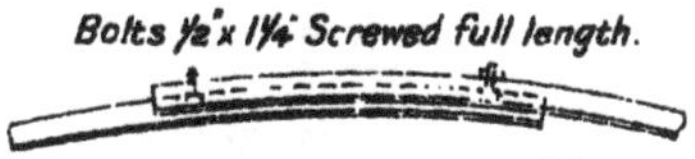
Bolts ½"x 1¼" Screwed full length.

DETAIL OF LAP "A"

Material required for each Shelter 17'·9" Long.
21 Sheets of large corrugated Shelter. (Each 2'·9" wide. 3" inch lap)
60 Bolts.
8 Bearers 10"x 6"x 4". 4 Pit Props.
160 Sq. Ft. of Boarding for floor.
Gas-proof double door frame complete.
60 Fascines Approximately.
Concrete material. Gravel 44 tons. Sand 22 tons. Cement 70 Barrels for each Shelter.
Concrete may be economised by grouping two or three shelters together.
Accommodation double bunked 12 men in each shelter.

8921. 11084. PP3115/387

Malby & Sons. Lith

Plate 144.

SPLINTER-PROOF SHELTERS.

INFANTRY TRENCH SHELTER BEHIND PARADOS.

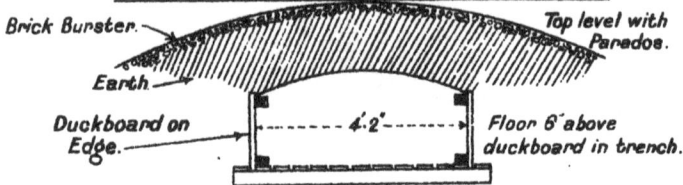

Brick Burster.

Top level with
Parados.

Earth

Duckboard on
Edge.

4'.2"

Floor 6' above
duckboard in trench.

SECTION

SKELETON SKETCH SHOWING CONSTRUCTION.

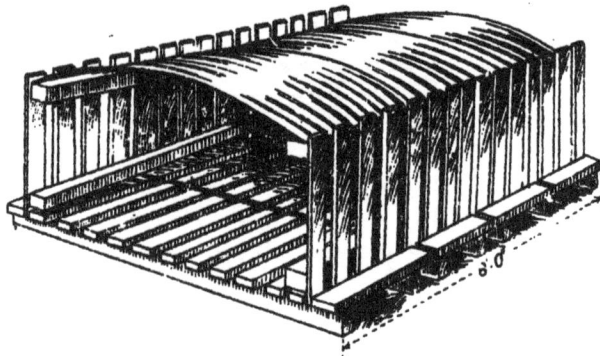

SKETCH SHOWING POSITION OF ENTRANCE ON SAFE SIDE.

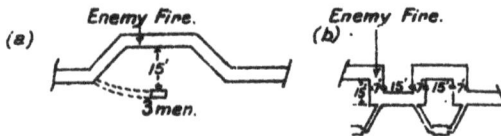

(a) Enemy Fire.

15'

3 men.

(b) Enemy Fire.

15'   15'

Small Elephant Shelters.

MATERIAL REQUIRED.

3. Curved sheets of small corrugated iron shelter.
2. Duckboards for sides.
4. Duckboards for floor.
2. Old corrugated sheets or boarding to form back.
   Brick rubble for top cover.
   Accommodation - 2 to 3 men.

NOTE :- Level of bottom of fire trench will regulate level of bottom of shelter, i e.
there will be a gradual down grade from bottom of shelter to bottom of
fire trench for drainage.

Malby & Sons, Lith

Plate 145.

CUT AND COVER DUG-OUT WITH TROUGHING PLATES.

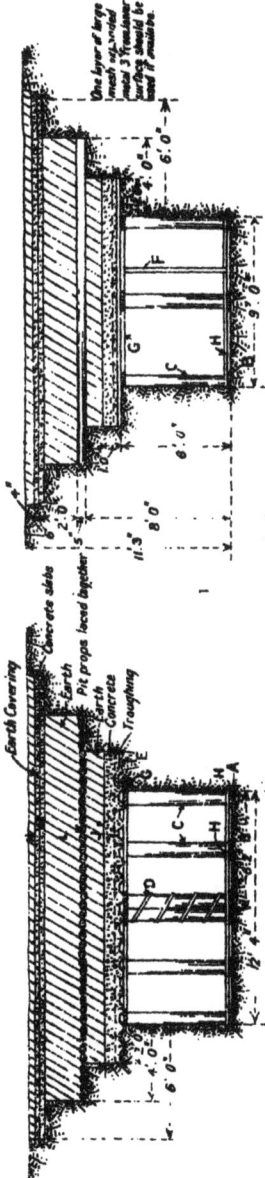

SECTION W.X.

SECTION Y.Z.

DETAIL OF UNIT "C"

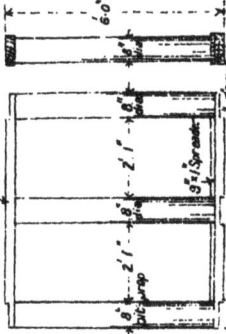

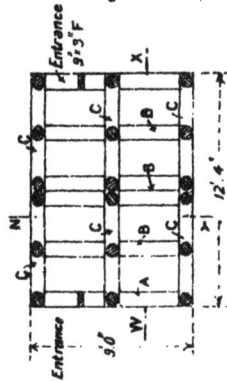

PLAN

No	MATERIAL		LETTER
2	Sills 9'-0"×3"×3".		A
2	9'-0"×3"×3".		B
6	Units Piecaces, heads and sills.		C
60	Ft. run Hoop Iron.		D
2	Troughing plates 3'-0"×3'.		E
2	Entrance frame 3'-6"×3"×3		F
4	Heads 3'-6"×3"×3 (turned at ends)		G
42	Lin. Ft. Stretchers 9'-1".		H
8	Cub. Yds. Concrete.		I
	" " Earth: returned.		J
7½	Lin. Ft. Pit prop. 5'dia		K
842	Lin. Ft. Pit prop. 3"dia		K
28	Cub. yds. Earth: returned.		L
680	Concrete Slabs 1'-6"×1'-0"×3"		M
5	Cub. Yds. Earth: covering returned		N
100	Sup. Ft. Flooring 7"×1"		Q

Mine Cases to answer also as in 151

Total Excavations 84 cub. yds.

Plate 146.

TIMBER CONSTRUCTION.

COMMON FAULTS.

FIG. 1.

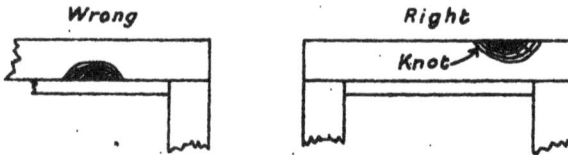

Wrong   *Right*   Knot

FIG. 2.

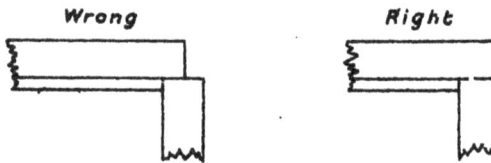

Wrong   *Right*

FIG. 3.

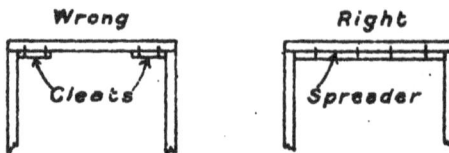

Wrong   *Right*

Cleats   Spreader

FIG. 4.

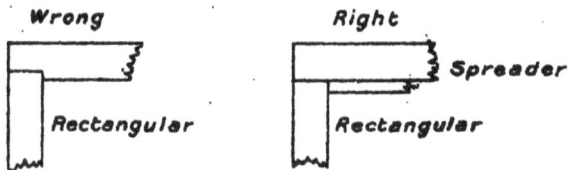

Wrong   *Right*

Rectangular   Spreader

Rectangular

8931. 11084 PP 3115/997.   Malby & Sons Lith

Plate 147.

TIMBER CONSTRUCTION.
COMMON FAULTS.

Fig. 1.

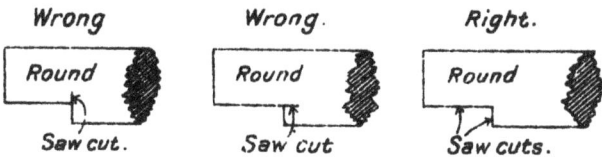

Wrong	Wrong.	Right.
Round	Round	Round
Saw cut.	Saw cut	Saw cuts.

Fig. 2.

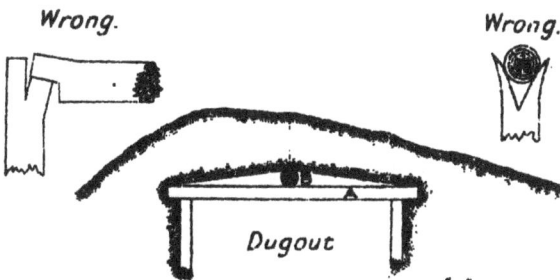

Wrong.

Wrong.

Dugout

If a roof beam is constructed with a log 'B' stretched across the centre of a beam 'A', the stresses in 'A' are due to the load on its centre i.e. concentrated and double what they would be if log 'B' were omitted, and the load distributed over the whole beam 'A'.

Fig. 3.

Wrong.

Wrong.

Right

Sandbags

Plate 148.

STRENGTH OF BEAMS.

CALCULATION OF LOAD.

FIG. 1.

Rails must be laid thus   not

Steel joists must be laid thus   not

Timber joists must be laid thus   not

FIG. 2.

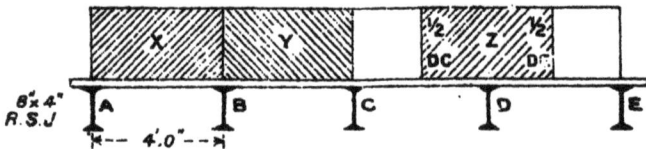

Beam "B" supports half "X" and half "Y".

Beam "A" supports half "X" only.

Therefore in a continuous roof with a load evenly distributed, estimate the load on any beam by taking
the cubic contents over half the distance between that
beam and those on each side of it, as at "Z".

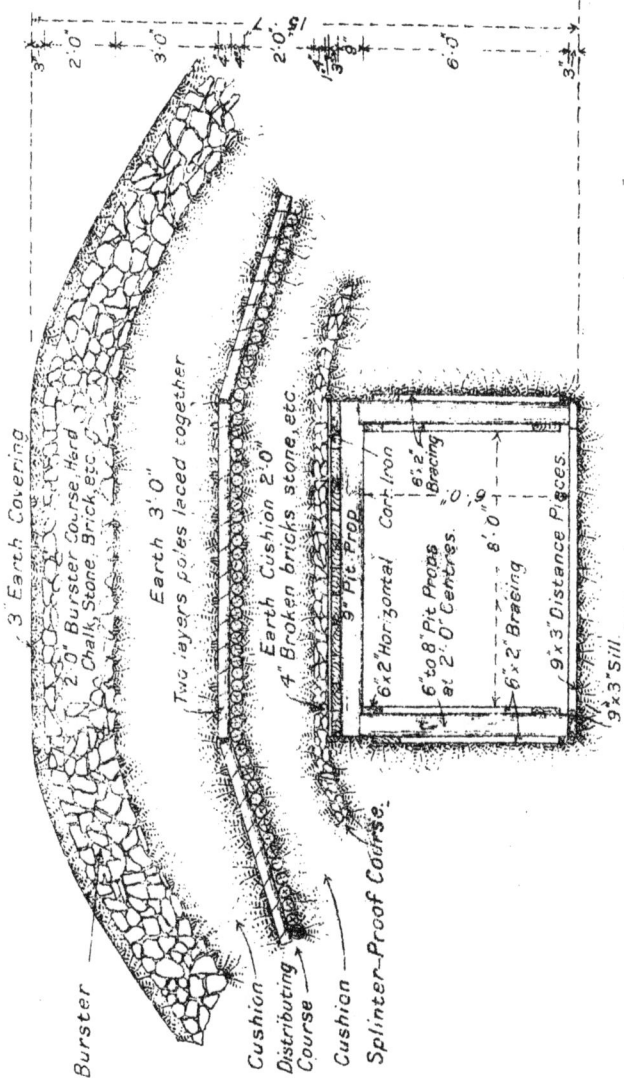

Plate 149.

SPECIMEN OF SUITABLE COVER.

PROOF AGAINST SHELLS UP TO 6 INCHES.

Burster

3' Earth Covering

2'0" Burster Course Hard
Chalk, Stone, Brick, etc.

Earth 3'0"

Two layers poles laced together

Cushion

Distributing Course

4" Earth Cushion 2'0"
Broken bricks stone, etc.

Cushion

Splinter-Proof Course.

9" Pit Prop

6"x 2" Horizontal Cart Iron

6":2"
Bracing

6" to 8" Pit Props
at 2'0" Centres.

6'0"

6" x 2" Bracing

8'0"

9 x 3" Distance Pieces.

9x3"Sill

CROSS SECTION.

3"
2'0"
3'0"
4"
2'0"
3"
6"
6'0"
3"

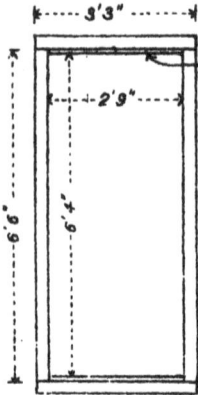

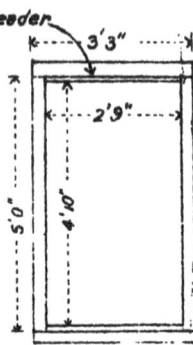

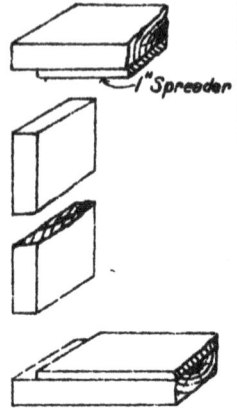

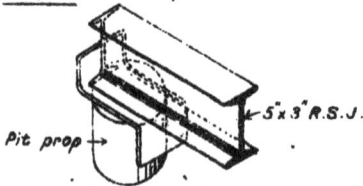

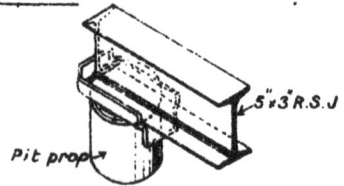

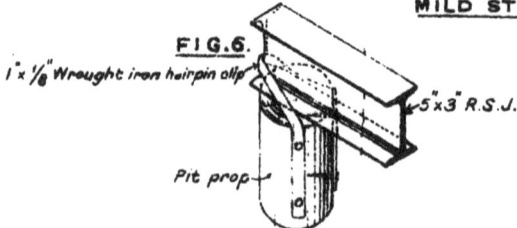

Plate 150.

STANDARD SETS & CLIPS.

3" TIMBER.

FIG.1.

3'3"

2'9"

6'6"

6'4"

1"Spreeder

FIG.2.

3'3"

2'9"

5'0"

4'10"

FIG.3.

1"Spreeder

ELEVATIONS

STANDARD JOINTS

FIG.4.

Pit prop

5"x 3" R.S.J.

DETAIL OF MILD STEEL Z SHOE

FIG.5.

5"x3" R.S.J

Pit prop

DETAIL OF
MILD STEEL SHOE Mᴷ II

FIG.6.

1"x ⅛" Wrought iron hairpin clip

5"x 3" R.S.J.

Pit prop

DETAIL OF HAIRPIN CLIP.

BCi. H984 PP 31 5/387

Malby & Sons Lith

Plate 151.

STANDARD DUG-OUT ACCOMMODATION FOR HEADQUARTERS.

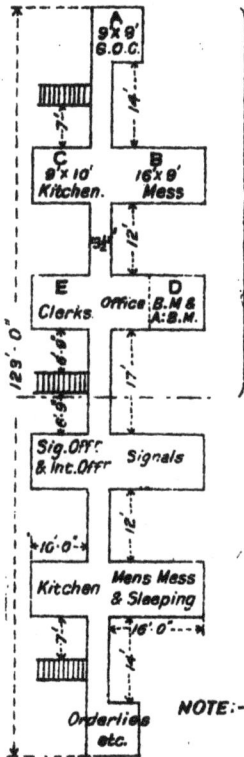

BRIGADE HEAD Q

A
9'x 8'
B.G.C.

7'

C
8'x 10'
Kitchen.

B
16'x 8'
Mess

8'

12'

E
Clerks.

Office

D
B.M &
A.B.M.

17'

129' 0"

Sig.Offr
& Int.Offr

Signals

12'

16' 0"

Kitchen

Mens Mess
& Sleeping

7'

16' 0"

Orderlies
etc.

14'

FOR BATT^N H.Q.

(Double Bn. H.Q.= One Bde H.Q.)

C.O.	Room A
Mess & Office	B
Kitchen	C
Adjt. & Clerks	E
Men & Orderlies	D

FOR H.A. BATTY H.Q.

B.C. Post	A
Wireless	E
Officers Bunks } Mess	B & C
Men 60 O.R. For 2 Guns	

FOR ARTY GROUP H.Q.

Same as for BATT^N

NOTE:- Wherever possible Bⁿ H.Q^{rs} should be arranged in pairs on the above lines, so as to be available as B^{de} H Q^{rs} in an advance.

8821.11084 PP3115/387.

Malby & Sons. Lith.

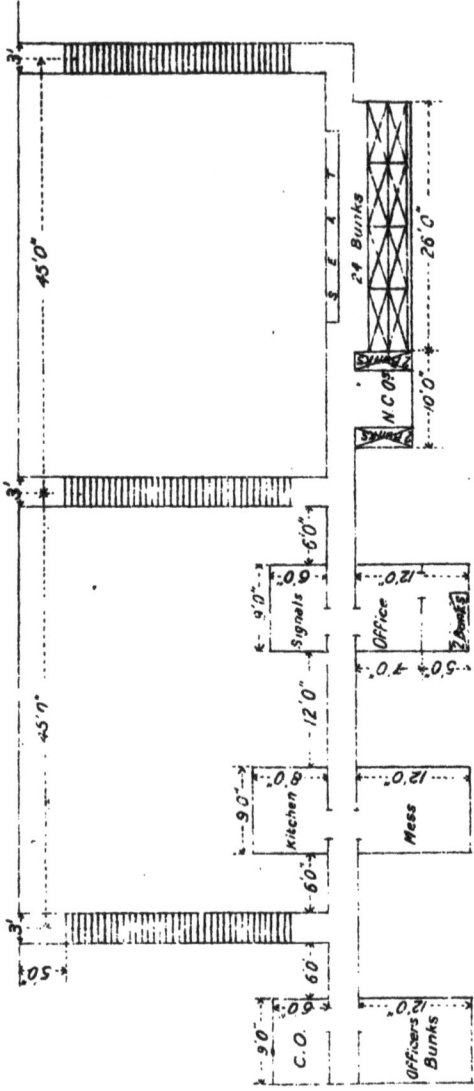

Plate 152.

BATTALION HEADQUARTERS

WITH ACCOMMODATION FOR 28 MEN.

24 Bunks

SEAT

Bunks

Bunks

N.C.O.s

26'0"

10'0"

45'0"

45'0"

Signals

9'0"

6'0"

12'0"

Office

12'0"

7'0"

5'0"

2 Bunks

Kitchen

9'0"

12'0"

6'0"

Mess

5'0"

6'0"

C.O.

9'0"

9'0"

Officers
Bunks

12'0"

8821.11084 PP3115/587

Malby & Sons. Lith

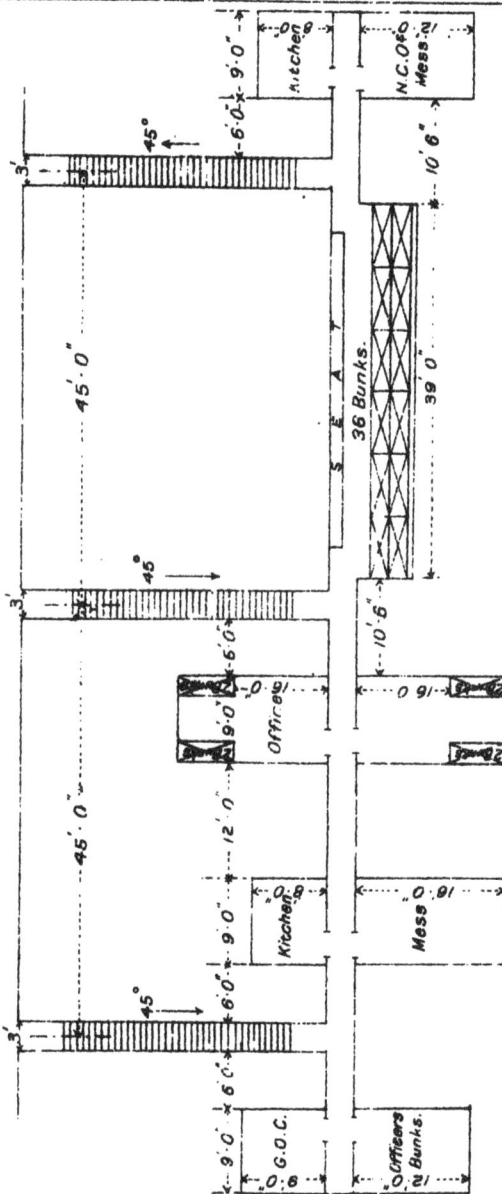

Plate 153.

BRIGADE HEADQUARTERS.

WITH ACCOMMODATION FOR PERSONNEL.

8521. 11084 PP 3115/387

Plate 154.

M.G. DUG-OUT & EMPLACEMENT.

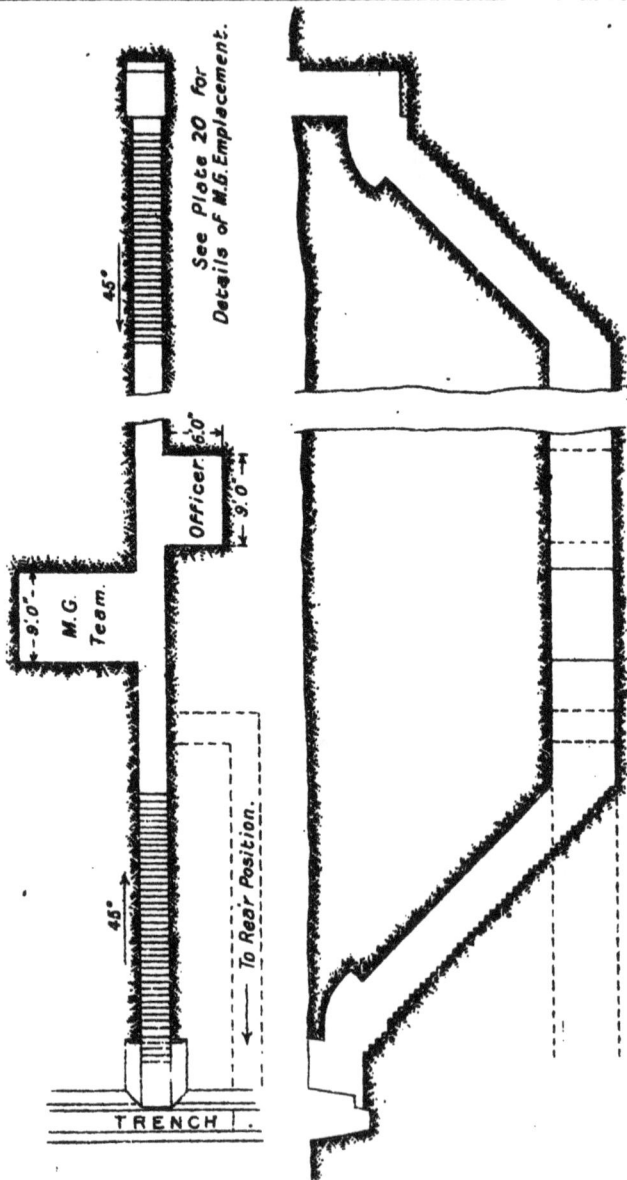

See Plate 20 for
Details of M.G. Emplacement.

45°

Officer. 6.0"

9.0"

9.0"

M.G.
Team.

45°

To Rear Position.

TRENCH

8921.11084 PP3115/387

Malby & Sons. Lith.

Plate 155.

DRESSING STATION.

As other end

9'. 0"

45'. 0"
4'. 6"
or
2'. 9"
13'. 0"

GENERAL ARRANGEMENT.
Fig. 1.

Entrance
Slope 1/2

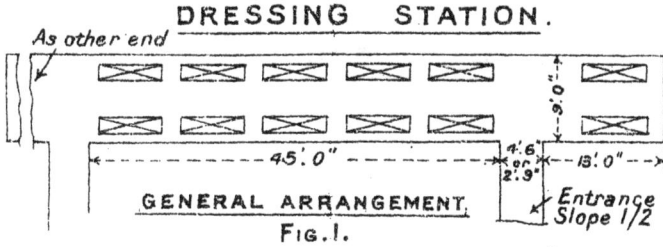

DETAIL OF BRACKETS.

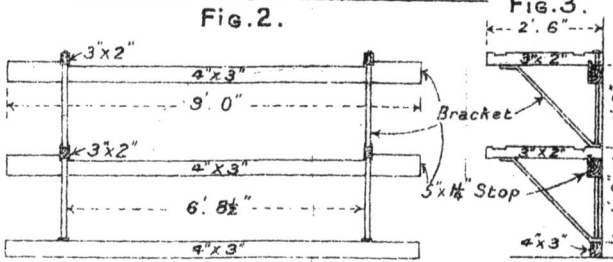

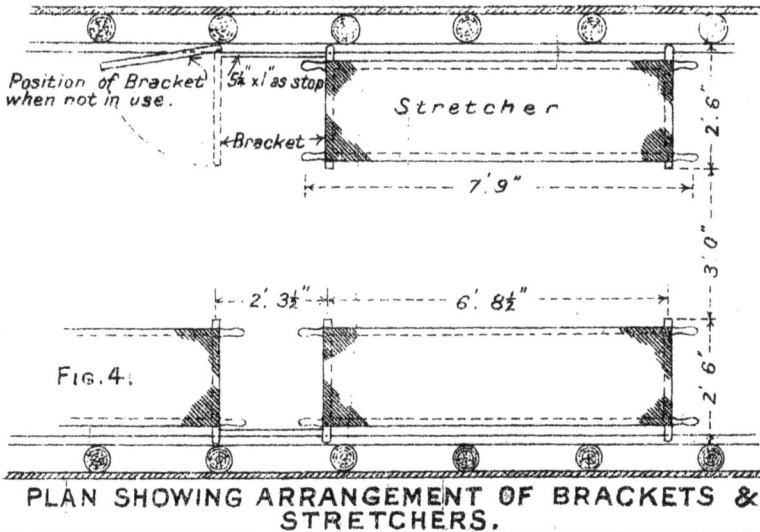

Fig. 2.

3"x 2"
4"x 3"
9'. 0"
3"x 2"
4"x 3"
6'. 8½"
4"x 3"

Fig. 3.

2'. 6"
3"x 2"
Bracket
1'. 9"
3"x 2"
5"x 1½" Stop
1'. 9"
4"x 3"

Position of Bracket
when not in use.
5¼" x 1" as stop
Stretcher
2'. 6"
Bracket
7'. 9"
3'. 0"

Fig. 4.
2'. 3½"
6'. 8½"
2'. 6"

PLAN SHOWING ARRANGEMENT OF BRACKETS & STRETCHERS.

8921. 11084·PP 3115/387.
Malby & Sons, Lith

Plate 156.

TYPES OF DUG-OUTS.

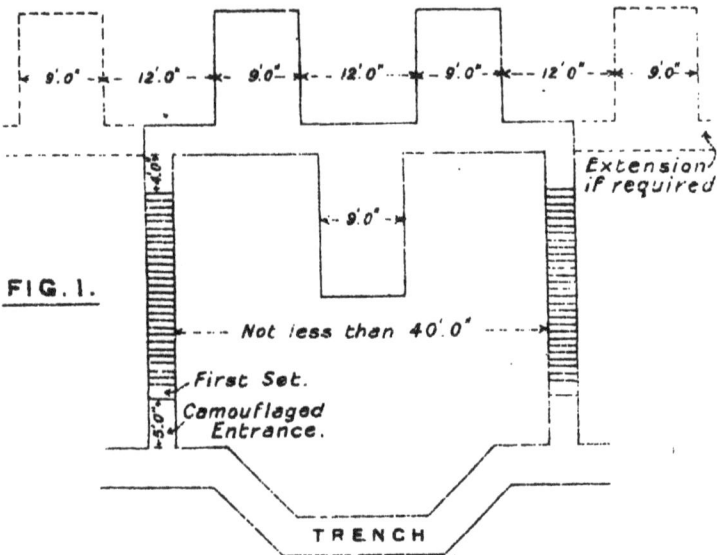

FIG. 1.

9'.0"   12'.0"   9'.0"   12'.0"   9'.0"   12'.0"   9'.0"

Extension
if required

4'.0"

9'.0"

- - - Not less than 40'.0" - - -

First Set.
Camouflaged
Entrance.

5'.0"

TRENCH

TYPE A.

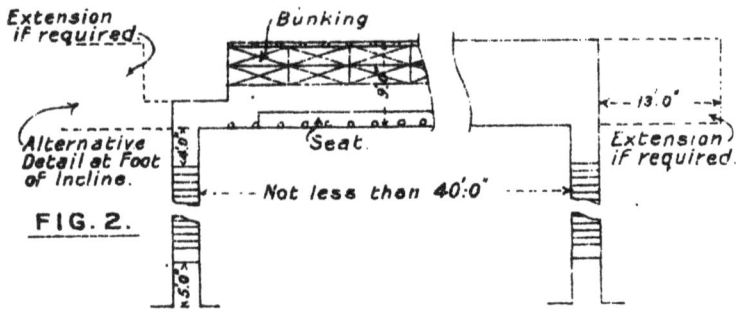

Extension
if required

Bunking

Alternative
Detail at Foot
of Incline.

FIG. 2.

4'.0"

Seat

- - - Not less than 40'.0" - - -

13'.0"

Extension
if required.

5'.0"

TYPE B.

     Malby & Sons. Lith

Plate 157.

ENTRANCES AND INCLINES.

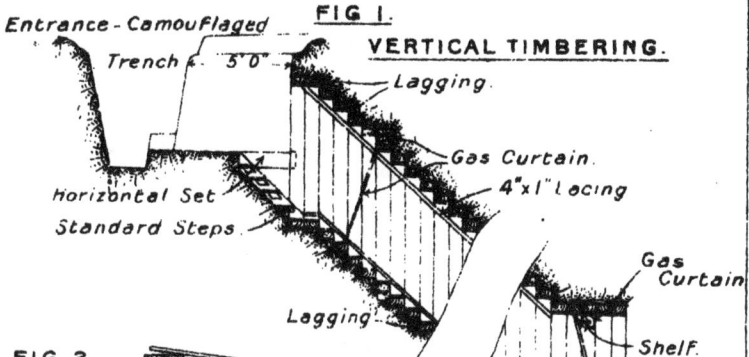

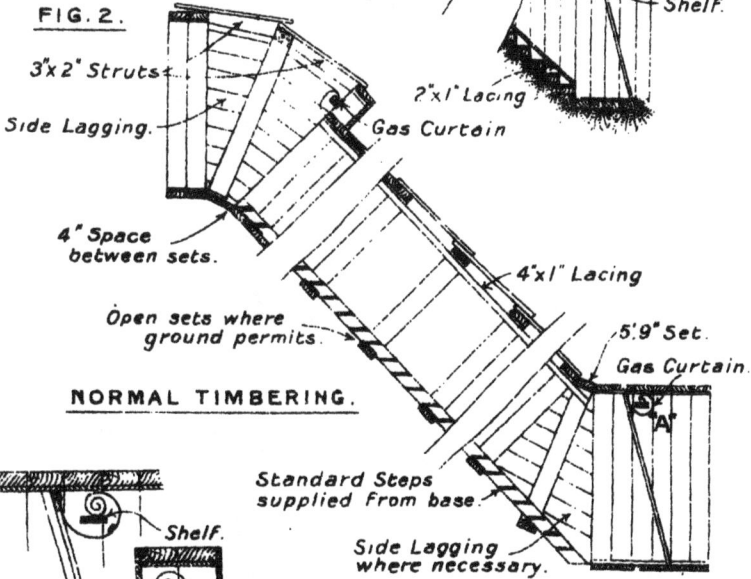

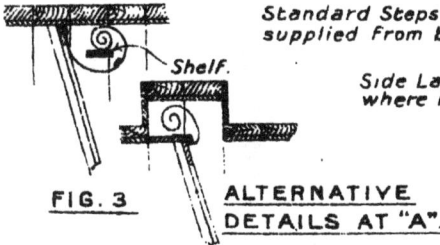

Entrance - Camouflaged

FIG 1.

Trench 5'0"

VERTICAL TIMBERING.

Lagging.

Gas Curtain.

4"x1" Lacing

Horizontal Set

Standard Steps

Gas Curtain

Lagging.

Shelf.

FIG. 2.

3"x2" Struts

2"x1" Lacing

Side Lagging.

Gas Curtain.

4" Space between sets.

4"x1" Lacing

Open sets where ground permits.

5'9" Set.

Gas Curtain.

NORMAL TIMBERING.

Standard Steps supplied from base.

Shelf.

Side Lagging where necessary.

FIG. 3

ALTERNATIVE DETAILS AT "A".

8921.11084. PP 3115/387.

Malby & Sons L'th

Plate 158.

S P I L I N G.

Whaling board

Space for driving forward lagging

Wood chock

Distance piece

4"x2" Whaling board removed later

A

Sills sunk in ground

4"x2" Whaling board

DETAIL AT A

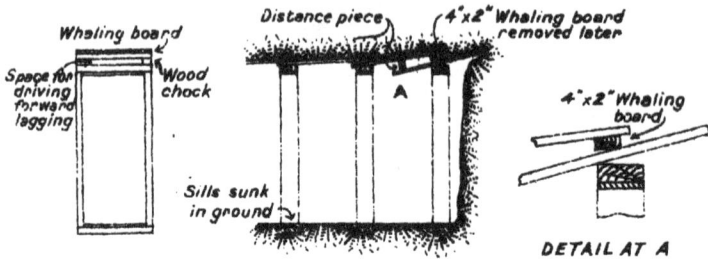

WITHOUT INTERMEDIATE SET

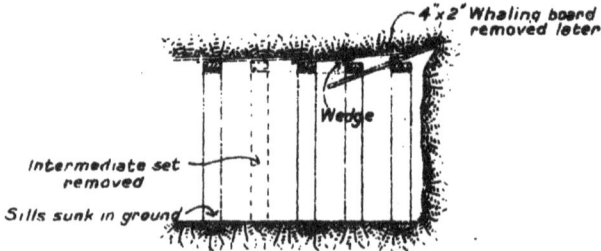

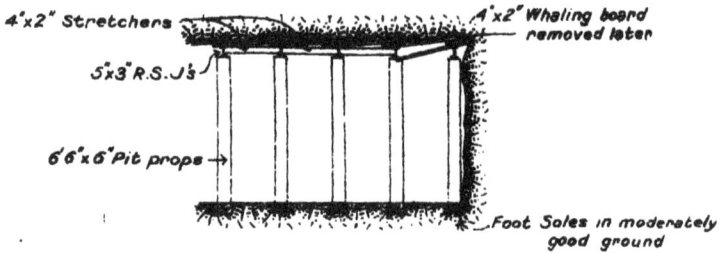

4"x2" Whaling board removed later

Wedge

Intermediate set removed

Sills sunk in ground

WITH INTERMEDIATE SET

4"x2" Stretchers

5"x3" R.S.J's

6'6"x6" Pit props

4"x2" Whaling board removed later

Foot Soles in moderately good ground

WITH PIT PROPS & R.S.Js.

Malby & Sons Lith

Plate 159.

FORWARD ROADS.

ROAD TEMPLET

Section showing camber of road from which the length of legs can be
calculated e.g. for a total rise of 8" at the centre. a 12'. 6" 11½" c. 9" d. 6".

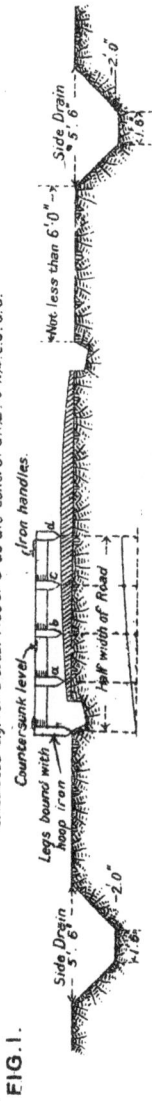

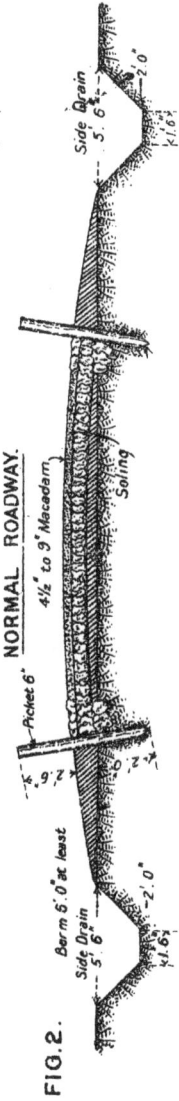

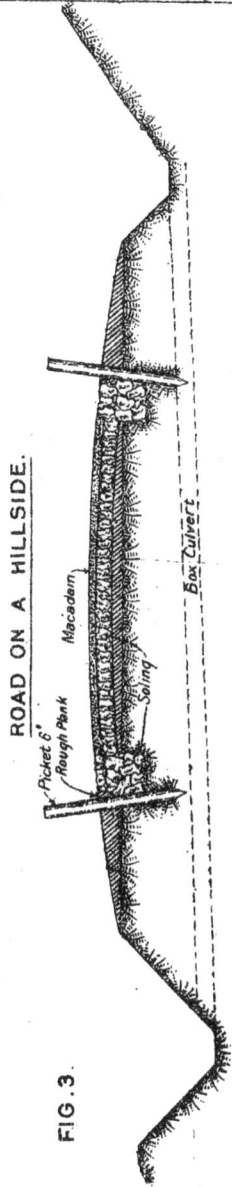

FIG. I.

Countersunk level.

Iron handles.

Legs bound with hoop iron.

Half width of Road.

Not less than 6'0"

Side Drain
5'. 6"

—2'.0"

Side Drain
5'. 6"

—2'.0"

NORMAL ROADWAY.

FIG. 2.

Picket 6"

4½" to 9" Macadam

Soling

Berm 6'.0" at least

Side Drain
5'. 6

—2'.0"

Side Drain
5'. 6"

—2'.0"

ROAD ON A HILLSIDE.

FIG. 3.

Picket 6"
Rough Pank

Soling

Macadam

Box Culvert

Plate 160.

FORWARD ROADS.

TYPICAL SECTION OF COUNTRY ROAD.

FIG. 1.

COUNTRY ROAD REMADE

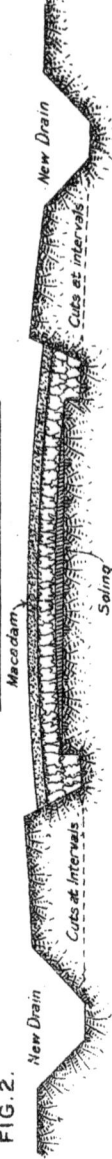

New Drain

Cuts at intervals

Macadam

Soling

Cuts at Intervals

New Drain

FIG. 2.

TREATMENT IN DEEP CUTTING.

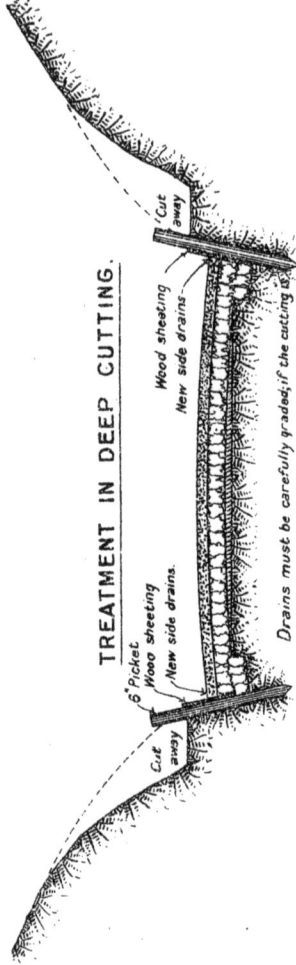

Cut away

Wood sheeting

New side drains

6" Picket

Wood sheeting

New side drains.

Cut away

Drains must be carefully graded; if the cutting is wide enough side drains clear of the roadway must be provided

FIG. 3.

Malby & Sons.

Plate 161.

FORWARD ROADS.

METHOD OF WIDENING PAVE ROAD.

Fig. 1.

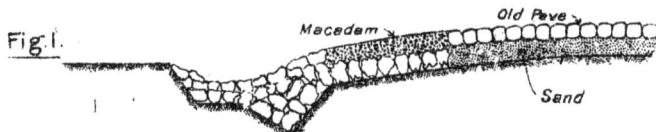

Old Pave

Macadam

Sand

Fig. 2. **SHELL HOLES.**

Cut the shell-hole square ramming the bottom of the hole.

Then fill the shell-hole with sandbags laid properly, on the top of the sandbags the roadway is now made in the usual way

If the shell-hole has water in it it must be pumped dry before it is filled.

Fig. 3. **ROAD REPAIR.**

Setion of road to be repaired Foundations completely disappeared
Mud and Sludge.

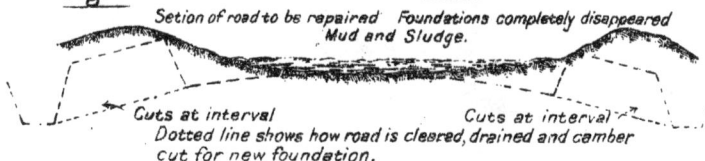

Cuts at interval   Cuts at interval
Dotted line shows how road is cleared, drained and camber cut for new foundation.

Fig. 4. **CORDUROY OR SLAB ROAD.**

Corduroy or slabs
9'·0" to 10'·0"

Berm 4'·0"

Riband 6"x 6"

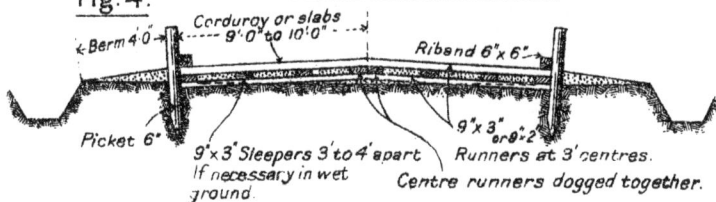

Picket 6"

9"x 3" Sleepers 3'to 4' apart If necessary in wet ground.

9"x 3" or 9"·2

Runners at 3' centres.

Centre runners dogged together.

8921.11084.PP9115/387.

Malby & Sons, Lith

Plate 162.

TANK CAUSEWAY.

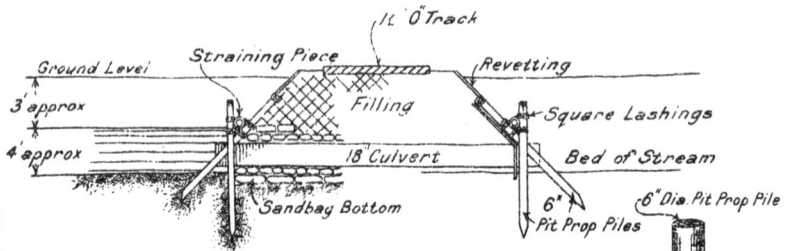

Ground Level

3' approx

4' approx

It O' Track

Straining Piece

Revetting

Filling

Square Lashings

18' Culvert

Bed of Stream

Sandbag Bottom

6"
Pit Prop Piles

SECTION.

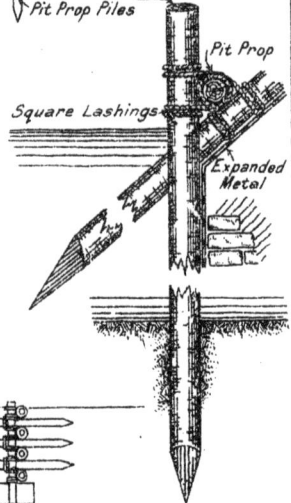

6" Dia. Pit Prop Pile

Pit Prop

Square Lashings

Expanded
Metal

DETAILS.

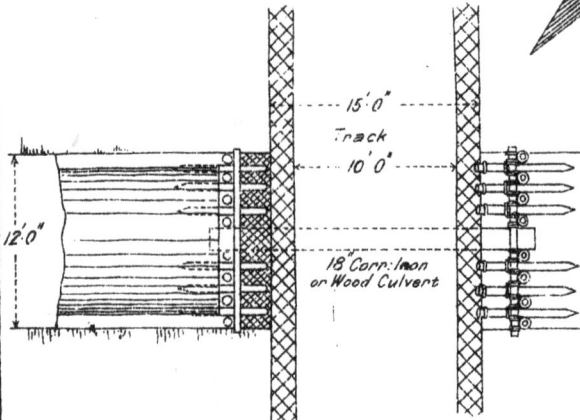

15' 0"

Track

10' 0"

12' 0"

18 Corr: Iron
or Wood Culvert

PLAN.

Malby & Sons. Lith

Plate 163.

9 – LB TRACK.

Fig. 1.

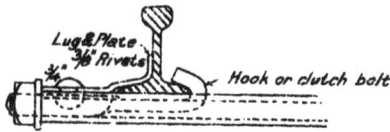

Lug & Plate
3/8" Rivets
3/4"
Hook or clutch bolt

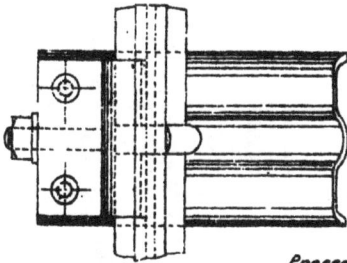

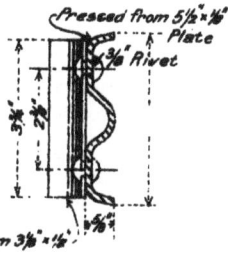

Pressed from 5½" x ⅞"
Plate
3/8" Rivet
3½"
2½"
5/8"
Pressed from 3½" x ⅛"

Fig. 2.

Joint
Plate

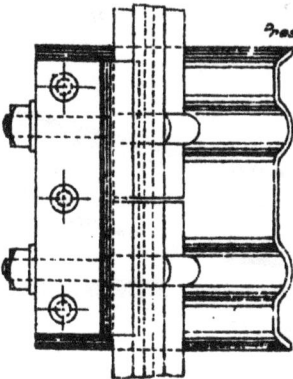

Pressed from 3½" x ⅛" Plate

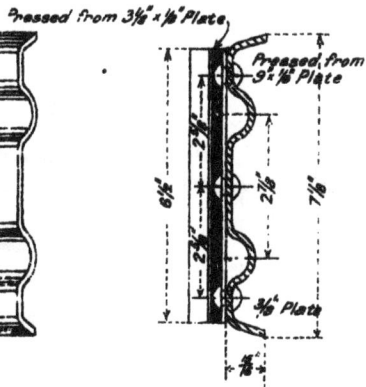

Pressed from
9" x ⅛" Plate
6½"
2½"
2½"
2½"
2½"
7½"
3/8" Plate
⅛"

Plate 164.

9-LB. TRACK.

FIG. 1.
5M STRAIGHT SECTION.

Length of rails 5m (16'·4⅞")

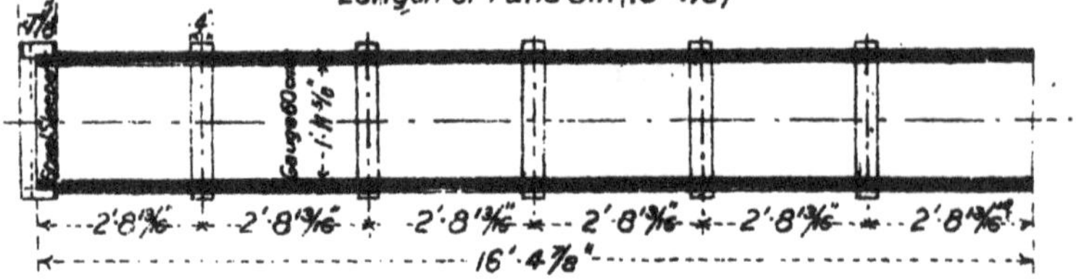

Gauge 60cm
2'·8 13/16"
16'·4⅞"

FIG. 2.
2.5M STRAIGHT SECTION.

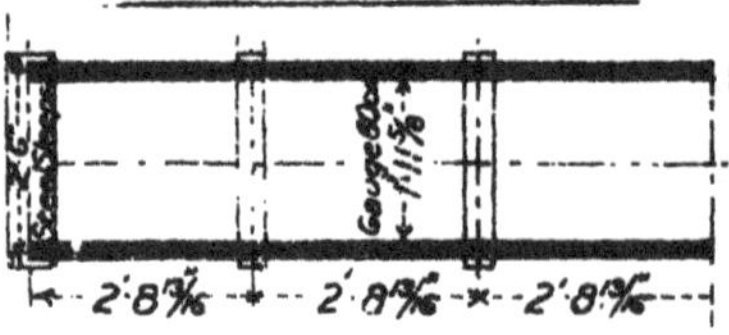

Gauge 60cm

FIG. 3.
2.5M CURVED SECTION (8'·2 7/16")

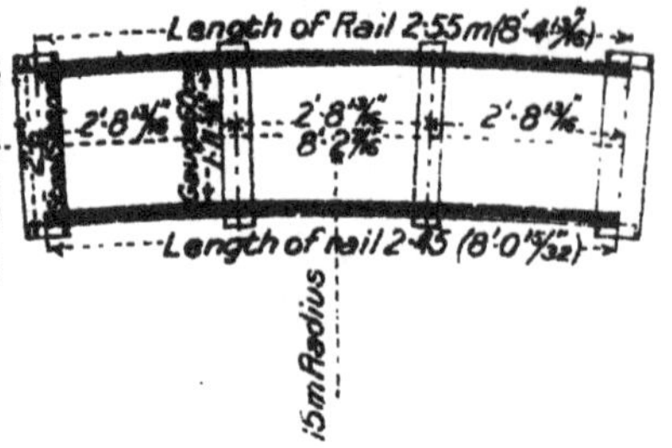

Length of Rail 2·55m
Length of Rail 2·45
15m Radius

FIG. 4.

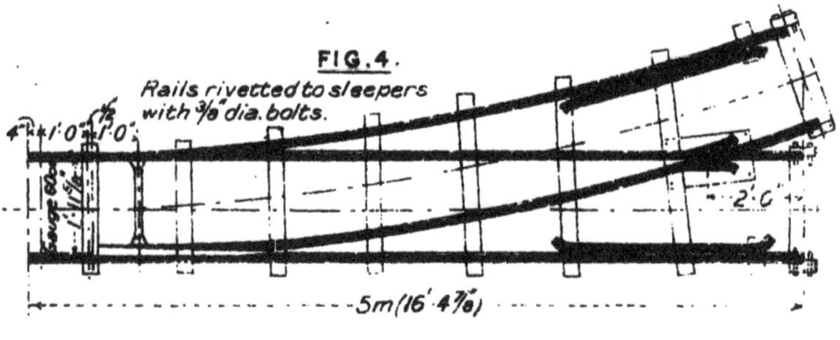

Rails rivetted to sleepers with ⅜" dia. bolts.
4"
1'·0"
1'·0"
Gauge 60cm
2'·6"
5m (16'·4⅞")

8921 11084. PP3115/387.

Malby & Sons Lith

Plate 165.

20-lb. TRACK.

FIG. I.
5 M. STRAIGHT SECTION (TYPE A)

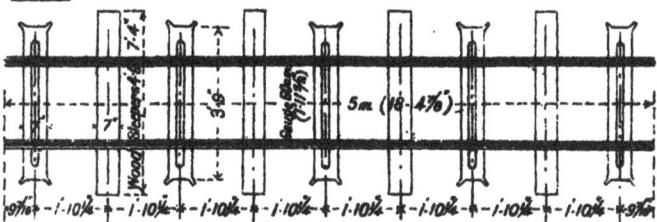

FIG. 2. 2·5 M. STRAIGHT SECTION (TYPE A).   **FIG. 3.** 2·5 M. CURVED SECTION (TYPE A)

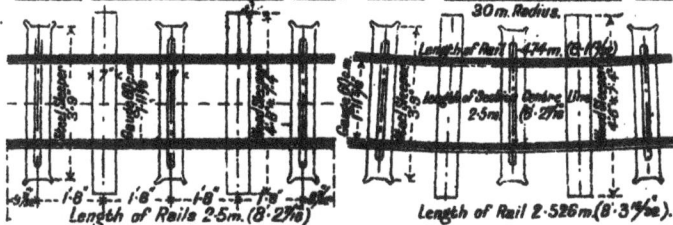

Length of Rails 2·5 m. (8·2⅞)   Length of Rail 2·526 m. (8·3⅞₃₂).

FIG. 4.   ### 5 M. CURVED SECTION (TYPE A).
30 m. Radius.

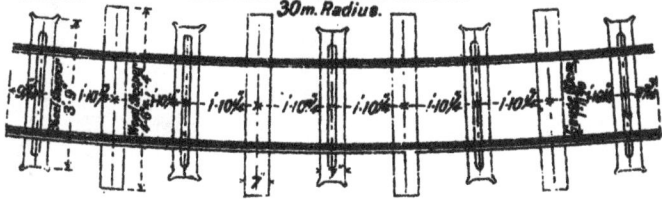

FIG. 5.   ### 5 M. STRAIGHT SECTION (TYPE B).

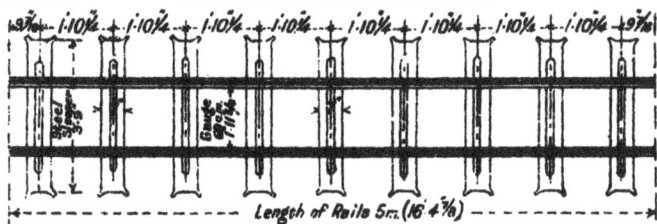

Length of Rails 5 m. (16 4⅞)

     Malby & Sons, Lith.

Plate 166.

20-LB. TRACK.

LONGITUDINAL SECTION THROUGH SLEEPER.

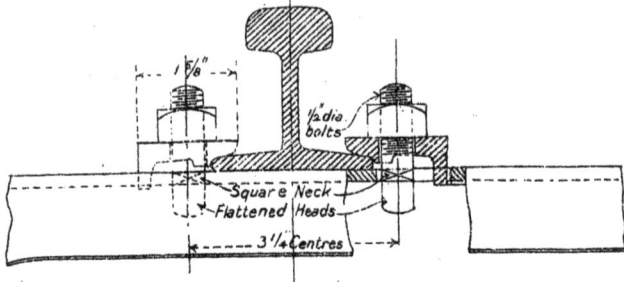

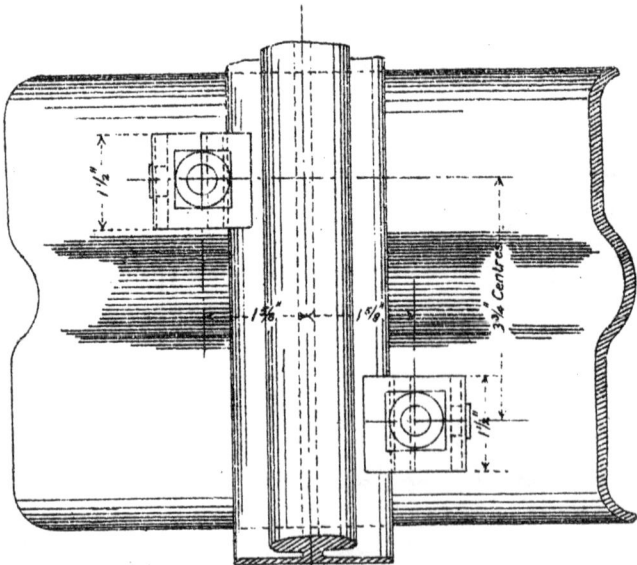

Malby & Sons. Lith

Plate 167.

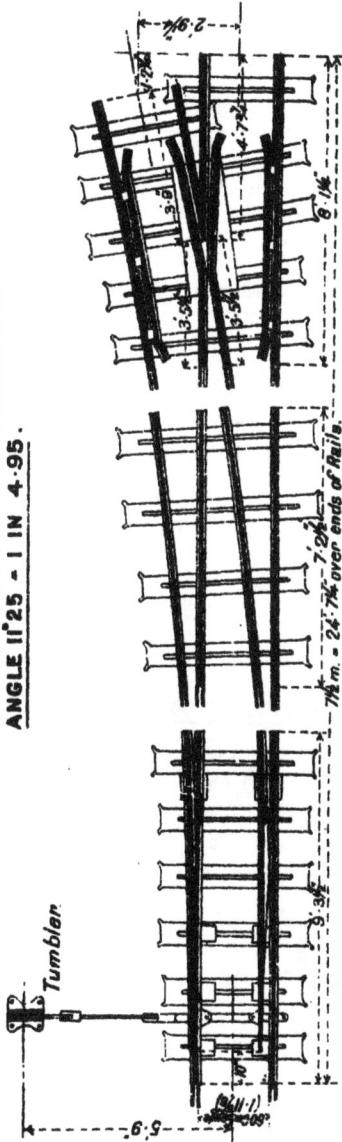

20-lb. TURNOUT.   30 m. RADIUS.

ANGLE 11°25′ = 1 IN 4·95.

Tumbler.

SWITCH SECTION.

CLOSURE SECTION.

FROG SECTION.

(A)

(B)

Main Line.

Siding or Branch Line.

RIGHT HAND TURNOUT.

Main Line.

Siding or Branch Line.

LEFT HAND TURNOUT.

Plate 168.

MULE WALK 20-LB. TRACK.

FIG.1

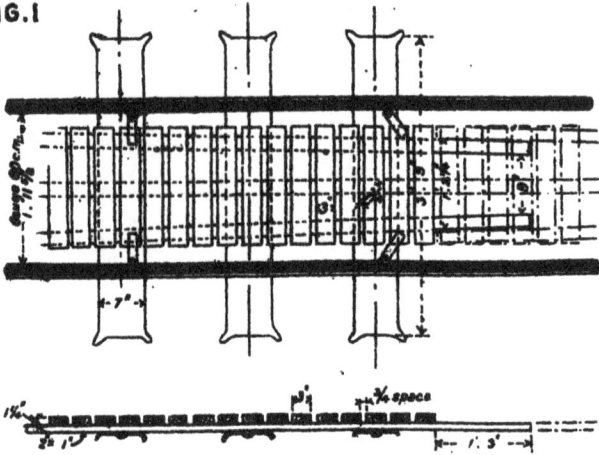

FIG.2.

DETAIL OF HOLDING DOWN CLIP.

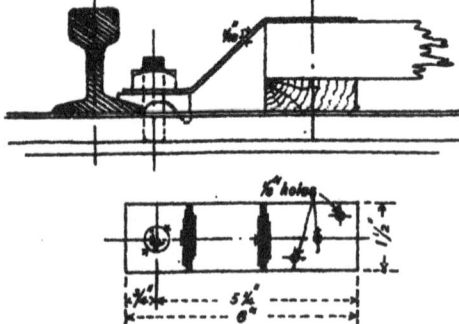

Can be made from Hoop Iron taken
from Corrugated Iron Binding.

Plate 169.

RAFT, TRACK 20-LB.

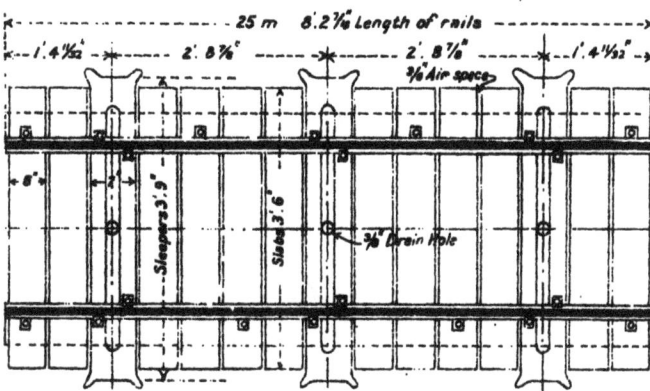

PLAN.

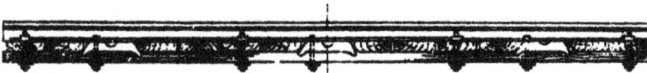

ELEVATION.

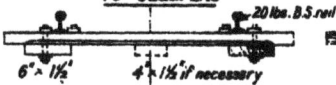

SECTION SHEWING ATTACHMENT TO SLEEPERS

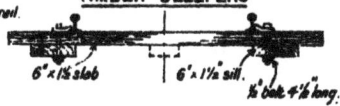

SECTION SHEWING ADDITIONAL TIMBER SLEEPERS

Malby & Sons. Lith.

Plate 170.

CROSS SECTIONS OF EARTHWORKS.

FIG. 1.

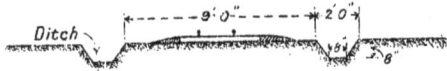

Ditch

FIG. 2.

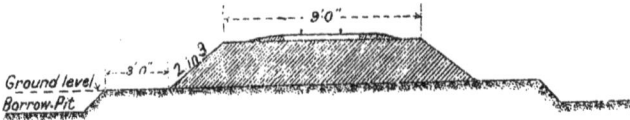

Ground level
Borrow Pit

FIG. 3.

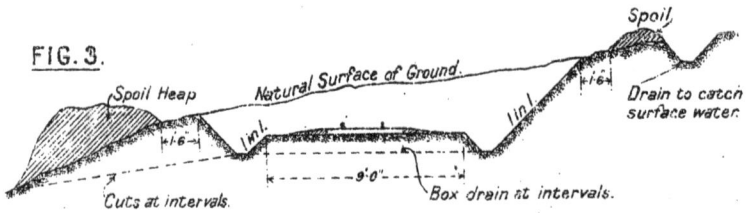

Spoil

Spoil Heap

Natural Surface of Ground.

Drain to catch
surface water.

Cuts at intervals.

Box drain at intervals.

Malby & Sons. Lith.

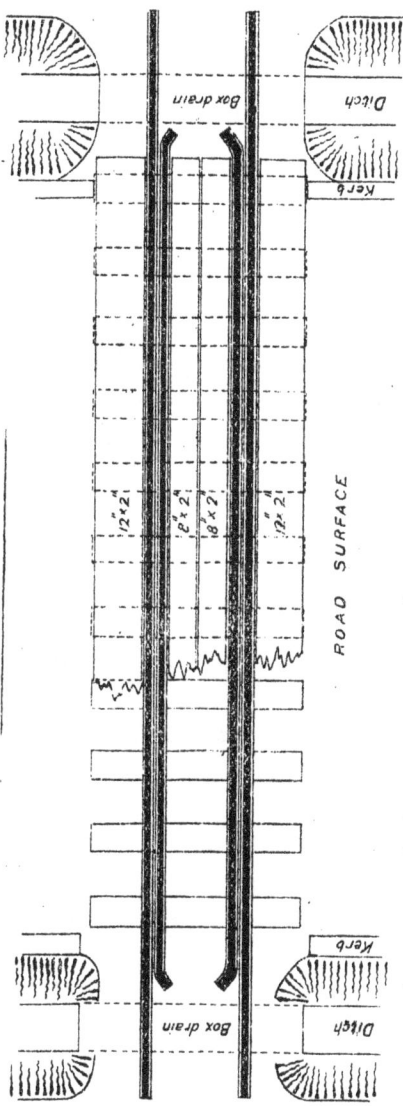

Plate 171.

LEVEL CROSSING.

Ditch

Kerb

Box drain

Kerb

Ditch

Box drain

ROAD SURFACE

Planks to be skew nailed with 5" spikes

Pave of Macadam

Dog spikes

Wood Sleepers

All sleepers on Level Crossings must be well packed
and if possible ballasted with brick or stone

Malby & Sons, Lith.

Plate 172

DEMOLITIONS. USE OF CORDEAU DETONANT.

FIG.1.

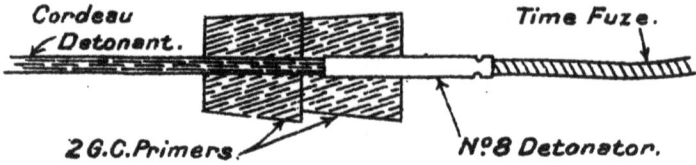

Cordeau Detonant.

Time Fuze.

2 G.C.Primers.

Nº8 Detonator.

FIG.2.

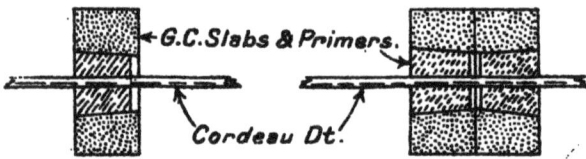

G.C.Slabs & Primers.

Cordeau Dt.

FIG.3.

FIG.4.

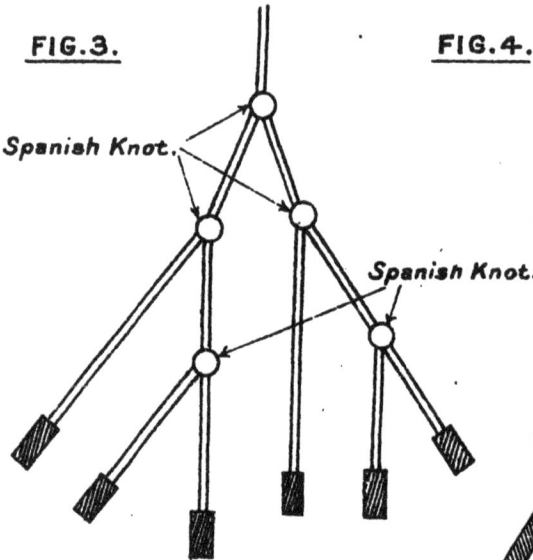

Spanish Knot.

Spanish Knot.

Spanish Knot.

Malby & Sons Lith

Plate 173.

DEMOLITIONS.

FIG.4.

FIG.3.

FIG.5.

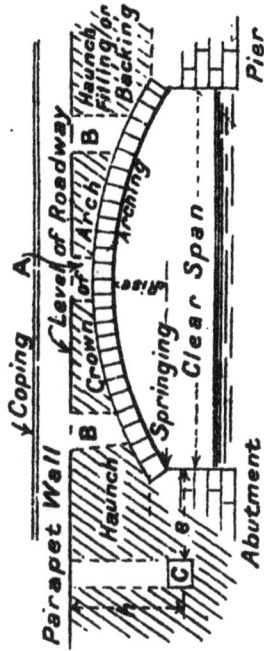

Coping
Level of Roadway
Crown of Arch
Haunch, Filling, or Backing
Arching
Springing
Pier
Clear Span
Springing
Abutment
Parapet Wall
Haunch
C
B
A
B

A. Charge at Crown.
B.B. Charges at Haunches.
C. Charge at Abutment.

FIG.2.

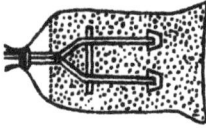

FIG.1.

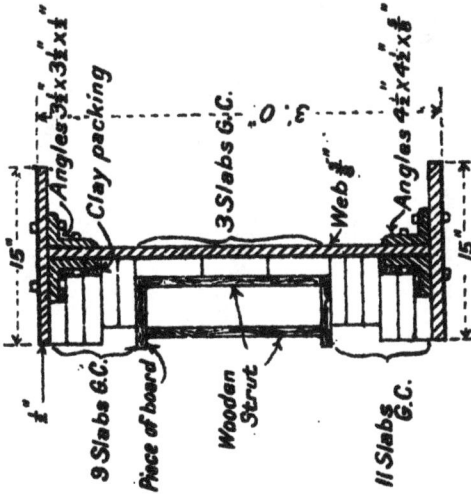

Angles 3¼×3¼×⅜″
Clay packing
3 Slabs G.C.
Web ⅜″
Angles 4½×4½×⅜″
3'. 0″
15″
15″
⅛″
9 Slabs G.C.
Piece of board
Wooden Strut
11 Slabs G.C.

Thickness of rivet heads ⅛″ throughout.

Malby & Sons, Lith.

Plate 174.

DEMOLITION OF RAILWAYS.

FIG.1.

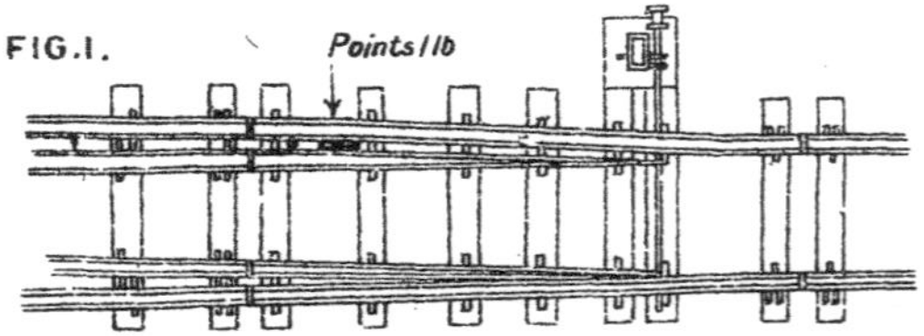

Points 1 lb

FIG.2.

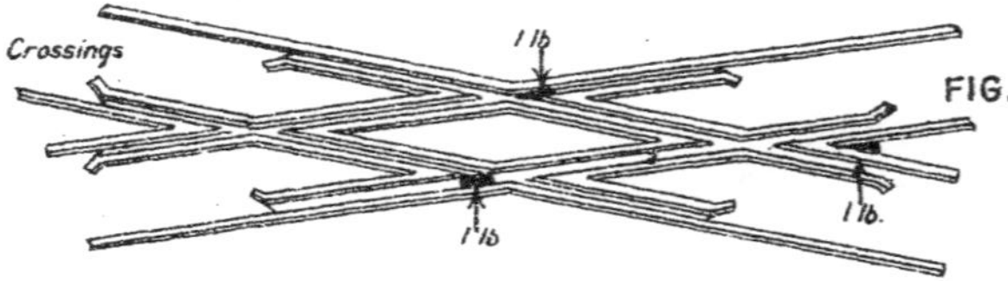

1 lb

1 lb

1 lb

1 lb

Crossings

FIG.3.

Probable Fracture 16 in.

Demolition of
Heavy Steel Rail
(105 lbs per yard)
I.G.C. Primer 1 oz.

Clay

5 11/16 in.

I.G.C. Slab, 15 oz.

String.

1 Detonator Nº 8.

Fuze

Slip Knot

Stick

Brickbat
or Stone

Fuze Safety

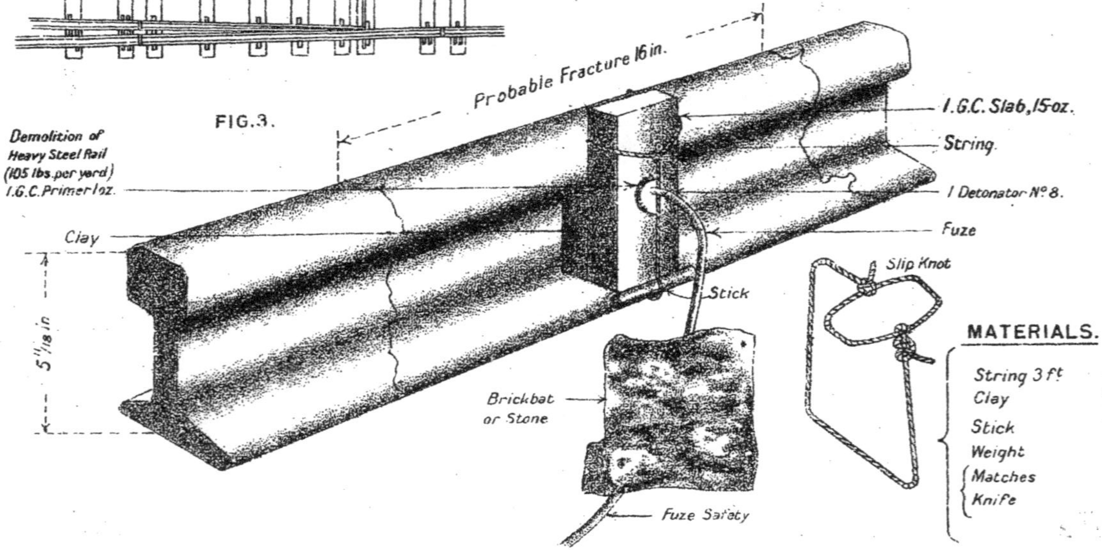

MATERIALS.

String 3 ft.
Clay
Stick
Weight
Matches
Knife

Plate 175.

ELECTRIC FIRING.

JOINTING WIRES.

FIG.1.

FIG.2.

FIG.3.

FIG.4.

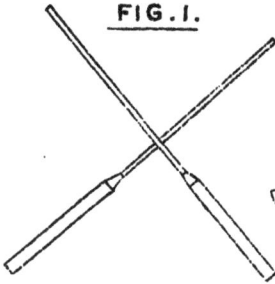

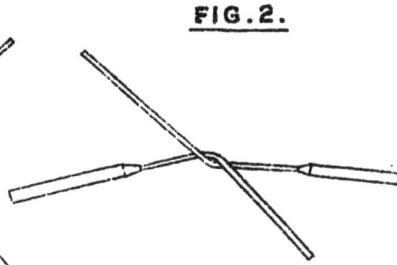

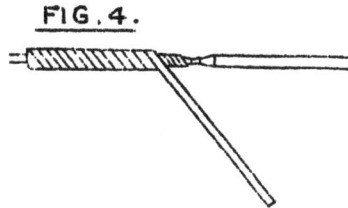

Connections for testing

Connections for firing

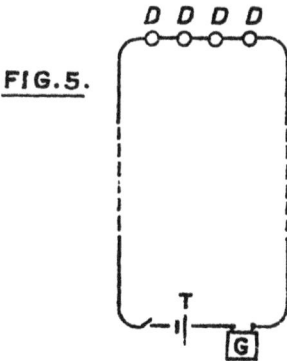

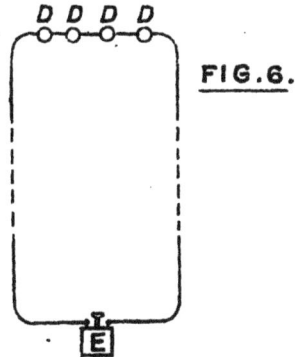

FIG.5.

D D D D

FIG.6.

D D D D

T

G

E

D.D.= *Detonators in charge.*
T = *Test Cell.*
G = *Q & I. Detector.*
E = *Exploder.*

Malby &Sons.Lith.

www.ingramcontent.com/pod-product-compliance
Lightning Source LLC
Chambersburg PA
CBHW031405180326
41458CB00043B/6619/J